POLYMER SCIENCE AND TECHNOLOGY SERIES

SERIES EDITORS: DR D M BREWIS AND PROFESSOR D BRIGGS

HANDBOOK OF POLYMER–FIBRE COMPOSITES

EDITOR:
F R JONES

DEPARTMENT OF ENGINEERING MATERIALS
UNIVERSITY OF SHEFFIELD

POLYMER SCIENCE AND TECHNOLOGY SERIES
SERIES EDITORS: DR D M BREWIS AND PROFESSOR D BRIGGS

I S MILES AND S ROSTAMI (eds), *Multicomponent Polymer Systems*
D E PACKHAM (ed.), *Handbook of Adhesion*
E A COLBOURN (ed.), *Computer Simulation of Polymers*
Forthcoming
D M BREWIS AND B C COPE (eds), *Handbook of Polymer Science and Technology*
H R BRODY (ed.), *Synthetic Fibres*
R N ROTHON (ed.), *Particulate Filled Polymer Composites*

Longman Scientific & Technical,
Longman Group UK Ltd,
Longman House, Burnt Mill, Harlow,
Essex CM20 2JE, England
and Associated Companies throughout the world

Copublished in the United States with
John Wiley & Sons, Inc., 605 Third Avenue, New York, NY 10158

©Longman Group UK Limited 1994

All rights reserved; no part of this publication may be reproduced, stored in a retrieval system, or transmitted in any form or by any means, electronic, mechanical, photocopying, recording, or otherwise without either the prior written permission of the Publishers or a licence permitting restricted copying in the United Kingdom issued by the Copyright Licensing Agency Ltd., 90 Tottenham Court Road, London W1P 9HE.

Trademarks
Throughout this book trademarked names are used. Rather than put a trademark symbol in every occurrence of a trademarked name, we state that we are using the names only in an editorial fashion and to the benefit of the trademark owner with no intention of infringement of the trademark.

First published 1994

British Library Cataloguing in Publication Data
A catalogue entry for this title is available from the British Library.

ISBN 0-582-06554-2

Library of Congress Cataloging-in-Publication data
Handbook of polymer fibre composites/editor, F.R. Jones.
p. cm. – (Polymer science and technology series)
Includes bibliographical references and index.
ISBN 0-470-23373-7
1. Fibrous composites. 2. Polymeric composites. I. Jones, F.R.
II. Series.
TA418.9.C6H343 1994   93–23652
620. 1'92–dc20   CIP

ISBN 0-470-23373-7 (USA only)

Set by 16JJ, in 10/12$\frac{1}{2}$ pt Times
Printed and Bound by Bookcraft (Bath) Ltd.

Contents

List of contributors		ix
Preface		xii

Chapter 1	Fibrous reinforcements for composite materials	1
	General introduction F R JONES	1
1.1	Acicular particulate reinforcement R ROTHON	4
1.2	Alumina fibres J E BAILEY	8
1.3	Aramid fibres L S PENN	12
1.4	Boron fibres J E BAILEY	15
1.5	Boron nitride and other covalent ceramic fibres J H SHARP and R J P EMSLEY	18
1.6	Carbon fibres from PAN – preparation and properties F R JONES	19
1.7	Carbon fibres from PAN – structure D J JOHNSON	24
1.8	Carbon fibres from PAN – surface treatment F R JONES	29
1.9	Carbon fibres from pitch B RAND and M TURPIN	34
1.10	Glass fibres – type and form F R JONES	38
1.11	Glass fibres – surface treatment F R JONES	42
1.12	Knitted reinforcements F K KO	48
1.13	Polyethylene fibres D L M CANSFIELD and I M WARD	54
1.14	Rigid rod polymer fibres R J YOUNG	58

	1.15	Silicon carbide fibres J H SHARP and R J P EMSLEY	62
	1.16	Silicon nitride and silicon carbonitride fibres R J P EMSLEY and J H SHARP	66
Chapter 2		Matrices	69
	2.1	Advanced thermoplastics J BARNES	69
	2.2	Conventional thermoplastics R S BAILEY	74
	2.3	Conventional thermoplastics – pultruded composite tapes G CUFF	79
	2.4	Epoxy resins F R JONES	86
	2.5	High temperature resins – thermosetting polyimides J N HAY	96
	2.6	High temperature resins – other thermosets J N HAY	101
	2.7	Phenolic resins – laminates P A SHEARD	106
	2.8	Phenolic resins – moulding compounds A SMITH	109
	2.9	Time–temperature–transformation diagrams for thermosets B ELLIS	115
	2.10	Unsaturated polyester resins R A PANTHER	121
	2.11	Urethane methylacrylates F R JONES	127
	2.12	Vinyl ester resins F R JONES	129
Chapter 3		Fabrication of polymer composites	133
	3.1	A guide to selection F R JONES	133
	3.2	Autoclave moulding F R JONES	138
	3.3	Centrifugal casting F R JONES	144
	3.4	Continuous fibre reinforced thermoplastic composites – shaping J BARNES	145
	3.5	Bulk moulding compounds and dough moulding compounds A G GIBSON	150
	3.6	Filament winding V MIDDLETON	154
	3.7	Hand lay T M GOTCH	160
	3.8	Injection moulding – thermoplastics M J FOLKES	165
	3.9	Injection moulding – thermosets A G GIBSON	168

3.10	Injection moulding – fibre management by shear controlled orientation P S ALLAN and M J BEVIS	171
3.11	Long glass fibre reinforced thermoplastic sheet P M JACOBS	176
3.12	Pultrusion A G GIBSON	180
3.13	Recycling of polymer fibre composites D W CLEGG	183
3.14	Reinforced reaction injection moulding (RRIM) F R JONES	187
3.15	Resin transfer moulding (RTM) C D RUDD	192
3.16	Resin transfer moulding (RTM) – practicalities A A ADAMS	198
3.17	Sheet moulding compounds A G GIBSON	200
3.18	Spray deposition T M GOTCH	205
3.19	Structural RIM (SRIM) P A SHEARD and M PARMAR	209

Chapter 4	Mechanical properties of composites – micromechanics	212
4.1	Continuous fibre reinforcement J E BAILEY	212
4.2	Continuous fibre reinforcement – Young's moduli J E BAILEY	215
4.3	Continuous fibre reinforcement – strength J E BAILEY	218
4.4	Continuous fibre reinforcement – off-axis properties of unidirectional laminae J E BAILEY	222
4.5	Hybrid effect M G BADER	225
4.6	Interfacial bond strength determination – critique I VERPOEST	230
4.7	Interfacial bond strength determination – fragmentation I VERPOEST and M DESAEGER	235
4.8	Interfacial bond strength determination – microdebond pull-out test L S PENN	238
4.9	Interfacial bond strength determination: pull-out test M J PITKETHLY	242
4.10	Laminates – angle plied S L OGIN	245

4.11	Laminates – crossply cracking F R JONES	249
4.12	Laminates – residual thermal and related strains F R JONES	254
4.13	Laminates – statistics of transverse cracking P W M PETERS	260
4.14	Short fibre reinforcement – direct measurement of fibre strain using Raman spectroscopy R J YOUNG	264
4.15	Short fibre reinforcement – stress transfer in discontinuous fibre composites M G BADER	269
4.16	Short fibre reinforced thermoplastics M G BADER	274
4.17	Short fibre reinforced thermosets F CHEN	278

Chapter 5   Mechanical properties – macromechanics   286

5.1	Damage accumulation – fracture mechanics S L OGIN	286
5.2	Delamination – fracture mechanics S L OGIN	289
5.3	Design considerations for aerospace applications G W MEETHAM	292
5.4	Failure criteria P A SMITH	296
5.5	Fatigue – aramid fibre reinforced plastics B HARRIS	299
5.6	Fatigue – carbon fibre composites P T CURTIS	305
5.7	Fatigue – glass fibre reinforced plastics B HARRIS	309
5.8	Fatigue – life prediction – role of materials and structure P IRVING	317
5.9	Fatigue – life prediction – damage calculation and assessment P IRVING	323
5.10	Impact performance – CFRP laminates G DOREY	327
5.11	Impact performance – residual compression strength G DOREY	330
5.12	Impact resistance – fibre reinforced polymers P HOGG	334
5.13	Knitted fabric composites – properties F K KO	341

	5.14 Laminate theory P A SMITH	347
	5.15 Non-destructive evaluation of composites B HARRIS	351
	5.16 Standard test methods for composite materials E W GODWIN	356
Chapter 6	Environmental aspects	362
	6.1 Mechanical properties under hygrothermal conditions E G WOLFF	362
	6.2 Moisture absorption – Fickian diffusion kinetics and moisture profiles T A COLLINGS	366
	6.3 Moisture absorption – anomalous effects F R JONES	371
	6.4 Osmosis and blistering in GRP F CHEN and A W BIRLEY	376
	6.5 Stress corrosion cracking of GRP F R JONES	379
	6.6 Thermal spiking/thermal cycling effects F R JONES	388

Glossary	392
Bibliography	395
Index of specific symbols	400
General index	403

List of contributors

A A ADAMS, formerly of Lotus Engineering, Hethel, Norwich, NR1 48
DR P S ALLAN, The Wolfson Centre for Materials Processing, Department of Materials Technology, Brunel University, Uxbridge, UB8 3BH
PROFESSOR M G BADER, Department of Materials Science and Engineering, University of Surrey, Guildford GU2 5XH
PROFESSOR J E BAILEY, Department of Engineering Materials, Sir Robert Hadfield Building, University of Sheffield, PO Box 600, Mappin Street, Sheffield, S1 4DU
DR R S BAILEY, Wilton Research Centre, ICI Materials, Rheology Centre, Wilton, Middlesborough, TS6 8JE
DR J BARNES, Wilton Research Centre, ICI Materials, Rheology Centre, Wilton, Middlesborough, TS6 8JE
PROFESSOR M J BEVIS, The Wolfson Centre for Materials Processing, Department of Materials and Technology, Brunel University, Uxbridge, UB8 3PH
PROFESSOR A W BIRLEY, IPTME, Loughborough University of Technology, Loughborough, LE11 3TD
D L M CANSFIELD, formerly The IRC in Polymer Science and Technology, University of Leeds, Leeds, LS2 9JT
DR F CHEN, Department of Engineering Materials, PO Box 600, Mappin Street, Sir Robert Hadfield Building, University of Sheffield, Sheffield, S1 4DU
DR D W CLEGG, School of Engineering, Sheffield City Polytechnic, Pond Street, Sheffield, S1 1WB
T A COLLINGS, 25 Fern Drive, Church Crookham, Fleet, GU13 0NW
G CUFF, formerly of Wilton Research Centre, ICI Materials, Rheology Centre, Wilton, Middlesborough, TS6 8JE
DR P T CURTIS, Materials and Structures Department, DRA Farnborough GU14 6TD
DR M DESAEGER, Department of Metaalkunde en Toegepaste, Materiaalkunde (MTM), Katholieke Universiteit Leuven, de Croylaan 2, B-3001 Leuven, Belgium

PROFESSOR G DOREY, Materials and Structures Department, DRA Farnborough, Farnborough, GU14 6TD

B ELLIS, Department of Engineering Materials, PO Box 600, Mappin Street, Sir Robert Hadfield Building, University of Sheffield, Sheffield, S1 4DU

DR R J P EMSLEY, School of Materials, University of Leeds, Houldsworth Building, Leeds, LS2 9JT

DR M J FOLKES, Department of Materials Technology, Brunel University, Uxbridge UB8 3PH

PROFESSOR A G GIBSON, Materials Division, Department of Mechanical Materials and Manufacturing Engineering, Herschel Building, The University, Newcastle Upon Tyne, NE1 7RU

E W GODWIN, Centre for Composite Materials, Imperial College of Science and Technology and Medicine, Prince Consort Road, London, SW7 2BY

T M GOTCH, 2 Lodge Close, Etwall, Derby, DE65 6NA

PROFESSOR B HARRIS, School of Materials Science, University of Bath, Claverton Down, Bath, BA2 7AY

DR J N HAY, Kobe Steel Europe Ltd, 10 Nugent Road, Surrey Research Park, Guildford GU2 5AF

DR P J HOGG, Department of Materials, Queen Mary and Westfield College, Mile End Road, London, E1 4NS

PROFESSOR P E IRVING, School of Industrial and Manufacturing Science, Cranfield University, Cranfield, Bedford, MK43 0AL

DR P M JACOBS, 3 Colne Close, Oadby, Leicestershire, LE2 4GA

PROFESSOR D J JOHNSON, Department of Textile Industries, University of Leeds, LS2 9JT

PROFESSOR F R JONES, Department of Engineering Materials, PO Box 600, Mappin Street, Sir Robert Hadfield Building, University of Sheffield, Sheffield, S1 4DU

PROFESSOR F K KO, Department of Materials Engineering, Drexel University, 31st and Markets Streets, Philadelphia, Pennsylvania 19104, USA

DR G W MEETHAM, 18 Short Avenue, Allertree, Derby, DE3 2EH

DR V MIDDLETON, Department of Mechanical Engineering, University of Nottingham, University Park, Nottingham NG7 2RD

DR S L OGIN, Department of Materials Science Engineering, University of Surrey, Guildford, GU2 5XH

R A PANTHER, Scott-Bader Company Ltd, Wollaston, Wellingborough, Northamptonshire, NN9 7RL

M PARMAR, PERA International, Technology Centre, Melton Mowbray, Leicestershire, LE13 0PB

PROFESSOR L S PENN, Department of Materials Science and Engineering,

University of Kentucky, 763 Anderson Hall, Lexington,
KY 40506-0046, USA
DR P W M PETERS, DLR, Institut für Werkstoff-Forschung, D-51140
Köln, Germany
DR M J PITKETHLY, Materials and Structures Department, DRA
Farnborough, Farnborough, GU14 6TD
PROFESSOR B RAND, School of Materials, University of Leeds, Leeds,
LS2 9JT
R ROTHON, Manchester Metropolitan University, Department of
Materials Technology, John Dalton Building, Chester Street,
Manchester, M1 5GD
DR C D RUDD, Department of Mechanical Engineering, University of
Nottingham, University Park, Nottingham, NG7 2RD
DR J H SHARP, Department of Engineering Materials, Mappin Street, Sir
Robert Hadfield Building, University of Sheffield, PO Box 600,
Sheffield, S1 4DU
DR P A SHEARD, Euro-Projects (LTTC) Ltd, Pocket Gate, Brook Road,
Woodhouse Eaves, Leicestershire, LE12 8RS
A SMITH, Perstorp Ferguson Ltd, Aycliffe Industrial Estate, Newton
Aycliffe, County Durham, DL5 6UE
DR P A SMITH, Department of Materials Science and Engineering,
University of Surrey, Guildford, GU2 5XH
DR M TURPIN, School of Materials, University of Leeds, Leeds, LS2 9JT
PROFESSOR I VERPOEST, Department of Metaalkunde en Toegepaste
Materiaalkunde (MTM), Katholieke Universiteit Leuven, de Croylaan
2, B-3001 Leuven, Belgium
PROFESSOR I M WARD, The IRC in Polymer Science and Technology,
University of Leeds, Leeds LS2 9JT
PROFESSOR E G WOLFF, Department of Mechanical Engineering, Oregon
State University, Rogers Hall 204, Corvallis, Oregon 97331-6001, USA
PROFESSOR R J YOUNG, Manchester Materials Science Centre, University
of Manchester and UMIST, Grosvenor Street, Manchester, M1 7HS

Preface

At the invitation of David Briggs, this book was conceived as an encyclopaedia along the lines of the *Handbook of Adhesion* by David Packham. However, in the planning stage it became apparent that the terminology and jargon of the discipline was rather specialized and would be difficult to retrieve from an alphabetical format without prior knowledge. Furthermore, the specialization has become so broad over the three decades since the discovery of high strength carbon fibres by William Watt that individual engineers and scientists have different major interests and needs. Therefore in this book the articles have been grouped into six important chapters which cover differing areas of interest: Fibres (Chapter 1); Polymer Matrices (Chapter 2); Fabrication (Chapter 3); Mechanical Properties (Chapters, 4,5) and Environmental Aspects (Chapter 6). In each chapter short contributions from leading international experts in the field are listed alphabetically to provide easy access to important issues within each area of the composites discipline. A further benefit of this format is that the titles of the individual articles have been carefully chosen so that there is a logic to their sequence and each chapter forms an introductory text in each of the main chapter headings.

To provide cross-referencing between articles a comprehensive index is provided. For example, if seeking the properties of a particular type of composite, this can be accessed through the individual component fibre and matrix. For detailed understanding of the nature of the damage and failure mechanisms which give rise to these properties, the reader is referred to Chapter 4 where the scientific principles are discussed. For a particular fabrication process, the information can be retrieved using the encyclopaedic listing in Chapter 3. To select an appropriate fabrication process for a particular application, guidance can be obtained by examining the relative costs provided on pages 136 and 137 and the relevant articles on selected manufacturing routes and individual materials.

Many of the applications of composite materials are in high performance structures employing continuous carbon (and other high performance) fibres. The important issues which dictate the design requirements are examined in Chapter 5.

It is not the intention of this book to provide a completely comprehensive text but to illustrate the science and engineering of these materials and provide sufficient background and bibliography for further research and scholarship. Each individual chapter could have been enlarged but specialist texts exist. The intense interest in composites began with the development of high modulus carbon and aramid fibres in the late sixties, and as a consequence encompassed the glass fibre reinforced plastics which had been available from the fifties and earlier. Recent times have seen the renaissance of phenolic materials, the first truly synthetic plastic (Bakelite) as glass fibre composites which have also been brought into the genre of polymer composites. Thus confusion often exists as to what a polymer composite is. The term now refers to a range of polymeric materials which employ particulate and fibre reinforcements. This text attempts to clarify these arguments with a series of articles which provide understanding on the range of polymer fibre-composites. More detailed information on particulate reinforcements can be obtained from the book of that name in this series by Roger Rothon and to a more limited extent from *Multicomponent polymer systems* by I S Miles and S Rostami, also in Longman's *Polymer Science and Technology Series*.

With 86 articles from 50 authors it proved difficult to unify the symbols (and units) employed without introducing too many errors. Where possible this has been achieved but an index of specific symbols has been included alongside a glossary to ensure that the differing use of the same or different symbol is clear. This arrangement provides a distinct advantage when using this text alongside other articles and books on composite materials.

Furthermore, for similar reasons, it was originally intended to provide five key references (i.e., a bibliography) with each article for access to the literature. In practice this has been restricted to a maximum of eight references, some of which are specific citations. It should be emphasised that together with the main bibliography, these are intended to provide a more comprehensive source of information. It does, however, mean that any references that are not cited in the text are intended as a bibliography for that topic.

I am indebted to the individual authors for their cooperation and patience which was needed for the preparation of the text. I would also like to thank former colleagues at Scott-Bader & Co. Ltd, the Department of Materials Science and Engineering, University of

Surrey, who stimulated my interests in polymers and polymer composites. This text would have not proved possible without the efforts of my present and former research students and research fellows at the Universities of Surrey and Sheffield, some of whom have contributed articles.

I have to thank Paula Turner and Kathy Hick of Longmans for their patient support during the preparation of the manuscript and, in particular, the painstaking subediting of the manuscript. My secretary, Carole Plant, who has typed most of my contributions and struggled with the indexes, is heartily thanked. My wife, Christine, is also thanked for typing some parts of the manuscript. My wife and family are also thanked for their patient support during the many evenings and weekends spent on the manuscript.

Finally, a text of this nature relies heavily on published literature and to the best of our knowledge, we have acknowledged and referenced all of these contributions for which we offer our thanks.

F R JONES
UNIVERSITY OF SHEFFIELD
MARCH 1994

Publisher's Note
While every effort has been made to trace the owners of copyright material, in a few cases this has proved impossible and we take this opportunity to offer our apologies to any copyright holders whose rights we may have unwittingly infringed.

CHAPTER 1

Fibrous Reinforcements for Composite Materials

General introduction

F R JONES

Composite technology utilizes the high stiffness and strength of filamentary materials. Common examples include straw reinforced clay bricks from biblical times and today's steel (fibre) braced radial tyre. For the layman, it is not fully understood that 'fibreglass' (for boat hulls, car bodies, etc.) refers to glass fibre reinforced plastic. The advantage of these materials lies in the high specific stiffness, which can be achieved, for example, with carbon fibres by molecular and crystal orientation along their fibre length, such that the chemical bonds carry a higher proportion of the load. Since the strength of a material is controlled by the population of Griffith flaws, it follows that a fine filament will have a lower density of critical (strength reducing) flaws[1]. As a consequence, high strength materials will be filamentary of fine diameter (typically 14 μm for glass and 7 μm for carbon fibres) and their measured strength will be gauge length dependent[2,3]. Furthermore, within a bundle of fibres individual strengths will differ and the bundle will fail under stress in a progressive manner. Similarly an embedded fibre, where good interfacial adhesion is maintained, will also fail progressively as a function of its strength statistics. Since the density of differing materials varies, the most efficient way of comparing performance is a comparison of specific properties as shown in Fig. 1. Specific strength is defined as strength/density, σ_{uf}/ρ, and specific modulus is defined as modulus/density, E_f/ρ, where E is Young's modulus (Pa), σ is stress (Pa), ρ is density (kg m^{-3}). The subscripts f and u refer to fibre and ultimate fracture respectively. Table 1 demonstrates the wide range of properties of various reinforcements and compares them to some typical isotropic materials, whereas Fig. 1 ranks the efficiency of reinforcement and demonstrates why carbon fibres and aramid fibres are important in high performance applications.

Table 1. Typical mechanical properties of reinforcing fibres in comparison to isotropic bulk materials

	Relative density	Young's modulus (GPa)	Tensile strength (GPa)	Failure strain (%)	Fibre diameter (μm)
Fibres					
E-glass	2.55	72	1.5–3.0	1.8–3.2	10–16
S-glass	2.5	87	3.5	4.0	12
S-2 glass	2.49	86	4.0	5.4	10
Carbon, mesophase pitch	2.02	380	2.0–2.4	0.5	10
Carbon, PAN					
High strength	1.8	220–240	3.0–3.3	1.3–1.4	7
High performance	1.8	220–240	3.3–3.6	1.4–1.5	7
High strain	1.8	220–240	3.7	1.5–1.7	7
Intermediate modulus	1.9	280–300	2.9–3.2	1.0	7
High modulus	2.0	330–350	2.3–2.6	0.7	7
SiCO continuous	2.5	200	3.0	1.5	10–15
SiC whisker	3.2	480	7.0	–	1–50
SiTiCO, continuous	2.35	200	2.8	1.4	8–10
Boron	2.6	410	3.4	0.8	100
α-Alumina (FP)	3.9	380	1.7	0.4	20
β-Alumina (Saffil)	3.3	300	2.0	0.5	3
δ/θ-Alumina (Safimax SD)	3.3	250	2.0	–	3.1
η-Alumina (Safimax LD)	2.0	200	2.0	–	3.2
Aramid					
High modulus	1.47	180	3.45	1.90	12
Intermediate modulus	1.46	128	2.65	2.4	12
Low modulus	1.44	60	2.65	4.0	12
Staple fibre (Poly m-phenylene isophthalamide)	1.4	17.3	0.7	22.0	12
Poly p-phenylene benzothiazole (PBT)	1.5	250	2.4	1.5	20
Polyamide 66	1.44	5.0	0.9	13.5	$\simeq 10$
Polythene					
Theoretical	–	> 200	–	–	–
Solution spun	1.0	100–120	1.0–3.0	–	–
Drawn (Tenfor)	$\simeq 1$	60	1.3	5	–
Bulk materials					
Steel	7.8	210	0.34–2.1	–	–
Aluminium alloys	2.7	70	0.14–0.62	–	–
Inorganic glass	2.6	60	–	–	–
Resins					
Phenolic	1.4	7	–	$\simeq 0.5$	–
Epoxy	1.2	2–3.5	0.05–0.09	1.5–6.0	–
Polyester	1.4	2–3.0	0.04–0.085	1–25	–
Thermoplastics					
Nylon 66	1.4	2.0	0.07	60	–
High density polyethene	0.96	1.3	–	–	–
Low density polyethene	0.91	0.25	–	–	–

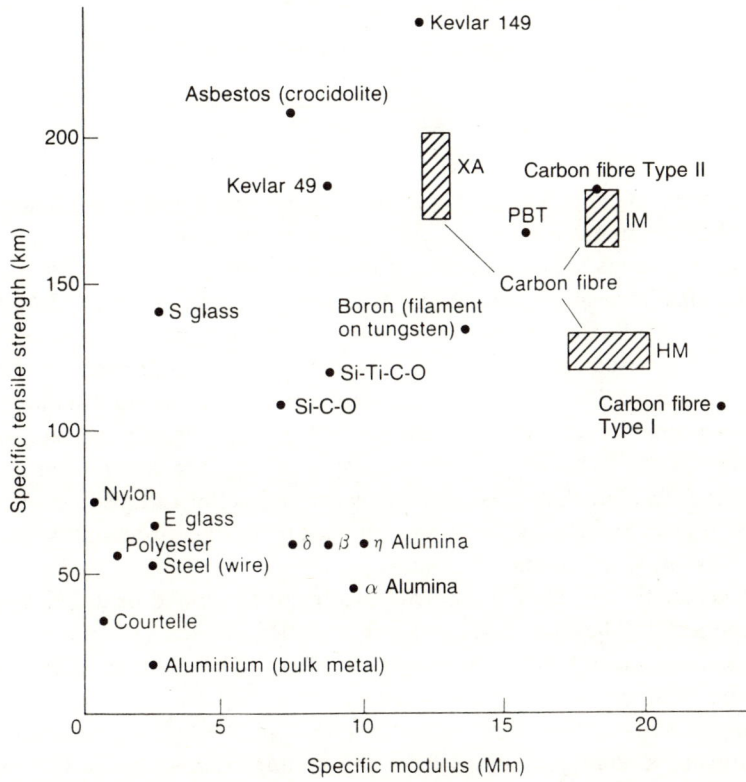

Figure 1 Specific properties of reinforcing fibres compared with other materials (after Jones[2])

References

1. A Kelly, N H MacMillan *Strong Solids*, 3rd Edn, Clarendon Press, Oxford, UK, 1986.
2. D R Lovell (Ch 3) and F R Jones (Ch 6) in *Composite Materials in Aircraft Structures*, ed. D H Middleton, Longman, Harlow, 1990.
3. *Handbook of Composites*, Vol. 1, Strong Fibres, eds W Watt, B V Perov, Elsevier, Amsterdam, 1985.

1.1 Acicular particulate reinforcement

R ROTHON

General aspects

There is considerable interest in using acicular inorganic particles as reinforcements in polymers. These particles are needle-like with diameters less than 5 μm and lengths generally below 100 μm. Some of them occur naturally, others are produced synthetically, usually by precipitation.

Acicular inorganic particles are stiff and relatively strong. Providing they have sufficient aspect ratio and can be bonded strongly to the polymer matrix, they can give good levels of reinforcement. The importance of these aspects is covered in more detail in Chapter 4. Their properties are often adequate but do not usually achieve the reinforcement levels of glass fibres, although they are easier to process with less fibre breakage, less machine wear and better surface finish than bulk fibres. Sometimes the chemical nature of the fibres gives extra benefits, especially flame retardancy.

Acicular fillers generally improve creep resistance and heat deflection temperature. They increase modulus and flexural strength more than non-acicular fillers but less than glass fibres. The main disadvantage compared to non-acicular fillers is higher viscosity and reduced impact strength. Milled glass fibres have somewhat similar effects, but contain significant amounts of debris from the milling process, hence they are generally less effective.

The fine needle shape of these fillers has given rise to some health concerns, especially after the asbestos experience. Fibre toxicity is complex. It depends on the chemical nature of the fibre as well as on the aspect ratio and the particle size. Information presented here is given in good faith and based on manufacturers' literature. The manufacturers should be contacted for more detailed and more recent information.

Natural products

Nowadays, only two natural products are of significant commercial interest, asbestos and wollastonite.

Asbestos is the classical microfibre, obtainable with diameters in the range 0.02–0.1 μm and with high aspect ratio. Various chemical forms exist but they are all hydrated silicates. The simplest form is chrysotile. It is the form most widely used in polymers and it is a hydrated magnesium silicate. Because of their chemical structure, chrysotile fibres are hollow. Chrysotile fillers provide the general benefits described earlier but they do lower the impact strength, are often difficult to process and can give rise to poor colour. The health problems associated with asbestos are well known and have severely curtailed its use. More information can be found in the review by Axelson.[1]

Wollastonite (calcium metasilicate) also occurs naturally in an acicular form but unlike asbestos is claimed to be of low toxicity. After extraction and beneficiation it is obtained as a fine white powder of varying aspect ratio. But the highest obtainable aspect ratio is a disappointing 20:1. This limits the obtainable reinforcement. A range of surface treatments has been developed to enhance particle/polymer adhesion and improve reinforcement. Despite its low aspect ratio, wollastonite can give a useful blend of properties at a reasonable cost, especially when blended with glass fibres. One of the principal uses appears to be in polyamides where a silane coupling agent is used to enhance bonding to the matrix. The main markets for wollastonite seem to be in the United States, presumably because this is the source of the purest deposits. There is more information on wollastonite in the review by Copeland.[2]

Synthetic products

A number of inorganic materials can be precipitated from aqueous solution as acicular particles and several are being offered commercially as polymer reinforcements. The principal forms offered to date are summarized in Table 2. The particles are generally single crystals and often are referred to by the manufacturers as whiskers. They generally contain more flaws than true whiskers grown by other processes but are still strong enough to exhibit considerable reinforcement.

From Table 2 it can be seen that the aspect ratios are between 20:1 and 100:1 and the obtainable reinforcement is generally claimed to be superior to wollastonite.

Aragonitic calcium carbonate has received surprisingly little attention but is now being offered by Maruo Calcium using its patented production method.[3] One of the problems of calcium carbonates has been poor bonding to polymer matrices but this has now been largely overcome.[4]

Another common mineral readily precipitated in acicular form is calcium sulphate and various forms of this have been marketed for several years. It is currently offered by US Gypsum under the name Franklin Fibre. Two forms are produced, hemihydrate and anhydrous. The latter is more thermally stable and less water sensitive and is thus of most interest in polymer applications. The fibres have quite a respectable aspect ratio but the strength of their interaction with polymers and the suitability of surface treatments is not clear. The product is claimed to be of low toxicity. More information can be found in the review by Milewski.[5]

Table 2. Principal synthetic acicular fillers

Property	CaCO$_3$	CaSO$_4$ anhydrous form	CaNa(PO$_3$)$_3$ Standard grade	NaAl(OH)$_2$CO$_3$	Mg(OH)$_2$	MgSO$_4$·5MgO·8H$_2$O
Crystalline or chemical name	aragonite	anhydrite	–	dawsonite	brucite	magnesium oxysulphate
Supplier	Maruo	US Gypsum	Monsanto	Alcoa	Kyowa	Ube
Aspect ratio (average)	40	30	20	~30	40	50–100
Average diameter (μm)	1	2	1–5	0.5	<1	<1
Surface area (m^{2}g^{-1})	n.a.	n.a.	1–2	15–17	n.a.	10
Hardness (Mhos)	3.5–4	~2.5	~4	n.a.	3	n.a.
Density (g cm^{-3})	2.9	3.0	2.9	2.4	2.4	2.3
Refractive index	birefringent ~1.68	1.59	1.57	1.53	1.56/1.58	n.a.
Thermal stability (°C)	~800	>800	>700	~300	~300	~250
Comments		Will transform to calcite at 520°C, effect of this not known	Crystals have fibrillar ends	Has a flame retardant effect	Has a flame retardant effect	

Monsanto have recently introduced a phosphate fibre with the formula CaNa(PO$_3$)$_3$. This contains long metaphosphate chains running parallel along the fibre axis. Although the aspect ratio is quite low, the crystals have fibrillar ends and this is claimed to improve reinforcement and give an effective aspect ratio nearer to 50:1. Further advantages claimed are the fibre's biodegradability and its low hazard to health. Little is known about its strength of bonding to polymers and the suitability of surface treatments. For more details, see the review by Crutchfield et al.[6]

Magnesium oxysulphate fibre has the highest aspect ratio of those listed. It was introduced some time ago by Ube Industries of Japan. The main limitation is its decomposition temperature of 250°C. This restricts its use in some thermoplastics. Its ability to bond to polymers is also relatively unknown. More details can be found in the review by Milewski.[5]

Finally, there are two acicular fillers that have been produced to combine reinforcement with flame retardancy. Both rely on endothermic decomposition and release of inert gases for their flame retardant action.

They are stable to about 300°C. Acicular magnesium hydroxide is available from Kyowa in Japan and dawsonite (sodium aluminium hydroxycarbonate) has been produced in acicular form by Alcoa in America. Again more information on dawsonite can be found in the review by Milewski.[5]

Unfortunately, fillers that produce flame retardancy in this way have to be used at high loadings (up to 60% w/w). This is very difficult to achieve with acicular particles alone. The best prospects for these products thus seem to be in blends with other flame retardants.

Conclusions

Despite the considerable efforts so far expended, acicular particulate reinforcements do not seem to have had the expected impact on the market. This is probably because no one product has yet been proven to have all the required features, among them high aspect ratio, high polymer interaction, chemical and thermal stability, low cost and low health hazard. Nevertheless, progress is being made and further interesting developments in this field can be expected.

References

1. J W Axelson in *Handbook of Reinforcements for Plastics*, eds J V Milewski, H S Katz, Van Nostrand Reinhold, New York, 1987, Ch. 9.
2. J R Copeland in *Handbook of Reinforcements for Plastics*, eds J V Milewski, H S Katz, Van Nostrand Reinhold, New York, 1987, Ch. 8.
3. H Katayama, H Skibata, T Fujiwara, European Patent Application 0406662 (1990).
4. J Hutchinson, J D Birchall, *Elastomerics*, **112** (7), 17 (1980).
5. J V Milewski, in *Handbook of Reinforcements for Plastics*, eds J V Milewski, H S Katz, Van Nostrand Reinhold, New York, 1987, Ch. 10.
6. M M Crutchfield, A R Henn, J A Hinkebein, B F Monzyk, in *Handbook of Reinforcements for Plastics*, eds J V Milewski, H S Katz, Van Nostrand Reinhold, New York, 1987, Ch. 7.

1.2 Alumina fibres

J E BAILEY

Introduction

On theoretical grounds alumina Al_2O_3 in the form of α-alumina (corundum), is one of the candidate materials for high strength. Strong multivalent covalent bonds give alumina crystals a high Young's modulus (460 GPa) and good thermal stability. In practice, polycrystalline α-alumina remains the dominant bulk engineering ceramic. It is available in polycrystalline form with varying grain size, porosity and composition. Addition of other oxides can ease fabrication, aid densification and control grain size. High strength whiskers of alumina were produced several decades ago in the search for practical systems.[1]

The development of continuous alumina fibre has been slow, probably for two reasons. Firstly alumina is comparatively dense (Table 1), therefore its specific strength and modulus are not very attractive for applications where low weight is a priority, e.g. space, aerospace. Secondly, alumina has no glass-forming characteristics, so it cannot be spun or drawn from a melt. Aluminosilicates can be spun and are produced in discontinous form for high temperature thermal insulation. For greater detail on the preparation routes and properties of alumina fibres, reference can be made to *Handbook of Composites*.[2]

Fabrication methods for polycrystalline alumina fibres

The various approaches to the production of polycrystalline continuous alumina fibres have in common the need to use a carrier material itself capable of being spun and subsequently thermally/chemically treated to convert it to an alumina fibre.

Extrusion and sintering of aqueous plastic suspensions or 'slips'

There have been several attempts to produce fibres by the development of extrudable highly plastic aqueous gel comprised of alumina hydrates (mainly boehemite related structures); the aqueous gel is sufficiently plastic to allow extrusion without the addition of a polymeric carrier material. A variant of this method is to use fibrillar colloidal boehemite gels with small quantities of lubricants to aid extrusion. Fibres down to diameters of $\simeq 100\,\mu\text{m}$ can be easily produced and the method can be regarded as a straightforward development of conventional ceramic firing methods. A critical aspect of the process is the dehydration and conversion of the gel to a metastable alumina oxide composition when the fibre is particularly weak. The reason for this is that hydraulic bonds are being destroyed and bonding resulting from sintering is only partially established.

Polymer precursor or 'relic' processes

In this process a carrier polymer fibre is loaded with fine oxide powder or impregnated with an inorganic salt solution, in this case containing alumina ions, normally by soaking and is then dried and fired in a similar manner to the extrusion process above. In a sense, the inorganic fibre so produced is a 'relic' of the original polymer fibre. Clearly during conversion to the ceramic there is a large emission of volatile products and a critical stage in the processing occurs when the fibre is weakened by decomposition at low temperature, $\simeq 400°C$, but the temperature is insufficient to promote substantial bonding by sintering. A particular example is the impregnation of water swollen, regenerated cellulose filaments with aqueous alumina nitrate or chloride solution. Excess salt solution is then removed and the filaments dried and slowly heated in an oxidizing atmosphere to 250°C. The salt is deposited in the fine voids of the polymer network and leads to an almost amorphous structure. At high temperature, the residual carbon char is removed and the salt oxidized. The induced microcrystalline nature of the oxide ensures relatively low sintering temperature.

These methods can produce strong fibres and a fine diameter, but the process is expensive because a preformed organic fibre is consumed and to date they have not been commercially exploited.

Spinning of solutions containing viscous alumina salt

The aim of this approach to the production of alumina fibres is the preparation of solutions of polymeric soluble precursors to the alumina oxide that provide the required viscous behaviour for spinning and extrusion. This eliminates the need for an expendable carrier polymer fibre. There is a range of potential solutions involving basic salts (such as aluminium oxychloride, formate, acetate), hydroflorosilic acid and colloidal alumina. The key requirement is the formation of a solution containing a high oxide equivalent concentration. For example, this comes about by the hydrolysis of the Al^{3+} ion in acidic solutions; the hydrated species need to bridge octahedral units, thereby producing polymeric networks containing a high density of alumina and oxygen ions. Rheology can be controlled by adding a small concentration of organic polymer.

Two commercially available fibre types, namely the Dupont F P polycrystalline continuous α-alumina fibre and the ICI Saffil staple polycrystalline transition alumina fibre are both produced by these methods.

The Dupont process[3] uses a suspension of alumina particles in an aqueous viscous phase containing an alumina precursor capable of forming a solution with alumina; whereas the resin spun Saffil filament involves mixed inorganic/organic viscous solutions.

Saffil is normally available in the form of a continuous random fibre blanket and is primarily used for the fabrication of metal matrix composites by squeeze casting. Fibre FP, however, is available as a continuous filament. Safimax has been introduced to provide a unidirectional reinforcement, which consists of a unidirectional mat with transverse stringers.[4] The properties of these fibres are included in Table 1.

Other methods

Alumina whiskers were produced quite early in the development of high strength filaments. The whiskers are produced by a vapour transport mechanism. Alumina is oxidized by steam and wet hydrogen at $\simeq 1400°C$. AlO and Al_2O sub-oxide species are formed and transported to cooler parts of the system where they condense to form fine whiskers, by the nucleation and growth of single crystal whiskers, if the correct thermodynamic conditions prevail. The whiskers normally have diameters of $\simeq 5\,\mu m$ and an aspect ratio ≈ 100. Growth occurs parallel to the <0 0 1> direction of the α-Al_2O_3 crystal with firm evidence for a dislocation nucleation and growth mechanism. Such whiskers are commercially available.

Large diameter α-Al_2O_3 filaments or rods, grown from molten alumina by crystal pulling methods, have achieved very high modulus and strength. But they are costly to produce in large volumes and remain of scientific interest only.

The mechanical properties of alumina fibres

The stiffness or Young's modulus of a polycrystalline ceramic fibre is determined essentially by the residual porosity of the fibres; ideally all porosity should be removed to achieve the highest value which will then depend upon the texture of the polycrystalline material if the structure of the crystal is anisotropic; for α-alumina this is a minor factor.

The presence of porosity has a marked effect upon the tensile strength as well as the modulus. Tensile strength is also affected by grain size. The ceramist seeking strength increases, is looking to decrease both grain size and porosity. The strength of brittle materials is determined by the presence of flaws introduced during fabrication. These flaws are often introduced at the surface of the material. They can be limited by grain size because cracks tend not to extend further than one grain; the grain boundary acts as a barrier. Surface roughness is an additional factor that can affect strength since a sharp discontinuity at the surface behaves as a crack in concentrating stress. Many fibres show an apparent dependence of strength upon fibre diameter. This has been documented for the ICI Saffil fibres.[1] Very careful production of glass fibres has shown these diameter effects are not an intrinsic property of glass fibres. The phenomenon is best understood in terms of the probability of damage to the fibre increasing with increasing diameter. Nevertheless, it can have a significant effect on the strength of commercially produced fibres, as shown in Fig. 2 for the Saffil filaments.

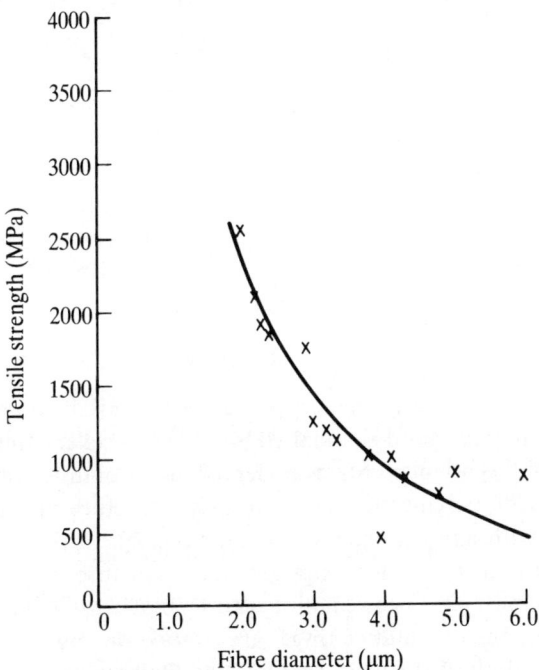

Figure 2 'Saffil' alumina fibre: variation in fibre tensile strength with fibre diameter. Redrawn from data given in reference 2.

Summary

Figure 2 indicates the strength levels presently achievable with continuous alumina fibres. The fibre stiffness lies in a range up to $\simeq 370\,\mathrm{GPa}$ and is dependent upon fibre structure.

Alumina fibres are available with high strength and stiffness mainly because alumina's stability is attractive for high temperature composites. Use of alumina fibres in reinforced plastics is more restricted because of their high density in relation to polymers, carbon and glass fibres. They may have some applications when abrasion and wear feature.

References

1. J E Bailey, H A Barker, *Chemistry in Britain* **10**, 465–70 (1974).
2. J D Birchall, J A A Bradbury, J Dinwoodie in *Handbook of Composites*, Vol. 1, Strong Fibres, eds W Watt, B V Perov, North-Holland Elsevier, Amsterdam, 1985, Ch IV pp. 115–54.
3. A K Dhingra in *New Fibres and their Composites*, ed. W Watt, Royal Society, London, 1980, pp. 411–417.
4. M H Stacey, M D Taylor, A M Walker, A new alumina fibre for advanced composites in *Sixth International Conference on Composite Materials, Second European Conference on Composite Materials* (ICCM6/ICCM2), eds F I Matthews *et al.*, Elsevier Applied Science, London, 1987, Vol. 5, pp. 5. 371–5. 381.

1.3 Aramid fibres

L S PENN

Aramids are high performance polymeric materials noted for their lightness in weight, good thermal stability and excellent toughness. The name 'aramid' is a United States Federal Trade Commission designation for a manmade polymer having the general structure of aromatic rings alternating with amide linkages (*aromatic amide*). In the United States, aramid is available in either a straight-chain structure (Fig. 3), known as the Kevlar family, or a bent-chain structure, known as Nomex, depending on whether the aromatic rings are para- or meta-disubstituted, respectively. Both of these are produced by Dupont. In the Netherlands, Akzo produces 'Twaron', which has a para-substituted structure.

Once synthesized, the aramid polymer is dissolved into a sulphuric acid solution, from which it is processed into fibres only a few microns in diameter by dry jet wet spinning.[1,2] The final morphology is achieved by treatment under tension at temperatures of 150–550°C. Bundles, or tows,

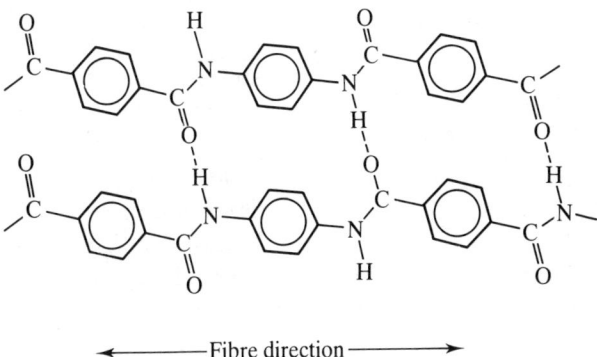

←————— Fibre direction —————→

Figure 3 Chemical structure of the straight-chain polymer comprising the Kevlar family of aramid. The polymer chain direction corresponds roughly to the fibre direction. In the transverse direction, neighbouring chains are held together by relatively weak hydrogen bonds and van der Waals' forces

Nomex, with bent-chain polymer structure, cannot reach the stiffness values of the Kevlars. Therefore its uses, e.g. thermal protective apparel, thermal and electrical insulation, and honeycomb core for aircraft interior panels, emphasize its thermal stability and lightness in weight. The Kevlars, with straight-chain polymer structure, are very stiff and can be processed into more than one high stiffness version, depending on the degree of polymer chain alignment introduced. The use of Kevlar as tyre cord, Kevlar 29 as underwater cable, and Kevlar 49 as reinforcing fibre in aerospace composites emphasizes the fibre strength and stiffness, perhaps more than its toughness. In recent years, an even stiffer version, Kevlar 149, has been introduced. Both Kevlar 49 and Kevlar 29 are used in ballistic protection for their toughness and ability to absorb energy.

Table 3 compares selected aramids with other common engineering fibres. In addition to the properties shown, important quantities for applications where lightness in weight is paramount are specific strength and specific stiffness, obtained by dividing the fibre strength and stiffness values by the corresponding fibre density. On this basis, the aramids in the table compare extremely well with the other fibres.

Aramid's yellow colour is derived from its extended conjugated π-electron system. This also means that the bare fibre is vulnerable to visible light (it absorbs in the range 300–400 nm), and deteriorates if exposed to sunlight.[4] For this reason, although the degradation products themselves screen the underlying polymer, Kevlar structures are coated or covered, or are buried within another part of the structure.

Table 3. Comparison of Kevlar aramid fibres with other engineering fibres

Fibre	Strength (MPa)	Modulus (GPa)	Density (g cm^{-3})
E-glass (Owens Corning Fiberglas)	3445	81.8	2.62
S-glass (Owens Corning Fiberglas)	4585	88.9	2.50
Carbon AS4 (Hercules)	4000	241	1.77
Carbon IM6 (Hercules)	4380	276	1.77
Kevlar 29 (Dupont)	3600	83	1.44
Kevlar 49 (Dupont)	4000	131	1.45
Kevlar 149 (Dupont)	3400	186	1.47

The conjugated π-electron system confers double bond character on most of the chemical bonds in the polymer structure. This in turn gives the aramids their excellent thermal stability. Thermal decomposition does not begin until above 400°C, a surprisingly high value for an organic polymer. In inert atmospheres (e.g. nitrogen) decomposition does not occur till almost 500°C.

The regular repeating structure and straight extended configuration of the polymer chains comprising the Kevlars allow high levels of crystallinity to be achieved. This level may be above 80%, a very high value for an organic polymer. Crystallographic studies have shown conclusively that the axes of the individual polymer chains are aligned fairly parallel to the fibre axis.[5]

The anisotropic polymer structure just described gives the fibre a very high tensile strength in the longitudinal direction, i.e. along the fibre axis. When load is applied, it is borne by the strong chemical bonds of the polymer chains. Neighbouring polymer chains in each crystallite are held together by hydrogen bonds and van der Waals' interactions, which are relatively weak and much easier to rupture than chemical bonds. Thus, the fibre is mechanically weak in the transverse (crosswise) direction.

This combination of strong bonds in the fibre direction (longitudinal direction) and weak forces holding the polymer chains together in the transverse direction creates interesting fibre behaviour. When the fibre is bent into a loop, it buckles on the inside of the loop and splits longitudinally on the outside of the loop. Furthermore, when loaded to breaking point, the fibre shows longitudinal cracking, or fibrillation, rather than a clean break.[5]

These fibre properties carry over to polymer matrix composites made with aramid (specifically, Kevlar) fibres. A unidirectional laminate, e.g. Kevlar/epoxy, is strong and stiff in the longitudinal (fibre) direction, but weak in the transverse direction. The compression strength is much lower than the tensile strength, and buckling under low compression loading is a problem. The composite exhibits a great deal of fibre fibrillation at failure, no matter whether the applied loading is shear or normal, tension or compression. These features are in contrast to brittle fibres, such as carbon and glass, whose composites show compressive behaviour nearly as good as their tensile behaviour, and whose failures are much cleaner.

Aramid fibres absorb moisture. This too is in contrast to glass and carbon fibres used in engineering applications. It is likely that moisture is attracted into small internal microvoids within each fibre,[5] rather than into the crystallites themselves. The transverse mechanical properties and the compressive properties, not the longitudinal tensile properties, are those most affected by the presence of moisture. Although the moisture content can reach 4.5% w/w, the polymer chains themselves are not degraded and all properties regain their previous high values upon drying. The fibres are dressed with spinning aids but surface treatments to promote adhesion are not effective so that fibre/matrix adhesion is generally poorer than for glass or carbon.

References

1. S L Kwolek, US Patent 3,671,542 (1972).
2. H Blades, US Patent 3,767,756 (1973).
3. C C Chiao, T T Chiao in *Handbook of Composites*, ed. G. Lubin, Van Nostrand Reinhold, New York, 1982, pp. 272–317.
4. L Penn, F Larsen, *J. Appl. Polym. Sci.*, **23**, 59 (1979).
5. M G Dobb in *Handbook of Composites*, Vol. 1, eds W Watt, B V Perov, Elsevier, Amsterdam, 1985, pp. 673–704.

1.4 Boron fibres

J E BAILEY

Introduction

Boron is one of the more attractive candidate materials for the production of high performance fibres, the light boron atom has a multivalent and high energy bond giving a high density of strong bonds per unit volume. Consequently solid boron can exhibit a high Young's modulus and low density and is therefore potentially of high strength. These factors are balanced by the difficulty in fabricating it in bulk by conventional methods, such as hot pressing and sintering, which has limited its use as a structural material. However, fibres in the form of continuous filaments were developed comparatively early in the evolution of high performance polymer composites and until recently were one of the major fibres; they are now being replaced by carbon and silicon carbide, which are cheaper. The specific properties of boron filaments are comparatively good and highlight their attractive performance in relation to other fibres.

Fabrication methods for boron filaments

The commercial production of boron filaments is based exclusively upon the chemical vapour deposition process (CVD) first reported in 1959. Boron is deposited from the vapour phase onto a fine clean refractory metal filament (normally tungsten) with a diameter $\simeq 12\,\mu$m which acts as the substrate for the deposition. Either boron halide or boron hydride are used to transport boron to the substrate filament. In the halide system, hydrogen is used to reduce the halide to boron. In the hydride system, thermal decomposition is used at reduced pressures. The process is outlined schematically in Fig. 4.

For the halide reduction process, the substrate wire temperature is normally in the range 1000–1300°C. The temperature chosen is a compromise between slow deposition rate at low temperatures and excessive crystal grain growth at high temperatures. Excessive grain growth reduces the filament strength. Precise process details, especially the deposition temperature, depend upon the purity of the reactant gases, which also affects the structure of the boron deposit. However, the concentration of the halide does not appear critical. Residence time in the reaction chamber is $\simeq 1$ min and production speed is several metres/min giving a filament of about 100 μm in diameter. Commercial production requires the equivalent of many hundreds of single reactions to produce tonnage output per annum. The hydride decomposition process is

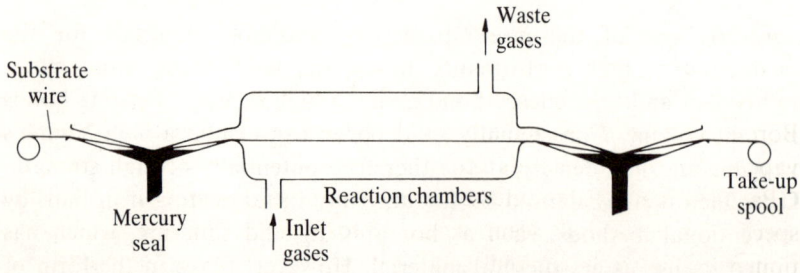

Figure 4 Boron fibre manufacture (schematic) (redrawn from Morley[1])

economically attractive but is inherently more hazardous and the strength properties are lower; it is no longer used[2].

Structure and mechanical properties of boron filaments

Boron deposits on the substrate wire can exist as one of several crystalline modifications dependent on reactant gas purity and temperature. Sections of the fibre show up a cone-like growth of boron grains. These cones are preferentially nucleated on the substrate surface, as evidenced by preferred growth from scratch marks on the tungsten wire. To obtain a homogenous boron structure, the tungsten surface needs to be clean and uniform. The growth of the cones gives a characteristic nodular appearance to the boron fibre surface. These cone/grain structures, typically 20 μm in size, themselves have a fine structure indicating a boron crystal size $\simeq$ 20 Å, normally the rhombohedral form. The fine structure and high density (low porosity) contribute to the high strength and stiffness of the filament.

During the reaction process, boron and tungsten react at the substrate wire surface to produce tungsten boride. Volume increases associated with boride formation give rise to internal stresses. These stresses form radial cracks in the interior of the fibre.

The surface roughness introduced by the nodular surface formation reduces the fibre strength and improved strength can be achieved by chemical polishing of the surface. For good quality boron filaments, surface flaws appear to be the strength controlling parameter.

Boron filaments exhibit good stability in normal gaseous environments and good thermal stability up to $\simeq$ 1000°C. Their hardness limits damage sensitivity. At high temperatures, boron reacts with molten metal, e.g. aluminium, and to overcome this problem, the boron is coated with a final layer of SiC by CVD. Some filament grades are produced like this.

Summary

Boron filaments with a diameter $\simeq$ 100 μm are produced by chemical vapour deposition methods and achieve tensile strengths in the range 2–4 GPa and a Young's modulus of 380 GPa. These filaments have attractive specific properties but seem expensive to manufacture. Future applications are likely to be limited.

References

1. J G Morley, *High Performance Fibre Composites*, Academic Press, London, 1987, pp. 38–45.
2. J-O Carlsson, *J. Mater. Sci.* **14**, 255 (1979).

1.5 Boron nitride and other covalent ceramic fibres

R J P EMSLEY and J H SHARP

The production of ceramic materials from organometallic polymer precursors has been possible since the early 1960s.[1,2] Popper and coworkers examined a range of polymers (unspecified) and their conversion into ceramics such as BN, AlN, Si_3N_4 and SiC. There is, however, no report of any attempt to produce ceramic fibres. Such fibres based on SiC are now commercially available (Section 1.15) and others based on the Si–C–N–O system or on silicon nitride are close to commercial exploitation (Section 1.16). There are several other covalently bonded ceramics which must, at least in theory, be capable of fabrication in the form of high temperature fibres but have not yet approached the stage of commercial development. The most likely candidates seem to be BN, B_4C and AlN; each of these systems will be discussed briefly.

Boron nitride is of interest as a potential ceramic fibre because of its close structural similarity to graphite. Early attempts to produce boron nitride fibres via the polymer precursor route involved the reaction of borazine derivatives and tristrialkylaminoboranes, $B(NHR)_3$, either together or individually.[3] Carefully controlled thermal treatment of these materials gave resins which could be melt spun then, on pyrolysis in an ammonia atmosphere, they yielded boron nitride fibres. Unfortunately, attempts to reproduce this work gave rise to infusible materials or boron nitride contaminated with carbon.[4] Paliorek *et al.*[4] attempted to design a polymer that would lose its side groups easily to yield pure boron nitride. β-triamino-N-tris(trimethylsilyl)borazine was chosen as the potential precursor, but it was found to undergo condensation on prolonged standing at room temperature and to form dimers and tetramers at 135°C. After various heat treatments in an atmosphere of ammonia, a pure white BN was formed. Fibres were produced by this route but had to be pyrolysed slowly over the temperature range 50–1000°C to avoid melting the fibres before crosslinking had fixed their shape.

The production of polymeric precursors for boron carbide, B_4C, has proved to be difficult. To date, boron carbide fibres have been made by only one method,[5] the reaction of decaborane, $B_{10}H_{14}$, with an organic diamine. Short fibres have been produced, which retained their shape on heating to 1000°C under argon to yield black ceramic fibres with a diameter of about 4 μm.

Few attempts have been made to produce aluminium nitride ceramics from polymer precursors and so far only one has been reported to yield ceramic fibres.[6] The polymer was synthesized by the reaction of ammonia with triethylaluminium warmed up with a further small amount of $AlEt_3$. Fibres were melt spun and cured with ammonia. Progress in this field is limited by the hazardous nature of organo-aluminium compounds.

References

1. F W Ainger, J M Herbert in *Special Ceramics*, ed. P Popper, Academic Press, New York, 1960, p. 158.
2. P G Chantrell, P Popper in *Special Ceramics*, ed. P Popper, Academic Press, New York, 1965, p. 87.
3. I Taniguchi, K Harada, T Maeda, Japan, Kokai (Patent), 76 53,000, 11 May (1976).
4. K J L Paliorek, D H Harris, W Krowe-Schmidt, R H Kratzer in *Ultrastructure Processing of Ceramics, Glasses and Composites*, Vol. III, eds D A Mackenzie, D Ulrich, John Wiley & Sons, New York, 1988, Ch. 5.
5. D Seyferth, W S Rees, *Mater. Res. Soc. Symp. Proc.*, **121**, 449 (1988).
6. T R Baker, J D Bolt, G S Reddy, D C Roe, R H Staley, F N Tebbe, A J Vega, *Mater. Res. Soc. Symp. Proc.*, **121**, 471 (1988).

1.6 Carbon fibres from PAN – preparation and properties

F R JONES

Preparation

Carbon fibres are manufactured from three different precursors: polyacrylonitrile or PAN, rayon and pitch. PAN was the original precursor for high strength carbon fibres[1] and is still the most important. In practice, PAN is typically a copolymer with 6 mole per cent methyl acrylate and 1 mole per cent itaconic acid. It is prepared by polymerization in aqueous sodium thiocyanate solution and carefully filtered. The dope is wet spun under 'clean room' conditions to produce the so-called special acrylic fibre (SAF) with a well controlled spin-stretch factor. The higher the spin-stretch factor, the higher the modulus of the final carbon fibre.[1] The production of carbon fibres involves a number of different stages, as illustrated in Fig. 5.

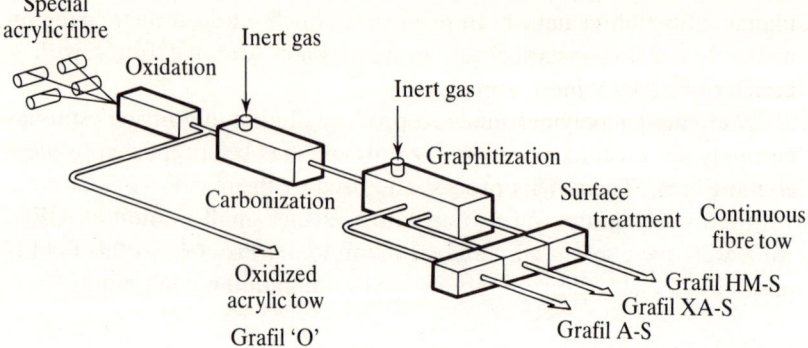

Figure 5 Carbon-fibre-manufacturing process from polyacrylonitrile (after Hysol-Graphil Co. Ltd)

Oxidation of the SAF to stabilize the fibre form during carbonization is carried out in an oxygen atmosphere at 200–220°C, under constraint, so that the molecular orientation induced during stretching is maintained. The reaction of polyacrylonitrile with oxygen is preceded by a self-induced thermal polymerization of the pendant nitrile groups. This forms ladder polymers. The itaconic acid comonomer acts as an initiator of the cyclization reaction whereas the methylacrylate comonomer reduces the glass transition temperature of the polymer, lowering the cyclization temperature and facilitating oxygen diffusion.

Since PAN is atactic, with short sequences of isotactic pendant nitrile groups, the cyclized sequences within the ladder polymer are limited to 5–15 groups. Furthermore where molecular orientation is incomplete, cyclization reactions involving syndiotactic groups will cause the ladder polymer to take on a curved rather than planar ribbon-like form. These cyclized rings are puckered and subsequent reaction with oxygen leads to aromatization and the formation of a planar aromatic structure. Oxygen-containing groups are also inserted into the rings. Together with the planar molecular structure, they promote the condensation of graphitic nuclei in the carbonization and graphitization stages. These stages are carried out in an inert atmosphere over the temperature range 400–2500°C. During polymerization into a graphitic structure, non-carbon elements are evolved as gases, such as H_2, N_2 and HCN, and oxygen-containing products. This process is illustrated below.

As can be appreciated from this mechanism, the degree of graphitization is determined by the extent of reaction which in turn is a function of the maximum temperature reached by the fibres. Several grades of carbon fibre have consequently been classified as a result of the moduli and strengths achieved at particular carbonization temperatures.

The fibres are surface oxidized (generally electrochemically) to promote adhesion to matrix resins and then coated with a sizing resin, usually an aqueous dispersible epoxy resin to protect them during handling.

The strength of PAN based carbon fibres

In the original work of Watt[1] and as shown in Fig. 6, the strength of the fibres was shown to reach a maximum at 1500°C – the so-called Type II or high strength (HS) fibre. Since the modulus increased continuously

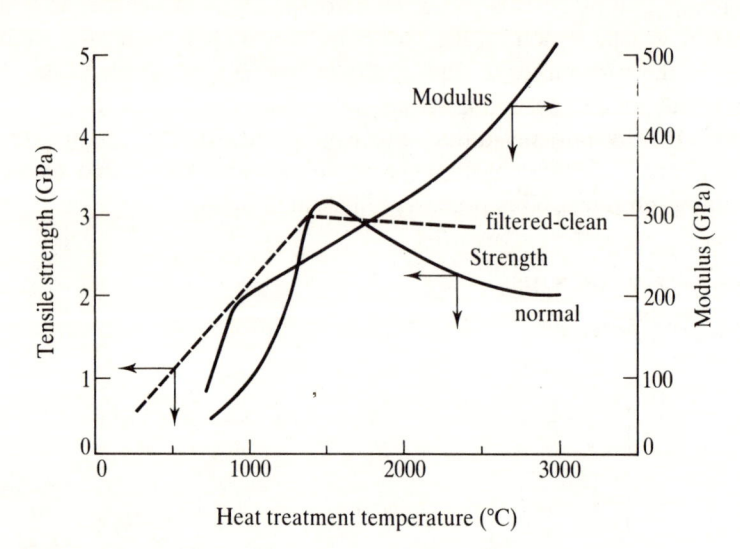

Figure 6 The effect of heat treatment temperature and clean conditions on the strength and modulus of carbon fibres

with heat treatment temperature, a high modulus (HM) or Type I fibre was designated. This had experienced a temperature in excess of 2500°C. Following the identification that the strength-limiting flaws resulted from 'particulate impurities'[2,3], the development of special acrylic fibre (SAF) led to increased strength for fibres subjected to 1000°C (Type A). These differences are illustrated by the work of Moreton.[3] As described in Section 1.7, these carbon fibres have a turbostratic form rather than a well-defined graphitic structure.

Johnson[2] has demonstrated that the critical flaw size for carbon fibres is greater than the crystallite dimensions determined by X-ray diffraction and TEM studies so that the misorientation of the graphite crystallite ribbons cannot account for the relatively low strength. The strength-controlling flaws have been shown to originate from catalysed graphitization around inclusions. This leads to sufficient continuity within the misoriented graphite crystal for the formation of a critical flaw. As a result, improvements in fibre strength have occurred continuously since their commercial inception in 1967[4].

With the recognition that flaw density dominated the strength of the fibres, a small degree of hot stretching was found to increase the strength and modulus as a result of improved orientation and a reduction in fibre diameter from 7 to 5 μm. These have been designated intermediate modulus (IM) fibres. Improvements in fibre performance can be seen in Table 4, which compares the strength/moduli of various commercial

Table 4. Typical properties of commercial PAN based carbon fibres

		Strength, σ_{uf} (GPa)	Modulus, E (GPa)	Failure strain, ϵ_{uf} (%)
Graphil (Mitsubishi)	XA-high strength	3.0–3.3	220–240	1.3–1.4
	XA-high performance	3.3–3.6	220–240	1.4–1.5
	XA-high strain	3.7+	220–240	1.5–1.7
	IM	2.9–3.2	280–300	1.0
	HM	2.5–3.1	350–405	1.3–1.4
Early	Type I (HM)	2.3–2.6	330–350	0.7
Early	Type II (HS)	2.7	250	1.0
Magnamite (Hercules)	XVHMS	1.9	440	0.43
	HMV	2.8	380	0.74
	HMS4	2.5	340	0.74
	IM7	4.7	280	1.7
	IM6	4.4	275	1.6
	AS6	4.1	240	1.7
	AS4	3.8	235	1.6
	AS2	2.7	230	1.2
	AS1	3.1	230	1.3
Torayca (Toray/Soficar)	M50	2.45	490	0.5
	M46J	4.2	450	0.93
	M46	2.35	450	1.1
	M40J	4.4	390	1.1
	M40	2.75	390	0.7
	M30	3.9	295	1.3
	T1000	7.1	295	2.4
	T800H	5.6	295	1.9
	T400H	4.4	250	1.8
	T300J	4.4	235	1.9
	T300	3.5	230	1.5
RK (RK Carbon)	HM	>2.3	330–350	>0.7
	IM	>2.5	280–300	>0.9
	35	>3.5	220–240	>1.6
	30	>3.0	220–240	>1.4
	25	>2.5	215–240	>1.2
Thornel (Amoco)	T50	2.4	393	0.6
	T40	5.65	290	1.95
	T650/42	5.0	290	1.7
	T500	4.0	245	1.6
	T300	3.65	230	1.6
Tenax (Akzo)	HM45	2.15	440	0.5
	HM40	2.25	390	0.6
	HM35	2.35	360	1.5
	IM500	4.8	300	1.6
	IM400	4.0	295	1.4
	HTA	3.5	240	1.5
	ST3	4.3	240	1.8

carbon fibres. The highest modulus fibre on record is 827 GPa for the graphitized pitch fibre Thornel P120 (Amoco Performance Products). The fibre with the highest reported strength (7 GPa) is Torayca T1000, which is PAN based.

One company, RK Carbon Ltd, specializes in utilizing textile acrylic precursor fibres for carbon fibre manufacture. In this way they are able to supply the lowest cost product at a small performance penalty. It is mainly marketed for industrial applications, e.g. carbon–carbon composites, rather than high performance aerospace applications where 20,000 filament tows have less consequence than 6–12,000 filament tows required for the high performance laminated composites.

Conclusions

PAN precursors give rise to a range of carbon fibres suitable as high strength/high modulus reinforcements for reinforced plastics. The properties of current commercial products are summarized in Table 4. For a more complete listing see reference 5.

References

1. W Watt in *Handbook of Composites*, Vol. 1, Strong Fibres, eds W Watt, B V Perov, Elsevier, Amsterdam, 1985, Ch. 9.
2. D J Johnson in *Chemistry and Physics of Carbon*, ed. P A Thrower, Marcel Dekker, New York, 1987.
3. W N Reynolds, R Moreton, *Phil. Trans. R. Soc. London*, **A294**, 451 (1980).
4. J-B Donnet, R C Bansal, *Carbon Fibres*, Marcel Dekker, New York, 2nd edn, 1989.
5. D R Lovell in *Composite Materials in Aircraft Structures*, ed. D H Middleton, Longman, Harlow, 1990, Ch. 3.

1.7 Carbon fibres from PAN – structure

D J JOHNSON

Introduction

It is now well understood that improvements in tensile strength of PAN based carbon fibres have come from continuing efforts to decrease flaws, arising from catalytic graphitization around inclusions originating in the precursor or from surface contamination, together with stringent control of the basic fibre structure, that is homogeneity, degree of orientation and crystallization.[1] It is now possible to identify the characteristics of structure needed for good properties, either predominantly for modulus

(HM types), for strength (HS types), or for an optimum combination of both modulus and strength (IM types).

Macrostructure

The macrostructure of any fibre is most easily observed in the light microscope (LM) or the scanning electron microscope (SEM); however, as carbon fibres do not transmit light unless sectioned, the SEM is used for routine examination of carbon fibres. PAN based carbon fibres have a highly convoluted surface and after failure in tension exhibit a typical 'brittle failure' fracture face with a smooth glass-like zone from which radiate striations in the so-called 'hackle' zone. Flexural and compressional failure give rise to buckling of the fibres and characteristic fracture faces with a clearly defined compression zone are observed.[2]

It is often stated that PAN based carbon fibres have a fibrillar macrostructure. In natural fibres such as wool and cotton, or high performance aramid, fibrils can readily be identified in situ or in breakdown products; this is not the case with PAN based carbon fibres. After failure in tension or compression, longitudinal splitting of aramid fibres is noticeable; this is not the case with PAN based carbon fibres, and we can state that these fibres do not have a fibrillar structure.

Although it is generally held that perfectly circular fibres give maximum tensile strength, multilobal fibres have been shown to improve both the tensile and compressive properties of carbon fibres. Control of the shape of PAN based carbon fibres may be more readily obtained following the introduction of a new melt spinning process using excess water.[3]

Microstructure

In order to observe the microstructure of carbon fibres it is necessary to prepare ultrathin specimens by sectioning or thinning techniques. Then, in the transmission electron microscope (TEM), it is possible to use bright field, dark field, and lattice fringe techniques to image the organization of the layer planes. In transverse sections (which are extremely difficult to prepare successfully) lattice fringe images reveal the complex morphology of the layer planes.[3] In HM fibres, layer planes follow the convolutions of the surface as shown in Fig. 7; there are no indications that the fibres are composed of stacks of parallel fibrils. The layer planes are highly folded forming complex interdigitating systems.

Longitudinally, the lattice fringe observations suggest a highly interlinked structure, particularly in the HM fibres. Fig. 7 is a good representation of the longitudinal structure.[1] Skin–core heterogeneity has been observed in HM fibres, but homogeneity is an essential element for

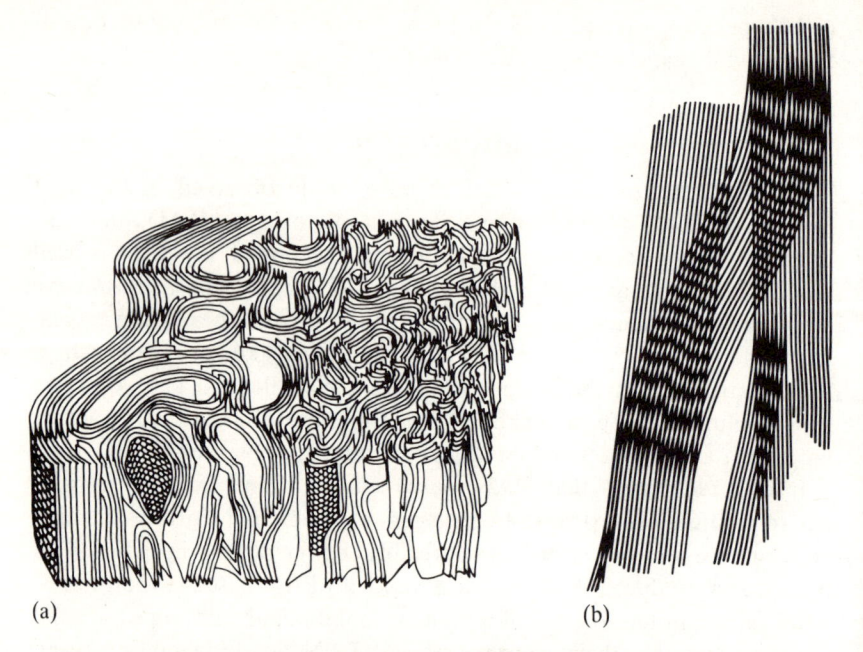

Figure 7 (a) Schematic three-dimensional representation of structure in HM· type PAN based carbon fibres. IM and HS types have a more disordered structure. (b) Schematic representation of interlinked longitudinal structure in HM type PAN based carbon fibres[4]

high strength, as the skin region can contain flaws from which cracks develop.[3] The basic interlinking nature of the layer planes ties together adjacent crystallites in the PAN based carbon fibres. When fibres are placed under tension, a build-up of shear stress develops in misoriented crystallites which cannot be relieved by slippage as the layer planes are pinned. Failure occurs by rupture of the layer planes with a crack spreading at right angles to the fibre axis as shown in Fig. 8. This is the Reynolds and Sharp mechanism of failure in PAN based carbon fibres.

If there is sufficient continuity of crystallites so that the crack exceeds the critical size, then catastrophic failure of the fibre occurs. Clearly, if the extent of the crystallites is limited, then a higher stress to failure can be sustained. Fibres containing graphitic inclusions caused by catalytic graphitization around impurity particles will fail at low levels of tensile stress because a crack, once initiated, can travel around the wall of the inclusion, thereby exceeding the critical size. Even when there are no obvious flaws in the fibre, it is necessary to limit the size of the crystallites. But if the crystallite size is made too small, the fibre will have poor orientation and an unacceptable modulus. The optimum structure is obviously a compromise.

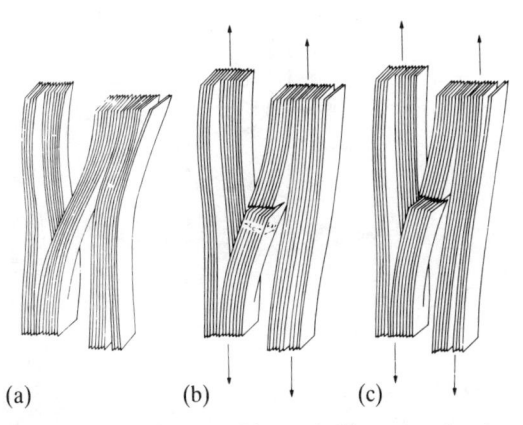

Figure 8 Representation of Reynolds and Sharp mechanism of tensile failure in HM type PAN based carbon fibres

Structural parameters

Wide-angle X-ray diffraction patterns of PAN based carbon fibres indicate that the roughly parallel layers of carbon atoms do not have perfect register with each other, they are said to comprise a turbostratic graphite structure. The difference between the perfect three-dimensional stacking of layer planes and the out-of-register turbostratic stacking is shown in Fig. 9. A measure of graphitization can be evaluated from the interlayer spacing $c/2$. If the value of $c/2$ for a perfect graphite is 0.335 nm, and for a very imperfect graphite is 0.350 nm, then the relative degree of interlayer (intracrystallite) disorder can be evaluated from this equation.

$$D_c = 1 - \left(\frac{0.350 - c/2}{0.350 - 0.335}\right)$$

A PAN based carbon fibre with very high strength Torayca T1000 shows little evidence of three-dimensional packing, $D_c = 86.7\%$, whereas the ultrahigh modulus, mesophase pitch based fibre Thornel P120 is almost perfectly graphitized, $D_c = 6.7\%$.

Values of the main characterization parameters of PAN based carbon fibres are provided by evaluation from wide-angle X-ray diffraction traces. The stacking size, L_c, the apparent crystallite length, $L_{a\parallel}$, and the apparent crystallite width $L_{a\perp}$, can be determined from the widths of the 0 0 2, 1 0 0 meridional and 1 0 0 equatorial peaks in the X-ray diffraction

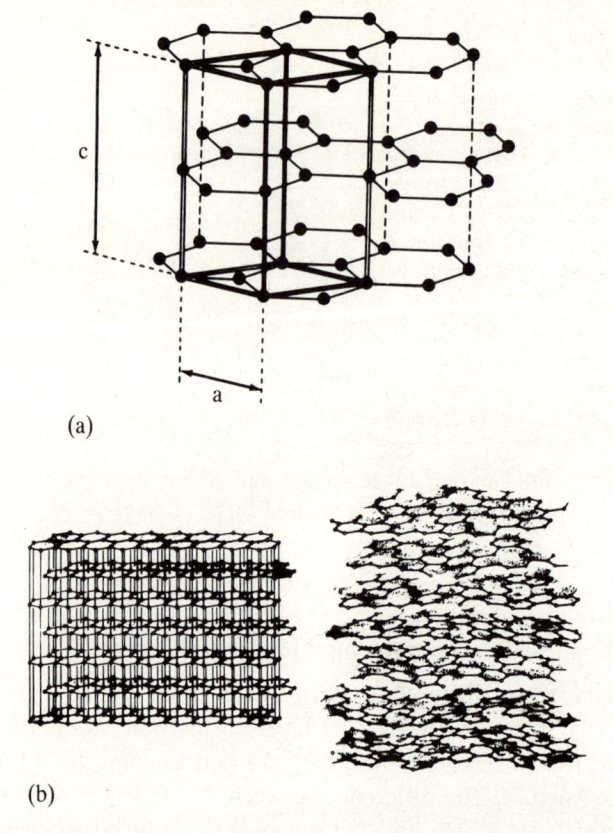

Figure 9 (a) Unit cell of graphite, a = 0.2456 nm, c = 0.6708 nm. (b) Comparison of 3-D graphite lattice with the turbostratic graphite structure

traces. A measure of orientation, Z, can be obtained from an azimuthal scan around the 0 0 2 arc.[5] Some typical values of characterization parameters for the main fibre types are given in Table 5; HM and IM are developmental fibres, HS is T1000.

Table 5. Characterization parameters of typical PAN based carbon fibres

Fibre	$c/2$(nm)	L_c(nm)	$L_{a\|}$(nm)	$L_{a\perp}$(nm)	Z (deg.)
HM	0.340	5.9	9.8	8.0	20.1
IM	0.345	1.9	4.8	7.2	29.8
HS	0.348	1.7	2.9	5.2	31.5

References

1. D J Johnson, *J. Phys. D: Appl. Phys.*, **29**, 286 (1987).
2. D J Johnson in *Introduction to Carbon Science*, ed. H. Marsh, Butterworths, London, 1989, Ch. 6.
3. D D Edie, E G Stoner in *Carbon-Carbon Materials and Composites*, eds J D Buckley, D D Edie, NASA Reference Publication 1254, Washington, DC, 1992, Ch. 3.
4. S C Bennett, D J Johnson, *Carbon*, **17**, 25 (1979).
5. D J Johnson, *Phil. Trans. R. Soc. London*, **A294**, 493 (1980).

1.8 Carbon fibres from PAN – surface treatment

F R JONES

Newly carbonized fibres do not adhere adequately to polymeric matrices to provide efficient reinforcement for optimum composite performance. An essential stage of fibre manufacture is a surface treatment to optimize fibre–matrix adhesion. While inferior interfacial adhesion can lead to poor shear properties, overtreatment can render the composite brittle.

Carbon fibres, from acrylic or PAN precursors, are defined by the heat treatment temperature used for carbonization. Thus Type I (high modulus, HM) fibres have experienced 2400°C and Type II (high strength, HS) 1500°C. With special acrylic fibre precursors and cleaner processing, the presence of strength limiting flaws could be reduced and lower carbonization temperatures (1000–1200°C) could be employed. In the latter Type A fibres, the graphitic structure is much less developed, therefore the surface microstructure is strongly dependent on the previous thermal history. For example, Type I fibres exhibit a core–sheath structure, with a high degree of order to the circumferential graphitic planes at the fibre surface. As a consequence, the conditions required to 'functionalize' the surface of the differing fibre types tend to be more vigorous for the HM fibres.

A number of surface modifications have been examined but anodic oxidation in an aqueous electrolyte (e.g. ammonium bicarbonate) is generally preferred. This is largely a result of the control which can be achieved with electrochemistry as much as with the convenience of adding an electrochemical bath to the production line. A typical unit is schematically given in Fig. 10. The degree of oxidation for a particular fibre type can be quantified by reference to the current density (typically 10–100 Cm^{-2}). However, much of the research is reported in terms of a fraction of the commercial surface treatment, which is set at 100%. Variation in surface treatment can be normalized accordingly by

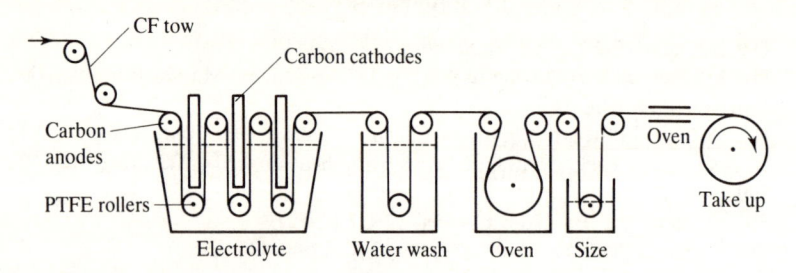

Figure 10 Typical arrangement for electrochemical surface oxidation of carbon fibres

adjusting the current density proportionally. Plasma oxidation is also employed but has not found general commercial acceptance.

Surface microstructure, microchemistry

Surface characterization of oxidized carbon fibres has been undertaken using various techniques. It is well established that the surface oxygen concentration increases, as shown in Fig. 11. The chemical state of the oxygen has been the subject of many papers but as shown in Fig. 11 a significant proportion exists as carboxylic acid groups. The remaining oxygen is in the form of adsorbed water, phenolic hydroxyl and/or

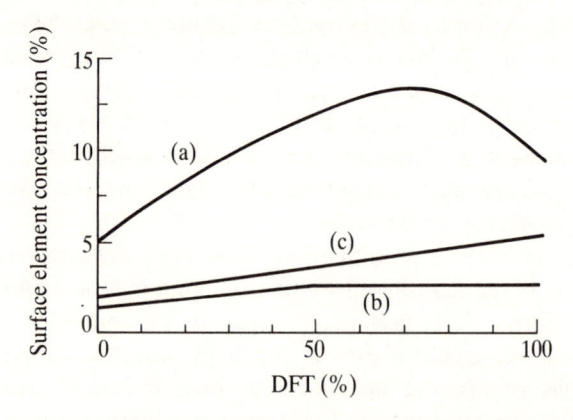

Figure 11 The surface concentration of a) oxygen atoms, b) oxygen atoms combined as carboxylic acid groups, c) adsorptive sites, as a fraction (in %) of 40,000 carbon atoms contained in a graphitic segment $\simeq 3.5$ nm thick DFT is the degree of fibre surface treatment[1]

ketonic groups. It is generally accepted that these functional groups are located at the edges of the basal planes in the graphite lattice. No correlation between surface chemistry and interfacial bond strength has been proven. Neither has a significant increase in surface area during surface treatment been detected. Denison and Jones[1] approached this problem by establishing a model segment for quantification of the surface derivatized groups. In this way it was possible to show that the surface of treated HS fibres was saturated with acid groups and their average separation was smaller than theoretically possible. As a result, they postulated that the functional groups were located within the micropores in the surface. Detailed study of HM fibres confirmed that the concentration of carboxylic acid groups rose to a maximum with surface treatment, while the dimensions of the micropores increased. It was demonstrated that thermal desorption of the oxygen-containing groups in vacuo at 1400°C did not always lead to a reduction in interlaminar shear strength, and that the measured acidity was not affected, i.e. the acid groups reformed by reaction with water during analysis. It was then concluded that active carbons, devoid of functional groups, could also be involved in adhesive bond formation.[2] It has also been observed that molecular-thin layers of epoxy resins can be chemically bound to the surface.[3,4] A schematic model of the surface of HM carbon fibres before and after surface treatment is given in Fig. 12. The spacing of these micropores is considered to be determined by the crystallite size distribution. The micropore dimensions of ≈ 0.7 nm demonstrate that they occur at the twist boundaries between these crystallites. Transmission electron microscopy of embedded fibres has also demonstrated that the fibre–matrix adhesion occurs predominantly at the edges of the exposed basal planes. The different surface microstructures for Type A and HM fibres have led to different interpretations, but it is clear the adhesion mechanism involves contributions from functional group chemistry, stereochemistry of the resin molecules and the presence of micropores whose dimensions must be large enough to allow access to the edges of the basal plane where the reactivity exists. For Type A fibres the presence of the graphitic skin is uncertain and consequently these fibres may have varying degrees of almost perpendicularly organized basal planes, exposed by surface treatment (Fig. 12b).

After surface oxidation the fibres are coated with a polymeric size, usually an epoxy resin. Whether this resin coating can be subsequently dissolved efficiently into the matrix is unclear. This aspect of the fibre–matrix interaction in carbon fibre composites is underresearched but the possibility of a matrix–size interface also needs to be considered. Recent arguments support the beneficial presence of an interphase.

(a)

6-8 nm

Axis

Skin

Core

$-CO_2$

(b)

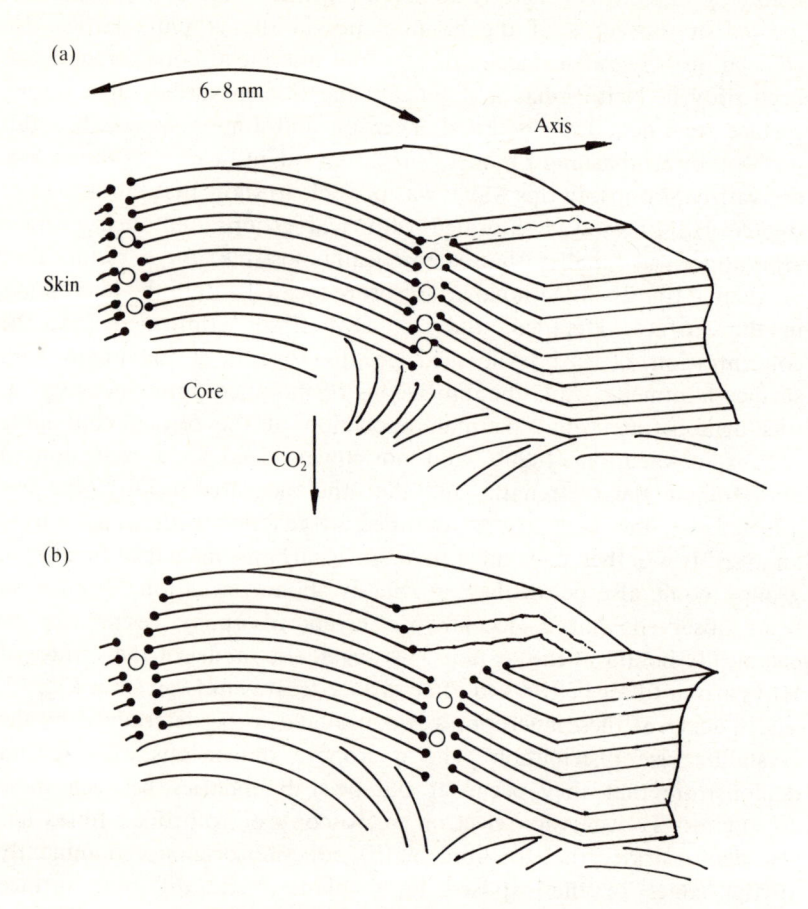

Figure 12 The schematic model of Type I (high modulus) carbon fibre surface (a) before and (b) after electrochemical oxidation, ● adsorptive sites which may or may not have functional groups attached, O adsorbate molecules (e.g. H_2O)

The effect of surface treatment on the mechanical properties of carbon fibre composites

Brittle fibres in a brittle matrix can have appreciable toughness because the cracks can get diverted along the fibre–matrix interface. If the bond is weak, the composite will not support loads in shear or compression, but when the bond is too strong, the material will be brittle.[5] These aspects are illustrated in Fig. 13 where it is seen that the interlaminar shear strength reaches a plateau but the notched tensile strength decreases monotonically with surface treatment. Optimization and careful control

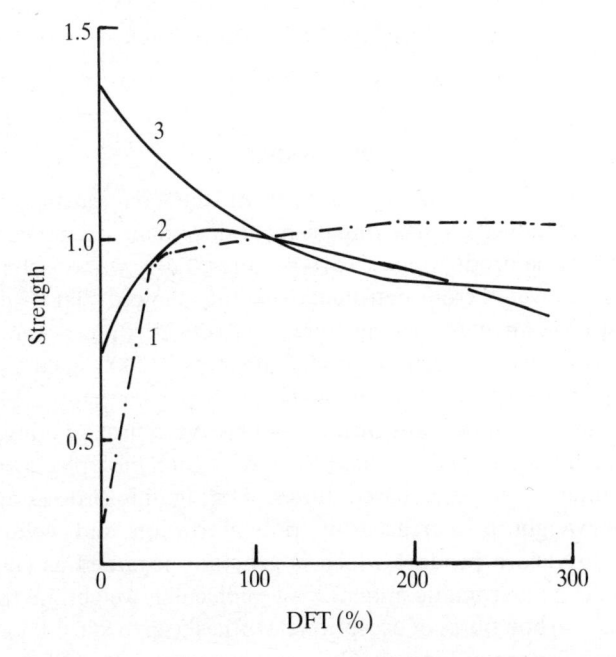

Figure 13 Effect of surface treatment on the mechanical performance of Type II carbon fibre epoxy composites, (1) interlaminar shear strength or a unidirectional composite, (2) impact strength, (3) notched tensile strength, of a $(0°/+45°/0°)s$ laminate. Redrawn from the data given by Dunford et al.[5]

of the interfacial bond strength is therefore essential for each fibre–resin system. It is also apparent from Fig. 13 that test methods other than fracture tests have to be used to accurately assess the interfacial bond strength.

References

1. P Denison, F R Jones, J F Watts, *J. Mater. Sci.*, **20**, 4647 (1985); *Surface and Interface Analysis*, **9**, 431 (1986); **12**, 455 (1988).
2. P Denison, F R Jones, A J Paul, J F Watts in *Interfacial Phenomena in Composite Materials*, ed. F R Jones, Butterworth, London, 1989, pp. 105–110.
3. P Denison, F R Jones, J F Watts in *Interfaces in Polymer, Ceramic and Metal Matrix Composites*, ed. H Ishida, Elsevier, New York, 1988, p. 77–85.
4. C Kozlowski, P M A Sherwood, *Carbon* **25**, 751 (1987).
5. D V Dunford, J Harvey, J Hutchings, C H Judge, RAE Technical Report TR 81096 (1981), HMSO, London.

1.9 Carbon fibres from pitch

B RAND and M TURPIN

Introduction

Pitch, a widely available by-product of the coal gasification and petrochemical industries, has long been considered an attractive, low cost precursor for the production of carbon fibres. Early work with isotropic pitches, as recovered from petroleum refining, showed that whilst fibres could be spun relatively easily, the fibre properties were poor – low strength and modulus – unless expensive, high temperature (2000–2800°C) tension processing conditions were adopted. The discovery in the mid 1960s that isotropic pitches can be transformed, by prolonged heat soaking, into an anisotropic liquid crystalline phase known as pitch mesophase has led to renewed interest in pitch based fibres. During this process the pitch molecular weight is increased by polymerization and volatilization processes to yield a product which is generally regarded as comprising large planar polyaromatic moieties of molecular weight in the range 800–2000.[1] Carbon fibres of exceptional stiffness (up to 900 GPa modulus) can be produced from this mesophase precursor as outlined below.

Fibre production

The mesophase pitch precursor is melt spun in a process shown schematically in Fig. 14. The molten precursor is extruded through a capillary spinneret, typically of 100–500 μm diameter, and is then drawn down and wound onto a rotating bobbin to produce fibres of 10–30 μm diameter. The shearing of the anisotropic mesophase pitch during spinning results in the orientation of the planar polyaromatics to lie along the fibre axis, this is shown greatly exaggerated in Fig. 14. It is also during the spinning process that the fibres develop characteristic transverse textures. These latter processing features are believed to result from vorticity in the threadline due to the radial velocity gradient associated with extensional flow.[2] The resulting radial textures can be classified into three main groups; random, radial and onion skin structures. They are shown schematically in Fig. 15. A wide variety of hybrid structures have been reported. These combine structural features of the three main groups. The vorticity during spinning, and by implication the fibre radial texture, has a marked influence on the mechanical properties of the fibres, but these effects are imperfectly understood at present.

'As spun' fibre is extremely brittle with low strength (typically 40 MPa). Oxidative stabilization is generally used to render it infusible. The fibre is

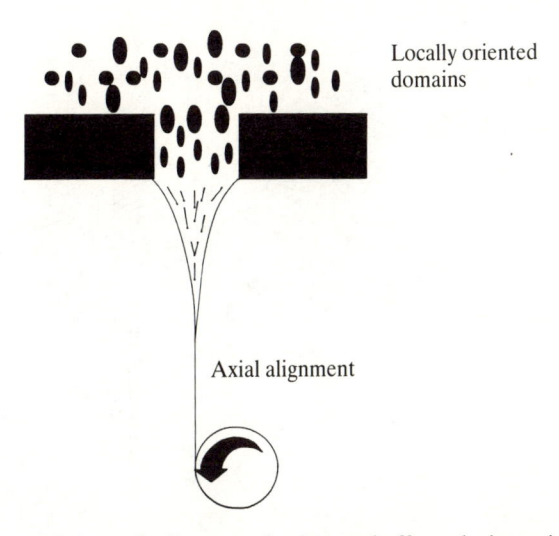

Figure 14 Schematic diagram of orientated effects during spinning

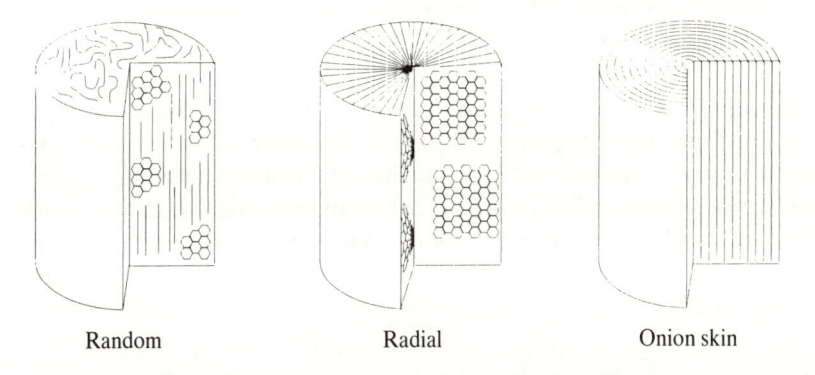

Figure 15 Typical radial textures of carbon fibres

heated to a temperature below the glass transition temperature in an oxidizing atmosphere, whereby oxygen is chemically incorporated into the fibre with resultant crosslinking. In this way the orientational order of the fibre produced during spinning can be retained during the carbonization step.

The fibre is carbonized by heating, generally rapidly, to a temperature in the range 1500–2800°C. At temperatures below 1000°C the predominant reactions lead to the removal of non-carbon atoms, particularly the oxygen incorporated during stabilization, and the

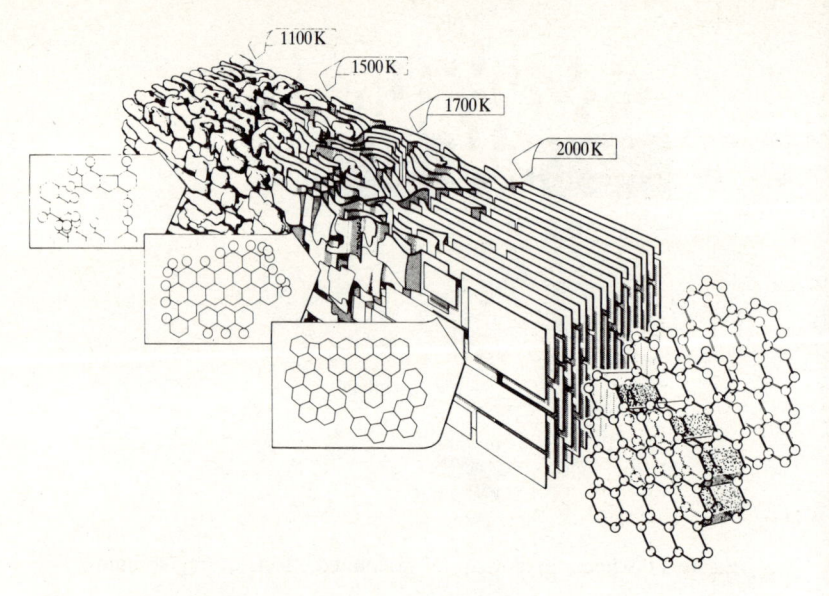

Figure 16 Schematic diagram of the structural changes occurring in mesophase during graphitization. Taken from Marsh and Henendez, 1989[3] by permission of the publishers, Butterworth-Heinemann Ltd, ©

growth of planar polyaromatic moieties. Above 1500°C the polyaromatic layers grow and become extensive. But they become distorted from the true graphitic structure by lattice defects and spatial dislocations parallel to the layer planes. This yields a so-called 'turbostratic' structure. Finally, above 2000°C true graphitization takes place, producing extensive regions of graphitic structure.[3] The carbonization process is shown schematically in Fig. 16.

Fibre properties

Unlike PAN based fibres, those derived from mesophase pitch are able to graphitize resulting in a product which may have a Young's modulus approaching that of graphite (1060 GPa). Indeed commercial fibres are readily available with moduli up to 894 GPa, in contrast to a maximum value of 588 GPa for a PAN based fibre.[4] At present, however, the high modulus of the fibres is offset by a reduction in strength, relative to the PAN based fibres. Presently, the strongest pitch based carbon fibres have strengths up to 4 GPa compared to 7 GPa for PAN. Additionally, pitch based carbon fibres have very poor compressive strengths, a feature that limits their use to tensile applications. The strength and moduli of commercial pitch based and PAN based fibres, available in 1991, are

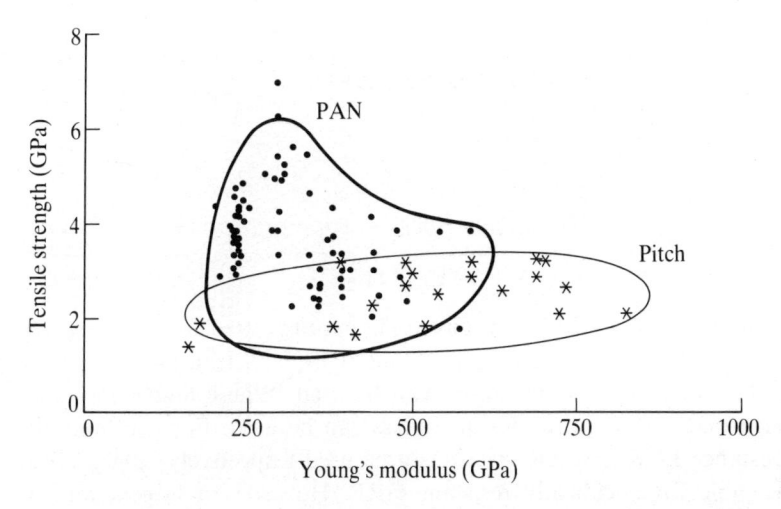

Figure 17 Comparison of the mechanical properties of PAN and pitch based carbon fibres (Data compiled from *Carbon and high performance fibres directory*, 5th edn, Chapman & Hall, 1991)

compared in Fig. 17. Pitch based fibres are generally favoured when stiffness is the desired material property, secondary to strength.

The graphitic nature of mesophase pitch based fibres results in properties that are considered desirable in the aerospace industry. In particular, the fibres have a high axial thermal conductivity, up to 640 W m^{-1}K^{-1} (almost twice that of copper), and are therefore of value in components where rapid heat transfer is important, in brake shoes for example. The fibres also have a negative axial thermal expansivity, typically -1.4 ppm °C^{-1}, which has led to interest in the development of composite materials of excellent dimensional thermal stability for non-terrestrial applications.[5]

At present, the main problem with high performance, pitch based carbon fibres is due to low demand coupled with high processing costs. This results in the price of the stiffest fibres being prohibitive for many applications.

References

1. B Rand, in *Handbook of Polymer Composites*, Vol. 1, *Strong Fibres* eds W Watt, B V Perov, Elsevier, Amsterdam, 1985, pp. 495–575.
2. D M Riggs, US Patent 4,504,454.
3. H Marsh, R Menendez in *Introduction to Carbon Science*, ed. H Marsh, Butterworth, London, 1989, p. 37.

4. *Carbon and High Performance Fibres Directory*, 5th edn, Chapman and Hall, London, 1991.
5. *Carbon and Graphite Fibers: Manufacture and Applications*, ed. M Sittig, Noyes Data Corporation, New Jersey, 1980.

1.10 Glass fibres – type and form

F R JONES

A number of glass compositions are commonly used for the production of glass fibres.[1-3] The most important is E-glass, whereas for specialist applications other compositions can be used. Where higher mechanical performance is required R- or S-glass can be used. For acid and alkali resistance ECR and AR glasses can be used respectively. C-glass is also available for chemically resistant GRP. However, it is generally only available as a veil for the reinforcement of a lining on a structural composite manufactured from E-glass or ECR continuous fibre roving. Some typical glass compositions for fibre drawing are given in Table 6.

The fibres are drawn from a bushing below a molten glass reservoir and immediately cooled in a spray of water. Since the presence of surface defects largely determines the strength of glass fibres, the fibres are immediately coated with a protective aqueous size solution. The size,

Table 6. Typical composition and properties of various glasses used for fibre formation

	Composition (wt %)						
	A	C	E	ECR	R	S	AR
SiO_2	72.0	64.6	52.4	58.4	60.0	64.4	61.0
Al_2O_3	1.2	4.1	14.0	11.0	25.0	25.0	0.5
CaO	10.0	13.4	17.2	22.0	9.0	–	5.0
MgO	2.5	3.3	4.6	2.2	6.0	10.3	0.05
Na_2O, K_2O, Li_2O	14.2	9.6	0.8	0.9	–	–	14.0
B_2O_3	–	4.7	10.6	0.09	–	–	–
BaO	–	0.9	–	–	–	–	–
ZnO	–	–	–	3.0	–	–	–
ZrO_2	–	–	–	–	–	–	13.0
TiO_2	–	–	–	2.1	–	–	5.5
Fe_2O_3	0.3	–	0.4	0.26	–	–	0.5
Specific gravity	2.45	2.45	2.56	2.6	2.58	2.49	2.74
Refractive index	1.512	1.520	1.548	–	–	1.523	1.561
Single fibre tensile strength (GPa)	3.1	–	3.6	3.4	4.4	4.5	2.5
Single fibre tensile modulus (GPa)	72.0	–	76.0	73.0	85.0	86.0	80.0
Softening point (°C)	700	690	850	900	990	1000	860

whose composition is the most commercially sensitive aspect of glass fibre manufacture, is applied by rubber roller before the fibres are wound up on to a bobbin to form a strand. The size or finish is crucial to the handleability of the fibres and their compatibility with matrix resins. This aspect is examined in detail in the following article.

A strand consists of a number of tows. A tow is the unitary element which represents the number of fibres drawn from a single bushing, typically $\simeq 200$. An assemblage of strands is called a roving. In new plants, this process occurs at the bushing and the products, called direct rovings, are used in applications where the incidence of a twist in the roving can be significant. Otherwise, the strands are assembled into rovings in a separate process. A slight twist is incorporated into the roving to improve its handleability. For continuous fibre composites, the selection of conventional or direct rovings depends on the nature of the fabrication process and the degree of perfection required in the fibre alignment. The number of filaments in a roving is defined by the tex, typically 600, 1200, 2400, etc. (1 tex = g/1000 m). Other textile processes are used to prepare woven rovings with differing texture (Fig. 18). These cloths[1,4] offer convenience of drape and multidirectional in-plane reinforcement. The rovings can also be chopped and deposited randomly to produce chopped strand mat (CSM), held together by an additional polymer binder. The binder[2] is chosen to provide controlled wet-out during resin impregnation so that the random orientation of the discontinuous fibres can be maintained. Furthermore, the binder should be selected to suit the application of the material. For example, the environmental durability of CSM based GRP can be strongly affected and for chemical resistance powder bound mat rather than general purpose emulsion bound mat should be chosen.

Continuous random mats (CRM) are also available in which continuous rather than chopped fibre in a random form is bound into a mat with superior drape, i.e. ability to be layed onto a curved mould without serious fibre distortion.

Chopped fibres at differing lengths are also available for incorporation directly into moulding granules and compounds. The chopped fibre length tends to be significantly longer for CSM (at > 20 mm) than those incorporated in moulding materials. Furthermore, the fibres are generally longer for dough moulding compounds (DMC) compared to moulding granules. Apart from the differing consistencies and processing requirements, the fibres in thermosetting materials are generally longer than those incorporated into thermoplastics.

Textile processing has recently become more popular for the production of three-dimensional reinforcements for use in structural RIM and resin transfer moulding (RTM). This aspect is discussed in

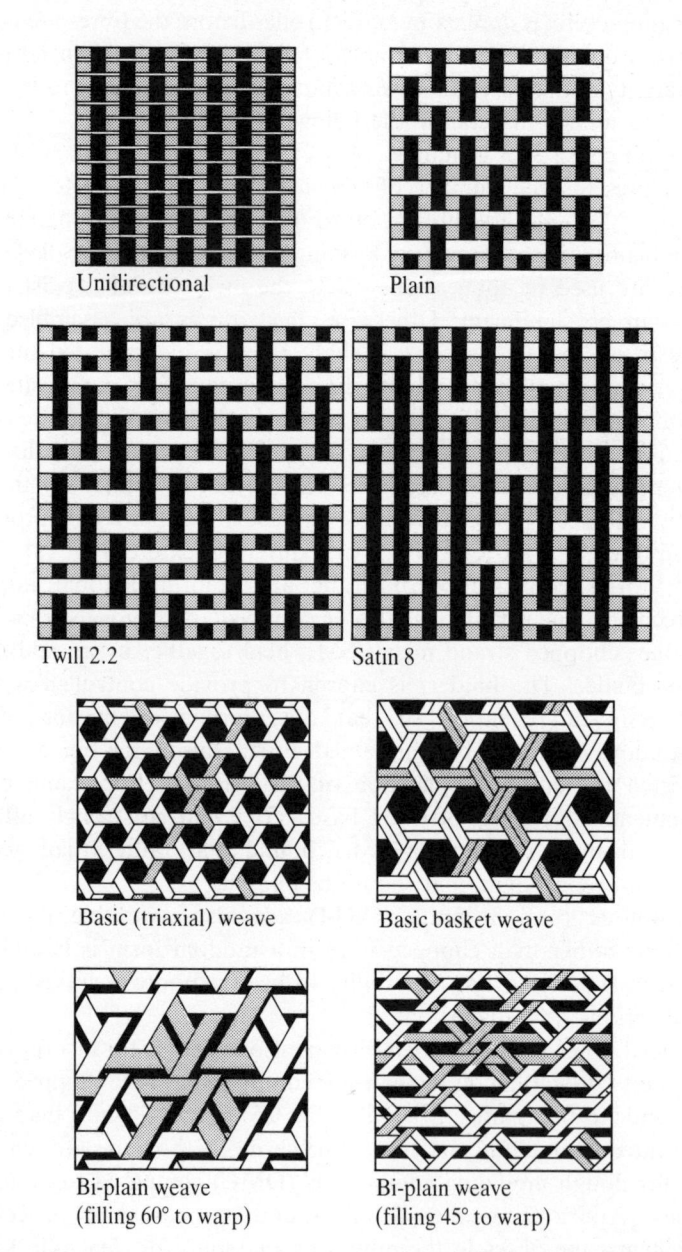

Figure 18 Schematic representation of differing reinforcing fabric weaves (after Lovell, 1990[4])

Glass fibres – type and form

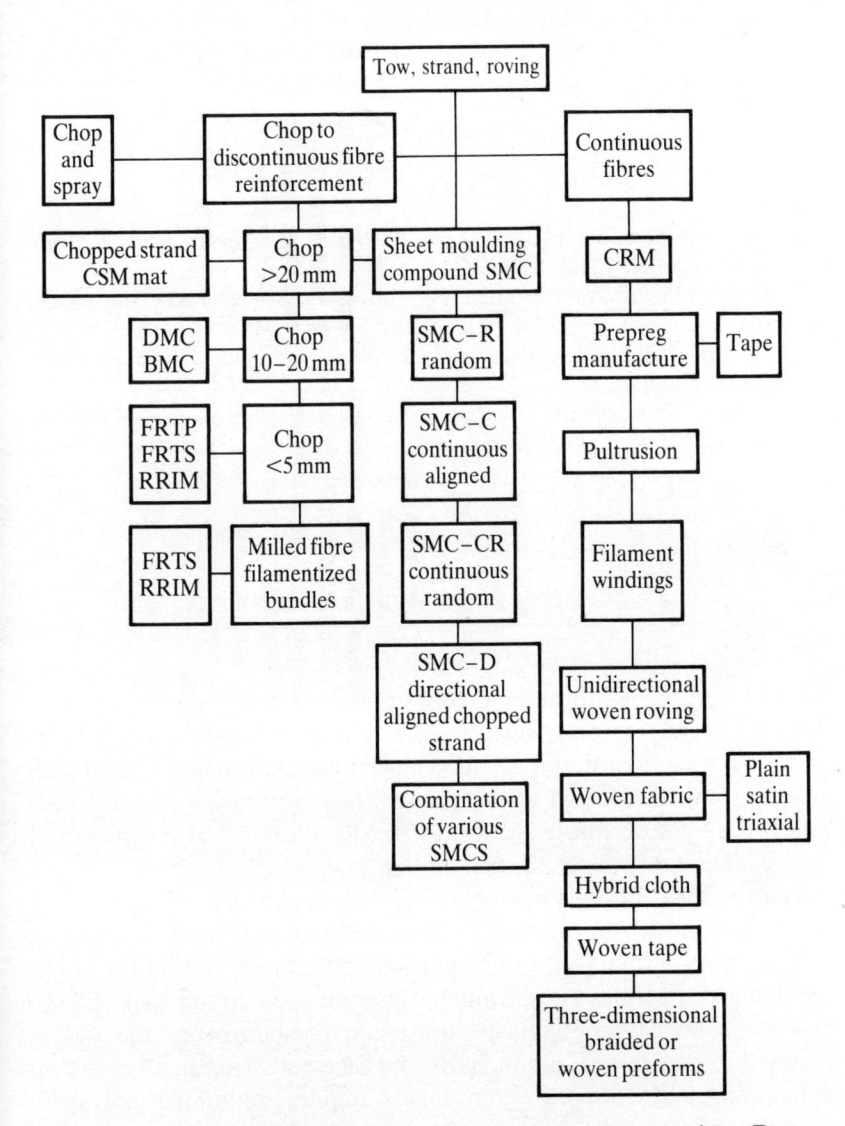

Figure 19 Types and forms of glass fibre products. R = random; D = directional; C = continuous; CR = continuous random, CRM = continuous (filament) random mat, CSM = chopped strand mat, DMC = dough moulding compound, BMC = bulk moulding compound, SMC = sheet moulding compound, FRTP = fibre reinforced thermoplastics, FRTS = fibre reinforced thermosets, RRIM = reinforced resin injection moulding

more detail in Sections 1.12 and 5.13. The nature and form of glass fibre reinforcements are summarized in Fig. 19.

References

1. R G Weatherhead, FRP Technology, *Fibre Reinforced Resin Systems*, Applied Science, London, 1980.
2. J Klunder, *Silenka Service Manual on Glass Fibres*, Silenka BV, Amsterdam, 1970.
3. K L Lowenstein, *The Manufacturing Technology of Continuous Glass Fibres*, Elsevier, London, 1973.
4. D R Lovell in *Composite Materials in Aircraft Structures*, ed. D H Middleton, Longman, Harlow, 1990, Ch. 3.

1.11 Glass fibres – surface treatment

F R JONES

While several glass-forming compositions are used to manufacture reinforcements for composites, E-glass is the most industrially important. The drawn fibres of $\approx 15\,\mu$m diameter are immediately cooled with a water spray and coated with an aqueous size in contact with a rubber roller. The 'size' or 'finish' is crucial to the handleability of the fibres and their compatibility with the resin matrix.

To cater for the variety of composite fabrication techniques, fibres may be formed into woven rovings, knitted preforms or continuous random mats (CRM) or chopped into lengths of >20 mm to <5 mm for use in moulding compounds or assembling into chopped strand mat (CSM). For these textile processes the polymeric component of the size is required to maintain strand integrity and filament strength. The random continuous and chopped strand mats require additional polymeric binders.

The glass fibre 'finish'

The 'finish' is applied as an aqueous emulsion. It consists of:

(i) adhesion promoter or coupling agent,
(ii) protective polymeric size,
(iii) additional polymeric binder (emulsion or powder),
(iv) lubricant.

Table 7. Typical sizing resins and binders for glass fibres

Glass type	Polymeric size	Polymeric binder	Application
E	PVAc	–	General purpose roving
E	PVAc	PVAc emulsion	General purpose CSM
E	Polyester	–	Environmentally resistant GRP
ECR	Polyester	–	Environmentally resistant GRP
E,ECR,S,R	Epoxy	–	High performance composites
E	Epoxy/ polyester copolymer	–	High performance composites with wide range of compatibility
E,ECR	Polyester	Various powder	CSM environmental resistant GRP CSM processing with controlled wet-out, CRM
E	Polyurethane	–	Roving for thermoplastics – short fibre moulding compounds (e.g. nylon)
C	–	Polyacrylate Polystyrene	Reinforcing veils for gel coats, chemically resistant barrier layers

PVAc = polyvinyl acetate, CSM = chopped strand mat, CRM = continuous random mat.

Items (ii)–(iv) impart good handleability and controlled wet-out kinetics with matrix resins and are therefore chosen for compatibility with the fabrication process. For specialist applications, such as environmental resistance, the chemical nature of the 'size' and binder are crucial and selected accordingly. Good economical design can be achieved by combining fibres with differing 'finish' in different laminae. Typical polymer sizings and binders are enumerated in Table 7.

Surface treatment for adhesion

The manufacturing process described in section 1.10 involves bringing hot glass into contact with water immediately before surface treatment. This results in a unique surface chemistry. A comparison of the bulk and surface chemistry of E-glass (Table 8) shows that the surface is silica rich. In addition, it is reported that a multimolecular layer ($\approx$ 20 monolayers) of water is adsorbed by hydrogen bonding through surface hydroxyl groups. The adhesion promoter therefore has the following function:

(i) to displace adsorbed water,
(ii) to create a surface which can be fully wetted with resin,
(iii) to develop strong interfacial bonds between the fibre and matrix

Table 8. Typical bulk (ICP) and surface (XPS) elemental composition for heat cleaned polished E'-glass plate* and as received non-treated, water sized fibres[†]

Element	Uncoated plate		Uncoated fibre	
	Bulk	Surface	Bulk	Surface
Si	22.3	25.1	22.8	22.4
Al	7.4	8.4	7.0	8.5
Ca	16.4	9.6	17.6	8.6
O	49.6	56.9	50.4	60.5
Mg	0.4	–	0.5	–
B	2.1	–	1.4	–
Fe	0.6	–	0.3	–

– Below the detectable limit for X-ray photoelectron spectroscopy (XPS).
* Cast from E'-glass marbles.
[†] Commercial fibres treated only with water during manufacture.
ICP = inductively coupled plasma technique.

which minimize the interfacial free energy, γ_{SL}, and maximize the work of adhesion, W_{SL}, as shown by the Young equation [1].

$$W_{SL} = \gamma_L (1 + \cos\theta) = \gamma_S + \gamma_L - \gamma_{SL} \quad [1]$$

where γ is the surface free energy of the solid (S), liquid (L) and θ is the contact angle. Thus the adhesion promoter has a primary role to interact with both the fibre surface and the matrix and is consequently called a coupling agent. The above requirements are best achieved for most matrices with organosilanes. However, chrome complexes and titanate coupling agents are also available (see Table 9).

The structure of the silane coating

The application of organosilane coupling agents has been described by Plueddemann.[1] More recently, the adhesion mechanisms have been reviewed by Jones.[2]

The silane coupling agents have a simplified structure

```
            R'
             \
              O
               \
    R'—O—Si—R
             /
            O
           /
          R'
```

Table 9. Typical coupling agents for resin based composites

Vinyl	$CH_2=CHSi(OCH_3)_3$
Epoxy	$\overset{O}{\overset{/\backslash}{CH_2CHCH_2OCH_2CH_2CH_2Si(OCH_3)_3}}$
Methacrylate	$CH_2=\underset{\underset{CH_3}{\mid}}{C}-COOCH_2CH_2CH_2Si(OCH_3)_3$
Primary amine	$H_2NCH_2CH_2CH_2Si(OCH_3)_3$
Diamine	$H_2NCH_2CH_2NHCH_2CH_2CH_2Si(OCH_3)_3$
Mercapto	$HSCH_2CH_2CH_2Si(OCH_3)_3$
Cationic styryl	$CH_2=CHC_6H_4CH_2NHCH_2CH_2NH(CH_2)Si(OCH_3)_3HCl$
Cationic methacrylate	$CH_2=\underset{\underset{CH_3}{\mid}}{C}-COOCH_2CH_2-\overset{Cl^-}{\overset{+}{N}}(Me_2)CH_2CH_2CH_2Si(OCH_3)_3$
Cycloaliphatic epoxide	(cyclohexane with epoxide)—$CH_2CH_2Si(OCH_3)_3$
Titanate	$[CH_2=C(CH_3)-COO]_3TiOCH(CH_3)_2$
Chrome complex	(chrome complex structure with $CH_2=C(CH_3)$, COO groups bridging two Cr centers with Cl, ROH, H_2O ligands)

where R' is C_2H_5 or CH_3 and R is a reactive or resin compatible functional group such as vinyl (for unsaturated polyester resins, epoxy or amino for epoxy resins). In the aqueous size, the $>Si-OR'$ group hydrolyses. This is often acid catalysed by adjusting the pH to ≈ 3.5 with acetic acid. As shown in [2] below further condensation of the silanol groups can occur, leading to a complex deposit.

$$R-\underset{OR'}{\overset{OR'}{Si}}-OR' \xrightarrow[-ROH]{+H_2O} R-\underset{OH}{\overset{OH}{Si}}-OH \xrightarrow{-H_2O} R-\underset{O}{\overset{OH}{Si}}-O-\underset{OH}{\overset{O}{Si}}-R \quad [2]$$

Since these reactions occur at the glass surface, there is a competition between surface silanol coupling agent condensation, step growth polymerization and cyclic oligomer formation.

The structure of the deposit has been the subject of chemical analysis over the last decade as sensitive analytical techniques became available. While individual coupling agents (varying R) polycondense to differing degrees, it is generally accepted that three distinct layers or components are deposited onto the glass surface as observed by differing hydrolytic stabilities: a strongly chemisorbed layer at the immediate glass surface with loosely chemisorbed and physisorbed overlayers. The last can amount to a significant proportion of the deposit, 80 monolayers out of 100 monolayers in the case of γ-methacryloxypropyltrimethoxysilane (γ-MPS),[3] and comprises cyclic and linear oligomers. The intermediate layer is a misnomer as its enhanced hydrolytical stability arises from the formation of a three-dimensional network polysiloxane, which is graded towards a higher density at the interface.[4] In the case of γ-MPS, the strongly chemisorbed layer, as well as being bound to the glass surface through siloxane bonds, may be homopolymerized through the functional group R, which is $CH_2=CCH_3-COO-$ in this case.

Mechanism of adhesion at the glass fibre resin interface

Several adhesive mechanisms have been proposed to account for the enhancement of interface dominated properties, such as retained tensile strength after environmental conditioning:

(i) chemical bonding theory,
(ii) deformable layer hypothesis,
(iii) surface wettability hypothesis,
(iv) restrained layer hypothesis,
(v) reversible hydrolytic bonding mechanism.

The simplistic chemical bonding theory is usually cited but cannot explain all the observations, not least the efficacy of γ-aminopropyltriethoxysilane (γ-APS) under conditions when chemical bonding through the amino group cannot be readily understood. Theory (v) allows for reversible reformation of chemical bonds on drying out of aqueous conditioned specimens.

It is now becoming clear that the adhesive bond arises from a combination of chemical bonding and the formation of a semi-interpenetrating network between the polysiloxane and matrix resin. The latter arises because the physisorbed oligomers dissolve into the resin leaving a molecularly porous structure with exposed functional groups into which the matrix resin can diffuse and copolymerize. This structure is schematically illustrated in Fig. 20. The presence of the so-called interphase has been confirmed by scanning secondary ion mass spectrometry[5] and C^{13}NMR.[4] The latter work also confirmed the

Glass fibres – surface treatment

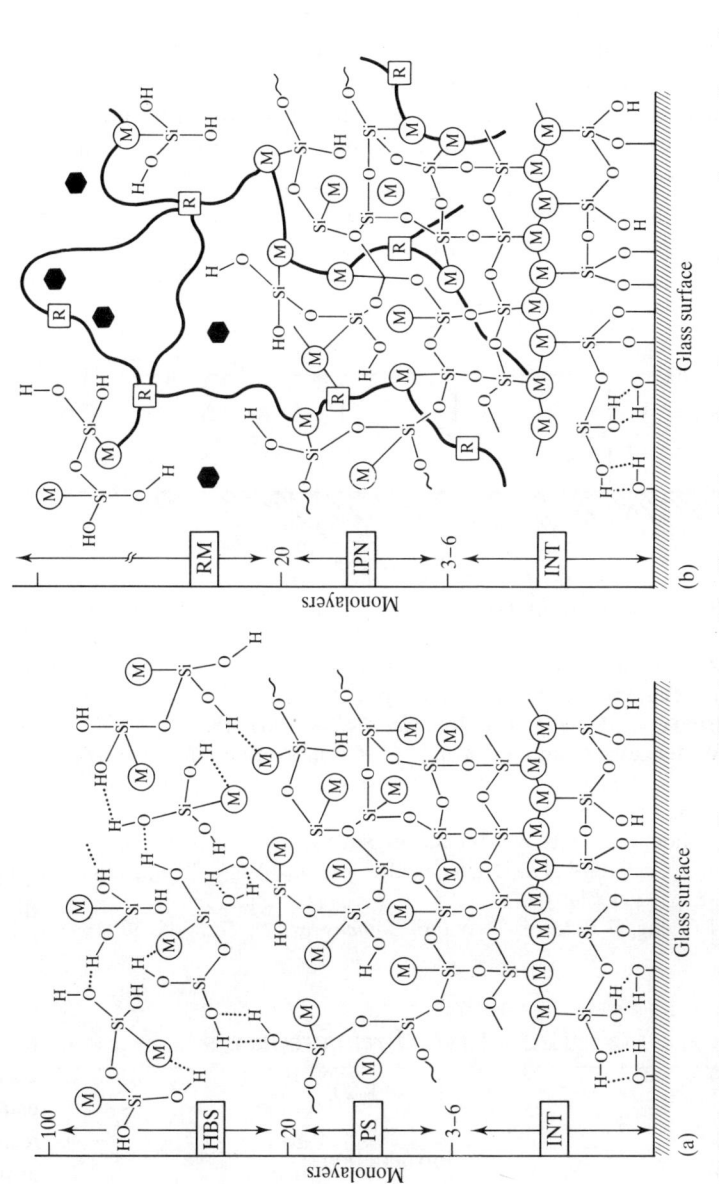

Figure 20 Schematic diagrams of (a) a polysiloxane deposit on glass fibre. M is methacryloxypropyl-trimethoxy silane, (b) composite interface. R-M is the interpenetrating copolymer with the resin matrix (R). ● is dissolved binder and/or size, PS = polysiloxane, HBS = hydrogen bonded oligomeric silanes, IPN = interpenetrating network, INT = interface.

reaction of the amino group in γ-APS with polyamide acid end groups. A consequence is that the concentration of silane in the 'size' emulsion can significantly influence the adhesion. More recently time-of-flight secondary ion mass spectrometry (ToF SIMS) in combination with XPS has demonstrated that the silane deposit can have an enriched aluminium concentration. It is further postulated that basic aluminium hydroxide sites are formed. These promote adhesion through acid–base interactions.[6] A schematic of the hydrolytically resistant silane interfacial layer is given below.

$$\begin{array}{ccccc} NH_2 & NH_2 & & & NH_2 \\ | & | & & & | \\ (CH_2)_3 & (CH_2)_3 & OH & & (CH_2)_3 \\ | & | & | & & | \\ HO-Si-O-(Si-O)_n-Al-O-Si-OH \\ | & | & & & | \\ O & OH & & & O \end{array}$$

References

1. E P Plueddemann, *Silane Coupling Agents*, 2nd edn, Plenum, New York, 1991.
2. F R Jones in *Interfacial Phenomena in Composite Materials '89*, ed. F R Jones, Butterworth, London, 1989, pp. 25–32.
3. H Ishida, J L Koenig, *J. Polym. Sci. Polym. Phys.* **18**, 1931 (1980).
4. L W Jenneskens and D Wang, F R Jones in *Interfacial Phenomena in Composite Materials '91*, eds I Verpoest, F R Jones, Butterworth-Heinemann, Oxford, 1991, pp. 11, 121.
5. J L Thomason, J B W Morsink in *Interfaces in Polymer, Ceramic, Metal Matrix Composites*, ed. H Ishida, Elsevier, New York, 1988, p. 503. See also ref. 2, pp. 171–180.
6. T-H Cheng, F R Jones, D Wang, *Composites Sci. Tech.* **48**, 89 (1993).

1.12 Knitted reinforcements

F K KO

Introduction

Knitted fabrics are traditionally identified with socks, underwear and sweaters. In the search for methods to reduce composite manufacturing costs, textile preforms including knitted structures are receiving increased interest in the composite industry. Knits differ from woven, braided and non-woven fabrics in that they are produced by the interlooping of yarns. While conformability and productivity are obvious attributes for knitted preforms, the availability of a broad range of micro- and macrostructural geometries has only recently been recognized. The non-linearity of knitting loops (and limited fibre packing density, resulting in the formation of resin pockets within a knitting loop) prevent knits from being considered for structural applications. The development of technology for the directional insertion of linear yarns in weft and warp knits greatly enhances opportunities for knitted preforms for conformable structural composites by combining the conformable foundation knit structure with directional reinforcement.

One of the earliest development of knitted structures for net shape fabrication of composite joints was carried out by Courtauld.[1] The most extensive use of knitted structures for composites is, perhaps, in Germany, supported by a well-established knitting machine manufacturing industry. A good example of the structural application of knitted composites that takes advantage of the high conformability of knits is the knitted composite helmet developed at MMB.[2] Knitted composites are used extensively in the marine industry in Europe as well as in the United States. Development programmes are also being carried out in the aircraft and the automotive industry. Sponsored by NASA in the United States, knitted structures are being developed for carbon–carbon composite engine housings and for carbon–epoxy aircraft structural components.[3,4] Ford Motor Company has also experimented with knitted structures for reinforcement of a composite crossmember for the Aerostar van.[5]

Classification of knitted structures

Knitted structures can be classified by basic loop formation mechanism into weft knits and warp knits. Weft knitting is carried out by sequential feeding of yarn and formation of loop on each knitting needle across or around a needle bed. The loops are joined in the horizontal (course) direction. In warp knitting the loops are joined in the vertical (wale) direction. Yarn feeding and loop formation in warp knitting are done simultaneously on each needle in the needle bar during the same knitting cycle. Fig. 21 shows the formation of weft knit and warp knit structures. A detailed description of the technology of knitting can be found in Spencer.[6]

50   Knitted reinforcements

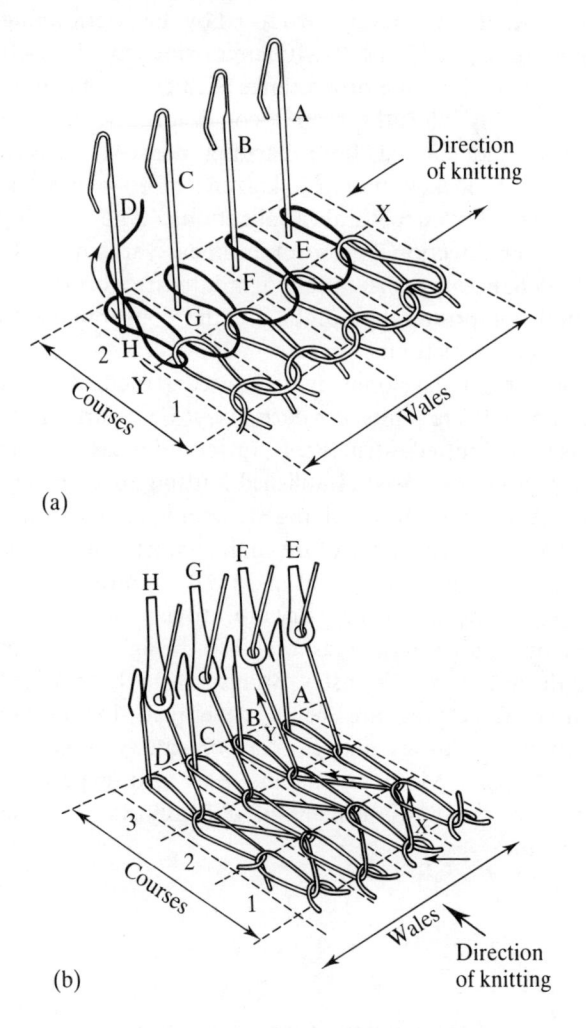

Figure 21 Formation of knit structures, (a) weft knit (b) warp knit

Weft knitted structures

There are four basic weft knit structures from which all the weft knitted fabrics are derived: plain, purl, rib and interlock. As shown in Fig. 22, a plain or 'Jersey' knit is produced by needles knitting as a single set to form a fabric having loops coming out from the technical back of the fabric. Plain fabrics are highly unstable. They tend to widen upon removal from the knitting machine and the fabric curls in a double curvature manner due to the recovery of residual processing energy developed during loop formation. If a yarn breaks, it will also cause the

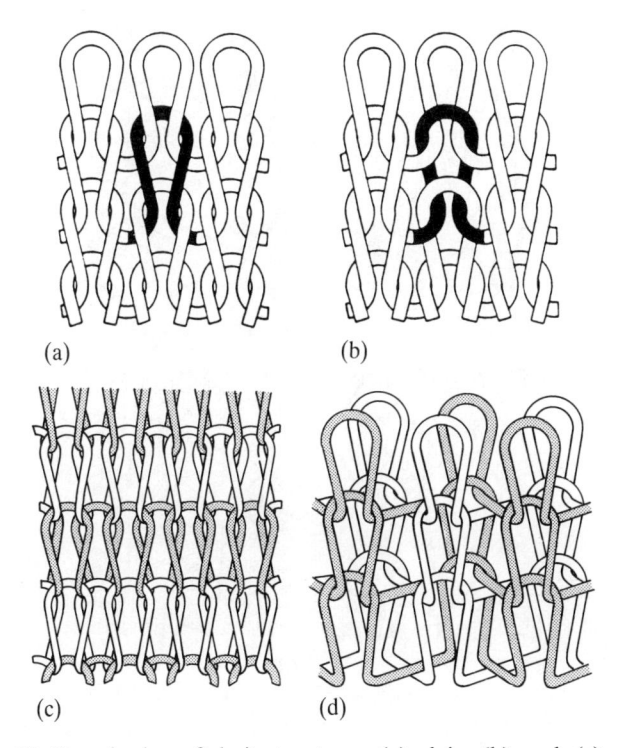

Figure 22 Four basic weft knit structures, (a) plain, (b) purl, (c) rub, (d) interlock

fabric to unravel. A purl knit is produced by the formation of alternate rows (courses) of face and reverse loops. The transverse extensibility of purl knits is equal to that of a plain knit, but the lengthwise extensibility is almost double. Rib knits are produced with two sets of needles opposite to each other to form a double thickness fabric. Because of the staggered nature of the columns of loops (wales), rib knits have high extensibility in the width direction. A variation of the rib knit is the interlock knit, wherein loop formation occurs in an alternate manner between the two needle beds, resulting in a stable structure of double thickness.

Warp knitted structures

Warp knits are produced by two basic types of machines: Raschel and Tricot. They use a wide variety of open or closed lapping motions around the needle to form chain and/or tricot stitch. By proper selection of the closed or open lapping motion and varying the lapping distances, a wide

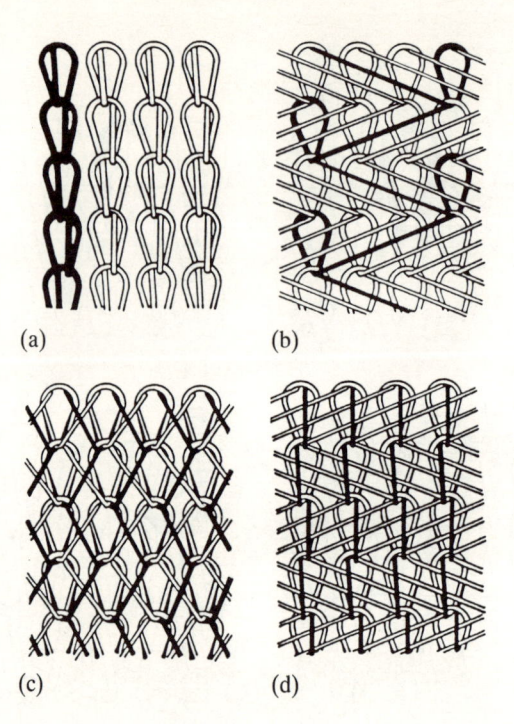

Figure 23 Different basic warp knit structures, (a) chain, (b) tricot, (c) locknit, (d) queenscord

range of structural stability and yarn segment orientation can be designed. Figure 23 illustrates the different basic warp knit structures.

Shape formation

Building upon the four basic knit structures, a wide variety of loop geometry can be produced by purposeful skipping or transferring of stitches. By changing the knitting construction, adjusting the length of yarn in a stitch and varying the number of needles in the knitting width, one can achieve planar shaping of a knitted fabric. The formation of complex shape, three-dimensional structures can be accomplished by a traditional forming process or by direct net shape knitting through selective transfer of stitches along with a variable take-down mechanism provided by the presser foot technology developed by Courtauld. Figure 24 illustrates the 'T-flange' and 'T-joints' produced in a presser foot knitting machine.[1]

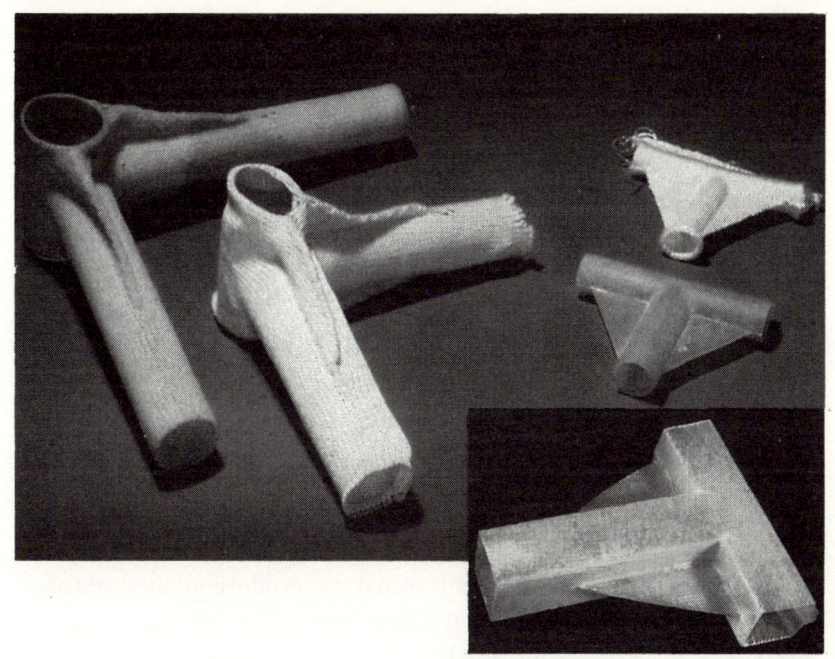

Figure 24 'T-flange' and 'T-joints' produced in a presser foot knitting machine (courtesy of Courtaulds Aerospace)

Directional reinforcement

Planar reinforcement

The interlooping nature of knitted structures provides excellent extensibility and conformability but strength translation efficiency is poor, due to non-linear loop geometry and low covering power. To overcome this deficiency, reinforcement yarns can be inserted in a linear or curvilinear fashion in the 0° (wale) and/or 90° (course) direction. Besides 0, 90 and 0/90 linear yarn insertion in warp knit structures, bias yarns with orientation ranging from 30° to 60° can also be incorporated into the fabric structure in both linear and curvilinear manners. The unique advantage of a planar curvilinear reinforcement is the possibility of tailoring toughness without creating stress concentrations at the yarn interlaces as in the case of woven fabrics. The most exciting development, however, is the multiaxial warp knits (MWK) with linear reinforcement insertions. Currently, there are three basic commercially available MWK systems: the Mayer RD 2S Raschel knitting machine, the Liba Copcentra multiaxial warp knit machine and the Malimo multiaxial knitting machine. The Mayer system is a more precise process with knit stitches wrapping around the well-placed insertion yarns, whereas the Liba and

Malimo systems are less precise in the knitting process. In these the insertion yarns are prone to be pierced by the knitting needles during stitch formation. The Liba system can insert up to eight layers of yarns with orientations ranging from 0° to ±60° to 90° plus a surface fibre mat. The Mayer system, on the other hand, is capable of inserting up to four yarn layers plus a surface fibre mat. With proper selection of the yarn orientation and placement sequence, MWK machines can produce fabric widths of over 2.5 m at a speed of up to 800 rpm, depending on yarn type and fabric construction.

Through the thickness reinforcement

Through thickness fibre reinforcement can be created by high-pile or plush knitting using a weft or warp knitting machine in addition to the use of higher strength stitching yarns in MWK fabrics. By proper selection of stitch geometry and loop length, different areal density and levels of pile height can be engineered to produce desired through thickness fibre volume fraction and thickness.

References

1. D Williams, *Advanced Composites Engineering*, **2** (2), 12–13 (1987).
2. F J Arendts, K Drechsler, MB6-Z-0356-RUB = OTN-033059, in *Mitteilungen aus dem Zentrallabor 22*, MBB Deutsche Aerospace, Munich, 1991 (Presented at the Third International Symposium, Advanced Composites in Emerging Technologies, University of Patras, Greece, 20–24 August, 1990).
3. A Taylor in *Proceedings of Fiber-Tex '88*, Clemson University, NASA CP-3038, 1988.
4. R Palmer, in *Proceedings of Fiber-Tex '92*, Drexel University, Philadelphia, NASA CP-3211, 1993.
5. C F Johnson, N G Chavka in *Proceedings of the 4th Annual Conference on Advanced Composites*, Sept. 1988, ASM International, pp. 535–565.
6. D J Spencer, *Knitting Technology*, Pergamon Press, Oxford, 1983.

1.13 Polyethylene fibres

D L M CANSFIELD and I M WARD

The low density of polyethylene means it will produce material of high specific strength (1.0–3.0 GPa) and modulus (60–120 GPa) when drawn to produce highly oriented fibre. The mechanical properties of the polyethylene fibres are therefore comparable to those of polyaramid and carbon fibres on a weight for weight basis. Combined with these properties is an extension to break of 3–4%. Although their principal application is currently marine ropes, the comparatively high extensibility also makes the fibres useful for textile applications and for fibre/resin composites where higher ductility and high energy absorption are required.

High modulus, high strength polyethylene fibres were first developed by a melt spinning/hot drawing route. This developed from fundamental studies on tensile drawing by Capaccio and Ward[1] at Leeds University on high density polyethylene polymers up to about 800,000 weight average molecular weight. Subsequently, ultrahigh molecular weight polymers ($\bar{M}_W = 1 \times 10^6$ to 3×10^6) were converted to high strength fibres by a solution spinning process, devised by Pennings,[2] and a process of gel spinning followed by hot drawing, developed by Smith and Lemstra.[3] These drawing processes have several important features.

(i) *For melt spun and gel spun fibres*, the initial modulus (stiffness) is a unique function of the draw ratio so that to obtain high modulus, the polymers must be drawn to very high ($> \times 30$) draw ratios (Fig. 25).

(ii) *For melt spun fibres*, the maximum draw ratio that can be obtained for a given draw temperature and strain rate depends on the polymer molecular weight and initial polymer morphology.

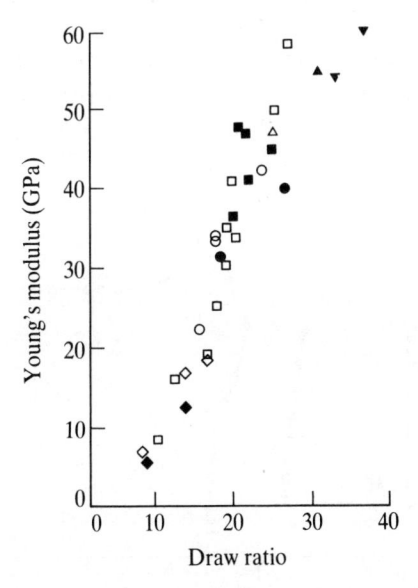

Figure 25 Modulus versus draw ratio for a variety of quenched (open symbols) and slow-cooled (solid symbols) linear polyethylene (LPE) samples drawn at 75°C. Sample 1(△), 2(□), 3(○), 4(◇) of increasing molecular weight

(iii) *Gel spun fibres* of ultrahigh molecular weight polymers can be drawn to give the highest specific strengths for any synthetic fibre.

The introduction of side chains in the polyethylene molecule was found to impose a severe limitation on the maximum draw ratio that could be achieved. The processes are therefore restricted to high density linear polyethylene and copolymers with less than five short side chains per 1000 carbon atoms.

A small pilot plant at Leeds University developed the basic principles for the continuous production of both multifilament and monofilament yarn by the melt spinning, high draw ratio route and this process has been licensed to SNIA Fibre in Italy and Bridon Fibres in the United Kingdom. SNIA Fibre have manufactured a multifilament yarn with a Young's modulus of 60 GPa, a breaking stress of 1.3 GPa and an extensibility of 5%. This was marketed under the brand name Tenfor. The melt spinning of fine fibres is restricted to comparatively low molecular weight polymer. This limitation is removed by the process of gel spinning and hot drawing of ultrahigh molecular weight polyethylene, a procedure devised by Smith and Lemstra and developed by DSM in Europe (Dyneema fibre) and Allied Spectra in the United States (Spectra fibre).

High modulus polyethylene (HMPE) fibres prepared by either the melt spinning or gel spinning routes are subject to creep under load. Creep restricts their effectiveness in some areas. This limitation has been markedly reduced as a result of an extensive series of investigations by Busfield, Woods and Ward[4] on the effects of irradiation of the drawn

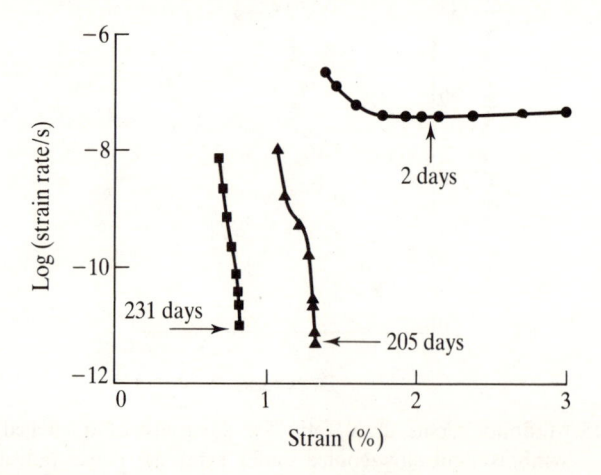

Figure 26 Creep response for crosslinked HMPE fibre at stress 0.17 GPa. ● 0 MRad 0% gel; ▲ 3 MRad 75% gel; ■ 6 MRad 82% gel

fibre by electrons and by γ-rays (Fig. 26). The significant discovery is that crosslinking, which restricts creep, can be appreciably increased by irradiation in an atmosphere of acetylene followed by annealing at a temperature of 80–120°C. This has been developed into a commercially viable process for producing creep resistant fibres.

Both melt spun and gel spun fibres are attractive as reinforcement for composites due to their high specific strength and good energy absorption. The other requirement for a good composite is a high shear strength in the fibre/matrix interface. This requires a firm bond between the two components. For polyethylene, this can be significantly improved by surface treatment with either oxidizing acids or exposure to a plasma. The development of these surface treatments has resulted from an extensive series of studies by Ward and coworkers.[5] Although treatment with acids gives some improvement in fibre/matrix adhesion, the optimum improvement results from plasma treatment in an atmosphere of oxygen. Plasma treatment affects the fibre surface in three stages.

(i) Short irradiation time oxidizes the surface, changing the surface energy and improving wettability.
(ii) Intermediate irradiation time develops crosslinking in the surface and removes the weak boundary layer that arises from segregation of low molecular weight polymer on the surface.
(iii) Long irradiation time results in a pitted surface which can be penetrated by the resin, improving adhesion by a mechanical interlock.

The improvement in adhesion between fibre and matrix is determined by measurement of the interlaminar shear stress (ILSS) of the composite. Composites incorporating untreated melt spun fibre (Snia, Tenfor) and gel spun fibre (Allied Spectra's Spectra) gave ILSS strengths of 16 and 8 MPa respectively. After oxygen plasma treatment these were raised to 28 and 14 MPa, an equivalent improvement in each case.

The combination of low density, high energy absorption, high strength, chemical inertness, and good matrix adhesion should enable high modulus polyethylene fibres to be used in the development of a range of composite materials with different properties from those available with glass or carbon fibre reinforced matrices. Hybrid fibre composites, incorporating glass or carbon and polyethylene fibres are of especial interest.[6]

References

1. G Capaccio, I M Ward, *Nat. Phys. Sci.*, **243**, 143 (1973).
2. A Zwijnenburg, A J Pennings, *J. Polym. Sci. Polym. Lett.*, **14**, 339 (1976).

3. P Smith, P J Lemstra, *J. Mater. Sci.*, **15**, 505 (1980).
4. D W Woods, W K Busfield, I M Ward, *Plastics Rubber Proc. Appl.*, **5**, 157 (1985).
5. N H Ladizesky, I M Ward, *J. Mater. Sci.*, **24**, 3763 (1989).
6. N H Ladizesky, I M Ward, *Composites Sci. Tech.*, **26**, 199 (1986).

1.14 Rigid rod polymer fibres

R J YOUNG

Introduction

Over recent years there have been several exciting developments in the field of high performance polymer fibres but, without doubt, the fibres with the most impressive mechanical properties have been the heterocyclic rigid rod polymer fibres produced through the USAF ordered polymer programme.[1-3] The best fibres so far have been obtained using molecules such as poly(*p-phenylene benzobisthiazole*), PBT; poly(*p-phenylene benzobisoxazole*), PBO; or poly(2,5(6)-benzoxazole), ABPBO. The chemical formulae of the molecules of PBT, PBO and ABPBO are given in Table 10. The fibres are made in a similar way to aromatic polyamide fibres (aramids) such as Kevlar by spinning from liquid crystalline solutions using aggressive solvents such as polyphosphoric

Table 10. Chain repeat units for various rigid rod and semiflexible chain polymers used in high performance fibres

Polymer	Chemical name	Repeat unit
PBT	Poly(*p*-phenylene benzobisthiazole)	
PBO	Poly(*p*-phenylene benzobisoxazole)	
ABPBO	Poly(2,5(6)-benzoxazole)	

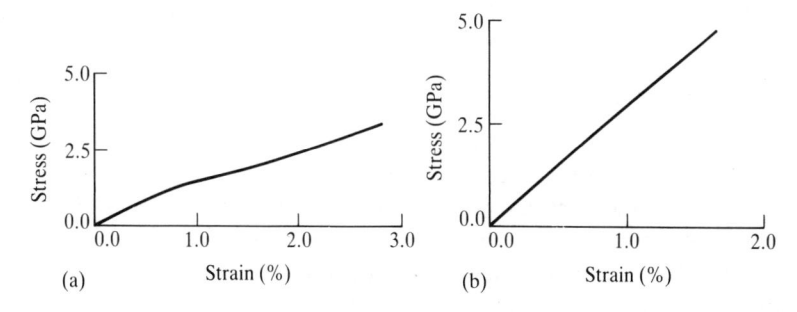

Figure 27 Stress/strain curves for single fibres of PBT, (a) as spun and (b) heat-treated

Mechanical properties

Fibres of PBT have a modulus of below 200 GPa in the 'as spun' condition but this can be increased to 320 GPa after heat treatment at elevated temperatures (up to 650°C), thought to improve the degree of molecular order in the fibres.[1,2] Stress/strain curves are given in Fig. 27 for both 'as spun' (AS) and 'heat-treated' (HT) fibres of PBT. It can be seen that the stress/strain curve for AS PBT shows a yield point at about 0.8% strain whereas the stress/strain curve for HT PBT is linear up to failure. More recent work has shown that heat treated fibres obtained from PBO can have even higher modulus values of up to 370 GPa.[3] Such fibres are also reported to have strengths of over 3 GPa and elongations to failure of up to 2%. Typical values of stiffness and strength for both as

Table 11. Typical mechanical properties of rigid rod and semiflexible polymer fibres

Polymer fibre	Tensile modulus (GPa)	Tensile strength (GPa)
AS PBT	190	3.0
HT PBT	250	4.0
AS PBO	145	4.6
HT PBO	260	3.4
AS ABPBO	95	2.1
HT ABPBO	133	2.2

spun and heat treated fibres are given in Table 11. Since these levels of stiffness and strength are combined with impressive oxidative and thermal stabilities and low specific gravities,[1] this means they are extremely promising as reinforcing fibres for high performance composites.

Although the fibres of rigid rod polymers are found to have excellent mechanical properties when deformed in tension, their performance under axial *compressive deformation* is generally poor and their compressive strengths can be a factor of 10 lower than their tensile strengths. This is a major factor that holds back their widespread use as many engineering applications impose both tensile and compressive deformation upon the fibres. The factors that control the compressive properties of high performance fibres are not yet fully understood but they may be inherent. To improve compressive strength, it may be necessary to introduce special solutions. For example, when they are used in composites, the fibres may be incorporated into hybrid mixtures with fibres of better compressive strength, e.g. glass.

Structure

The excellent mechanical properties of the fibres stem from the inherent stiffness of the polymer backbones. The molecules of PBT and PBO are true rigid rods. The aromatic rings in the molecules (Table 10) lie along a linear chain and the molecules have no flexibility. They are only soluble in a few solvents and the molecules in solution adopt rigid rod conformations. In concentrated solutions the molecules form liquid crystalline domains and when they are spun into fibres and coagulated, high degrees of molecular alignment are obtained. In the 'as spun' state, the fibres have a good degree of molecular alignment but no well-defined crystal structure. The structure is improved by heat treatment without significantly affecting the level of molecular alignment. It is found that heat treated PBT does not have a three-dimensional crystal structure in the conventional sense.[1] It has a crystal structure in which there is regular sideways packing of the molecules but no register in the chain direction due to the smooth profile of the ribbon-like molecules. In contrast, the crystal structure of PBO has a more three-dimensional character.

As well as leading to high degrees of molecular ordering, the rigid rod character of the molecules makes them inherently very stiff when deformed in tension parallel to the chain axis. Since there are no kinks or zig-zags in the chains, axial deformation is either by stretching of the covalent bonds or by distortion of the stiff aromatic ring structures. Hence it is possible to realize the high inherent levels of stiffness of the covalent bonds as in the case of diamond. This, combined with good orientation, leads to fibres with very high levels of stiffness and strength.

The molecule ABPBO is not a true rigid rod polymer but is a semi-flexible chain polymer due to there being a slight kink in the polymer chains (Table 10). This means that the material is rather more easy to process but the mechanical properties are somewhat inferior to the true rigid rods (Table 11).

Molecular deformation

It is possible to follow molecular deformation in the rigid rod polymers during macroscopic deformation using Raman spectroscopy.[4] Well-defined Raman spectra can be obtained from fibres of PBT, PBO and ABPBO and it is found that the positions of the Raman bands shift generally to lower frequency under the action of tensile stress or strain. This has been shown to be due to the macroscopic deformation causing direct stretching of the covalent bonds along the polymer backbones and so the strain induced Raman band shifts are a direct measure of the level of chain stretching in the fibres. In low modulus isotropic polymers only very small strain induced Raman band shifts are found, as most of the deformation on the molecular level is by bond rotation and molecular uncoiling. For the rigid-rod polymer fibres,[4] it is generally found that more chain stretching occurs in the higher modulus HT fibres, due to their more perfect microstructure.

Molecular composites

An exciting application of rigid rod polymers is in molecular composites.[5,6] This is, in principle, a simple concept whereby, instead of forming the molecules into fibres then incorporating them in a matrix of flexible polymer molecules, the rigid rod molecules are dispersed directly in the matrix. In order to obtain good mechanical properties, it is essential that the reinforcing molecules are well dispersed, otherwise the clusters of molecules will act as particles rather than fibres, giving a lower level of reinforcement. Previous attempts to make molecular composites have met with problems due to clustering, but it has been demonstrated[5] that it is possible to make molecular composites from carefully prepared blends of PBT with the semi-flexible molecule ABPBI. Films of the material are found to have an in-plane modulus of the order of 80 GPa.[5] In addition, it has been demonstrated using Raman spectroscopy[6] that mechanical deformation of the molecular composite film leads to stress transfer to the reinforcing PBT molecules, as would be expected for a material acting as a molecular composite.

References

1. S R Allen, A G Filippov, R J Farris, E L Thomas, *Strength and Stiffness of Polymers*, eds A E Zachariades, R S Porter, Marcel Dekker, New York, 1983, Ch. 9.
2. S R Allen, R J Farris, E L Thomas, *J. Mater. Sci.*, **20**, 2727 (1985).
3. S J Krause, T B Haddock, D L Vezie, P G Lenhert, W-F Hwang, G E Price, T E Helminiak, J F O'Brien, W W Adams, *Polymer*, **29**, 1354 (1988).
4. R J Young, P P Ang, *Polymer*, **33**, 975 (1992).
5. S J Krause, T B Haddock, G E Price, W W Adams, *Polymer*, **29**, 195 (1988).
6. R J Young, R J Day, P P Ang, *Polym. Comm.*, **31**, 47 (1990).

1.15 Silicon carbide fibres

J H SHARP and R J P EMSLEY

Engineers require high performance ceramic fibres stable at elevated temperatures, for the reinforcement of metal and ceramic matrices. Carbon fibres have excellent mechanical properties but are vulnerable to oxidation, while oxide fibres do not have the requisite strength and toughness for the most demanding applications. In recent years a small but slowly increasing number of fibres based on other ceramic systems, notably silicon carbide and silicon nitride, have become available. These materials can be conveniently divided into two main groups, those fibres made by chemical vapour deposition (CVD) and those produced by the polymer precursor route (PPR).

CVD fibres

There are two commercially available silicon carbide fibres based on CVD, which share similar fabrication technology. Their physical properties are summarized in Table 12 (Section 1.16) in which they are compared with those of other ceramic fibres. The production of a CVD fibre requires an electrically conducting substrate fibre, usually carbon (SCS) or tungsten (Sigma), resistively heated in the presence of a volatile silane, such as methyltrichlorosilane (CH_3SiCl_3), methane and hydrogen. Argon may be used as an inert carrier gas. For the most part, the balance of these gases is such that silicon carbide is produced, but the mixture of gases may be varied so that carbon-rich or silicon-rich fibres are produced. This ability to vary the chemical composition across the diameter of the fibre allows the production of fibres with interfaces tailored to meet a particular application.

At present, the major application of silicon carbide fibres produced by the CVD route is in metal matrix composites (MMCs), where their high strength and modulus (see Table 12) and their resistance to chemical attack is greatly valued. Other applications are in ceramic matrix composites (CMCs) and to a lesser extent in polymer matrices. The main disadvantages of these fibres are their expense and relatively large diameters. Their thickness determines the number of fibres that can be used per unit volume of the composite and also restricts the radius of curvature of the artefact, making it difficult to fabricate complex shapes.

PPR fibres

Ceramic fibres produced by the PPR route have a common general fabrication route involving the synthesis of a polymeric precursor. This is spun into fibres then undergoes a curing operation to crosslink the polymer molecules and render the fibre infusible. The cured fibre is then pyrolysed in an inert atmosphere at the required temperature. The advantage of this route is that it leads to small diameter fibres, which cannot currently be obtained using alternative processes.

Although the production of ceramic materials from organometallic precursors has been possible since the early 1960s, it was not until Yajima produced a carbosilane polymer in the late 1970s that this technology could be used for the production of silicon carbide based fibres. The ceramic fibre produced by Nippon Carbon and sold under the name of Nicalon is a direct result of the research and development work of Yajima and his collaborators.[1,2] Further development has led to the production of Tyranno fibres by Ube Chemicals[3] and, at the time of writing, these two Japanese fibres are the only PPR silicon carbide fibres commercially available.

Yajima's early polycarbosilane (a polymer having a silicon–carbon backbone) was produced by cleavage of the dodecamethylcyclohexasilane ring, $[(CH_3)_2Si]_6$, in an autoclave. Fractionation of the product gave a polymer fraction that could be dry spun into fibres. Yajima[1] considered this route to be 'technically difficult, expensive and time consuming'. An alternative, simpler procedure was developed in which dichlorodimethylsilane, $(CH_3)_2SiCl_2$, was reacted with molten sodium in xylene to yield polydimethylsilane. This was then either heated in an autoclave or reacted in the presence of a catalyst (polyborodiphenylsiloxane) to yield the polycarbosilane, which was fractionated to remove the volatile and insoluble material. The melt spun fibres were subjected to a controlled oxidation to fix their shape and prevent them from remelting during subsequent heat treatment. This pyrolysis of the organometallic polymer to the inorganic ceramic could theoretically yield pure silicon carbide:

'cured polycarbosilane' → SiC + CH$_4$ + H$_2$O + H$_2$

but in reality the products of the pyrolysis are more complex:

'cured polycarbosilane' → SiC$_x$O$_y$, + CH_4 + H_2

Hence the final product contains some silicon dioxide and excess carbon as well as silicon carbide.

Early Nicalon fibres, known as standard grade, had a nominal composition of Si, 52.6%; C, 30.4%; O, 14.2%; and had only modest resistance to oxidation at temperatures greater than 1000°C. Improved processing resulted in the production of ceramic grade Nicalon fibres, which have a nominal composition of Si, 58.4%; C, 31.2%; O, 10.0%, equivalent to SiC, 70.8%; SiO$_2$, 18.8%; C, 10.0%. These fibres have much improved chemical stability at high temperatures in both oxidizing and inert atmsopheres.

Although methods of non-oxidative curing[4] have been used in the laboratory and improved fibre properties have been obtained, the commercially available fibre is still based on oxidative curing. The incorporation of significant levels of oxygen in the fibre leads to degradation at high temperatures even in inert atmospheres. The strength and Young's modulus of Nicalon fibres (Table 12) show little change up to 1000°C, but above this temperature they both decrease, especially the strength. On heating in air, the ceramic grade fibres rapidly take up oxygen to a limit of about 20%. The carbon content is reduced and a thin silica sheath is formed around each fibre. These processes are accompanied by grain growth, which almost certainly contributes to the observed loss in strength.

Three forms of Nicalon ceramic fibres are currently produced. In addition to the ceramic grade fibres, there are HVR and LVR grades, which have high and low volume resistivities of > 10^6 ohm cm and 0.5–5 ohm cm, respectively. All three grades have high strength and modulus which can be maintained for 1000 h at temperatures up to 1000°C. Nicalon is available as either continuous fibre, chopped fibre or woven cloth and may have a variety of surface treatments applied, e.g. different sizes to meet the requirements of the matrix, whether metal, ceramic or polymer, into which they are to be placed. A carbon coated version of ceramic grade Nicalon is available with a 10–20 nm coating designed to provide a weak fibre–matrix interface in ceramic and glass–ceramic matrix composites.

Although Nicalon fibres are the most widely used small diameter non-oxide ceramic fibres, extensive research and development continues in the search for fibres which retain their properties to higher temperatures. These include Tyranno fibre (Ube Chemicals), and MPS and MPDZ fibres (Dow-Corning Celanese) and related Si–C–N–O fibres to be discussed in Section 1.16. To produce Tyranno fibres, polydimethylsilane is mixed with a titanium alkoxide in the approximate ratio 10:1 and heated at 340°C in nitrogen for 10 h. The resultant polytitanocarbosilane is spun into fibres and oxidatively cured in a manner similar to that used to produce Nicalon. The molecular weight of the polymer is low, around 1500. This precursor fibre is converted into an amorphous ceramic fibre by heating in air at 180°C, followed by heat treatment in nitrogen at about 1000°C. The final product contains about 2% by weight of titanium, which is said to inhibit crystallization, and similar amounts of SiC, C and SiO_2 as Nicalon. The fibres, which have a diameter of about 9 μm, have good mechanical properties (Table 12). They retain their strength up to 1200°C, at least for short periods, and eventually crystallize into β-SiC and SiO_2.[4] By changing the amount of titanium incorporated into the fibre and by modifying the processing conditions, fibres having a range of resistivities can be produced for various electrical applications.

There are undoubtedly several commercial systems under development, although manufacturers are concerned about the high development costs and the long delay before these can be recouped from such a specialized market. To obtain a fibre that retains its mechanical properties to very high temperatures, it will be necessary either to avoid crystallization or to inhibit grain growth of the crystallites. As explained above, the present fibres are non-stoichiometric and contain silica and carbon as well as silicon carbide. Future developments are expected to produce fibres with an Si:C ratio close to 1 and a much reduced oxygen content.

References

1. S Yajima, Y Hasegawa, J Hayashi and M Iimura, *J. Mater. Sci.*, **13**, 2569 (1978).
2. Y Hasegawa, M Iimura and S Yajima, *J. Mater. Sci.*, **15**, 720 (1980).
3. T Yamamura, T Hurushima, M Kimoto, T Ishikawa, M Shibuya and T Iwai, *High Tech. Ceramics*, ed. P Vincenzi, Elsevier, Amsterdam, 1987, p. 737.
4. K Okamura, M Sato, T Sequchi and S Kawanishi, in *Controlled Interphases in Composite Materials*, ed. H Ishida, Elsevier, New York, 1990, p. 209.
5. A R Bunsell, *J. Appl. Polym. Sci. Appl. Polym. Symp.*, **47**, 87 (1991).

1.16 Silicon nitride and silicon carbonitride fibres

R J P EMSLEY and J H SHARP

Since Yajima produced the first silicon carbide fibres, there has been considerable interest in the development of fibres capable of withstanding still higher temperatures and more chemically aggressive environments. In addition to the Nicalon and Tyranno fibres developed in Japan (see Section 1.15) several other fibres, based on the Si–C–N–O system, have been or are being developed to the point of commercial exploitation. Some of these are essentially based on silicon nitride, whereas others are more correctly considered as silicon carbonitrides. They are, however, conveniently classified together and are not greatly different, since they are all based on covalent bonding.

The precursor for the hydridopolysilazane (HPZ) fibre (Dow-Corning Celanese) was produced from the reaction product of trichlorosilane and hexamethyldisilazane.[1] Care was taken to exclude oxygen from the precursor. Fibre spinning was carried out in an inert atmosphere and the fibre was cured by exposure to a multifunctional chlorosilane ($RSiCl_3$). Pyrolysis of the cured fibre resulted in a loss in mass of 30–35% accompanied by the evolution of ammonia, chlorosilanes, methane and other volatiles. The resulting fibre, which is believed to be completely amorphous, had the chemical composition, Si, 60.0%; N, 32.6%; C, 2.3%; and O, 2.2%, whereas stoichiometric Si_3N_4 has the composition, Si, 60.06%; N, 39.94%.

The fabrication of an almost pure silicon nitride fibre from perhydropolysilane has recently been described by Tonen.[2] The polymer was synthesized by the reaction of ammonia with a stable adduct of dichlorosilane and pyridine:

$$n\,SiH_2Cl_2 \cdot 2C_5H_5N + 3n\,NH_3 = (SiH_2NH)_n + 2n\,C_5H_5N + 2n\,NH_4Cl$$

After the ammonium chloride had been filtered off, the filtrate was saturated with ammonia and heated to yield a polymer with an average molecular weight of 12,000. The polymer was dry spun into fibres, nitrided in ammonia at temperatures below 1000°C and finally pyrolysed by heating above 1200°C. The chemical composition of the product was Si, 59.8%; N, 37.1%; C, 0.4%; and O, 2.7%, which is close to that of Si_3N_4. The tensile strength and Young's modulus of this fibre are given in Table 12. They compare well with the mechanical properties of other ceramic fibres. After heat treatment at 1200°C in air for 12 h, about 85% of the original strength was retained, although slight surface oxidation of the fibre was noted.

A silicon carbonitride ceramic fibre, called 'Fiberamic',[3] has been developed by Rhone-Poulenc by the reaction of ammonia with

Table 12. Properties of selected ceramic fibres

Fibre type	Source	Density (kg m^{-3})	Diameter (μm)	Tensile strength (GPa)	Young's modulus (GPa)
†SCS	Textron Speciality Materials	3000	140	3.45	400
†Sigma	BP Metal Composites	3400	100	3.9	390
*Nicalon	Nippon Carbon	2550	15	2.75	193
*Tyranno	Ube Industries	2370	9	2.94	176
*HPZ	Dow-Corning Celanese	2350	10	2.2	150
*Tonen	Tonen	2500	10	2.5	300
*Fiberamic	Rhone-Poulenc Chemie	2400	15	1.8	220

*PPR; †CVD.

chlorosilanes containing silane (Si–H) and vinylsilane (Si–CH=CH$_2$) groups. Careful control of the reaction conditions enabled the production of a polymer precursor having the desired molecular weight distribution to be suitable for either melt spinning or dry spinning of the precursor fibres. The fibres were then cured and their shape fixed by either thermal or catalytic hydrosilylation reactions between the groups Si–H and Si–CH=CH$_2$:

$$-Si-H + -Si-CH=CH_2 \rightarrow -Si-CH_2-CH_2-Si-$$

The cured fibre was then pyrolysed at a temperature between 1200 and 1400°C to yield a ceramic with typical composition, Si, 57%; N, 22%; C, 13% and O, 8%. The product was 90–95% amorphous Si–C–N–O, but also contained 5–10% free carbon, which existed as small clusters. Fiberamic has good mechanical properties and the amorphous nature was retained up to 1400°C. After ageing in an oxidizing atmosphere, the tensile strength decreased to 85% of its initial value after 100 h at 1000°C and after 10 h at 1200°C.

The properties of these fibres are compared with those of silicon carbide fibres in Table 12. It is clear that there is scope for further development in these systems and new commercial fibres, perhaps with better physical properties or with better resistance to crystallization at high temperatures, are likely to be developed.

References

1. G E Legrow, T F Lim, J Lipowitz, R S Reaoch, *Mater. Res. Symp. Proc.*, **73**, 553 (1986).

2. T Isoda in *Controlled Interphases in Composite Materials*, ed. H Ishida, Elsevier, New York, 1990, pp. 21–24.
3. P Poirier in *New Materials and Their Applications*, Proceedings of the 2nd International Symposium, ed. D Holland, Institute of Physics, Bristol, 1990, pp. 157–162.

CHAPTER 2

Matrices

2.1 Advanced thermoplastics

J BARNES

Introduction

Whilst the field of continuous, fibre reinforced composites is now well established, and such materials are an everyday part of our lives there is still room for change to the established order. This is aptly demonstrated by the relatively recent trend towards the inclusion of high performance fibres such as carbon and glass in what we may term 'engineering' thermoplastics based on aromatic polymers.

The key difference between such composites and the more traditional thermoset based materials (i.e. epoxy, polyester, phenolics, etc.) lies in the behaviour of the matrix during processing. The thermoplastic matrix is not required to undergo a 'cure' process to achieve its final mechanical properties – all that is required is melting, shaping and subsequent solidification. In consequence, such materials can offer the potential for extremely rapid forming.

Types of polymer

Within the general class of thermoplastic polymers suitable for use in composite form, we can draw a distinction between two families, those which are amorphous and those which are semicrystalline. Amorphous polymers may best be considered as those which on a molecular level consist of a collection of random coils with little local ordering. Such polymers are severely affected by temperature and lose most of their mechanical performance suddenly when raised to above their glass transition point, T_g. The effect of common organic solvents is usually also severe, causing effects such as crazing, swelling and dissolution.

Table 13. Mechanical properties (manufacturer's data) of two thermoplastic polymers: poly ether ether ketone, PEEK, and poly ether sulphone, PES

Property	PEEK (semicrystalline)	PES (amorphous)
Relative density	1.32	1.37
Glass transition temperature (°C)	143	210
Typical level of crystallinity (%)	35	0
Failure strain at 23°C (%)	50	40–80
Tensile strength at 23°C (MPa)	92	84
Tensile strength at 180°C (MPa)	17	41
Flexural modulus at 23°C (GPa)	3.6	2.6
Impact strength (Izod, 23°C) (kJ m^{-2})	8	8
Maximum continuous service temperature (°C)	250	180

Polymers in which the molecule contains regular repeat units and few bulky side chains may have a capacity for local ordering; such materials are usually termed semicrystalline. They contain islands of regular (commonly thought to be chain folded) regions separated by areas of random, amorphous material. As such polymers are raised above the T_g, the effects of thermally induced loss in stiffness occur only in the amorphous regions, hence the decay in structural performance is less marked in semicrystalline materials. Additionally, the crystalline components tend to limit the capacity for ingress of solvents, and semicrystalline polymers are in general more solvent resistant than their amorphous counterparts; some of the semicrystalline materials are virtually insensitive to common solvents.

Table 13 shows typical properties of two polymers. Table 14 is a list of polymers available in composite form. The effects of common solvents on PEEK/carbon fibre composites are shown in Table 15. (PEEK is poly ether ether ketone.)

Impregnation strategies

The variation in resistance to solvent attack in the two polymer types leads to differences in the types of impregnation strategy which may be used to provide a pre-impregnated sheet of fibres (or 'prepreg') for lamination (see Cogswell[1] for a more complete discussion). Many of the amorphous polymers can be completely dissolved in common solvents, which can therefore be used to provide low viscosity solutions for impregnation. Whilst this can be a route to providing a tacky prepreg by the inclusion of residual solvent at the end of the process, there are costs associated with ensuring that all the solvent is eliminated. Environmental effects of residual solvent are also a concern.

Table 14. Thermoplastic composites (after Cogswell[1])

T_g (°C)	Approximate processing temperature (°C)	Semicrystalline matrices	Amorphous matrices	Composite suppliers
260	320–350		Avimid KIII	Dupont
260	360–390		Torlon C	Amoco
250	350–370		Radel C	Amoco
230	315–360		Victrex ITA	ICI Fiberite
230	315–360		PES	Specmat, BASF
220	315–360		PEI	Cyanamid, Ten Cate
220	340–360		Ryton PAS-2	Schriner
220			Radel X	Phillips
175	390–410	Victrex ITX		Amoco
156	360–370	PEKK		ICI Fiberite
145	320–340		J Polymer	Dupont
143	380–400	PEEK		Dupont, ICI Fiberite, BASF, Tech Textiles
95	330–350		Ryton PPS	Asahi, Nittobo, Texxes, Phillips

PEKK = poly ether ketone ketone, PEEK = poly ether ether ketone, PES = poly ether sulphone, PEI = poly ether imide, PPS = polyphenylene sulphide.

Table 15. The effects of common solvents on carbon fibre/PEEK APC-2/AS4[5]

Environment	Immersion conditions	Weight change (%)	90° Flexural strength (MPa)
Control	23°C/50%RH/15 days	+0.02	130
Water	100°C/15 days	+0.23	127
MEK	23°C/15 days	+0.03	143
Skydrol LD4	70°C/15 days	−0.08	139
Ethylene glycol	70°C/15 days	+0.02	141
Methylene chloride	23°C/15 days	+2.41	131
Jet fuel JP4	82°C/14 days	+0.10	152
Jet fuel JP5	82°C/14 days	+0.03	152

A number of alternative methodologies involve the incorporation of fine polymer powders within the fibres or interweaving the reinforcement with fibres spun from the matrix polymer. It is common for such tows to be presented to the end user in the form of a dry woven mat, usually at least partly drapable. The systems are usually not tacky, but localized welding is used during the lay-up process to maintain a degree of handleability. However, each of these impregnation routes requires that the wetting of the fibres by the matrix takes place during the process of forming, which requires both high pressure and significant time. Since one of the primary advantages of thermoplastic composites is their potential for rapid forming, such materials have been found to be increasingly unattractive.

An alternative means to impregnate beds of fibres is to use melt impregnation. It may be argued that this is the preferred route since it requires no pretreatment of the polymer to form powders or fibres and eliminates the need for binders or solvents. However, difficulties arise due to the high viscosity of most useful grades of thermoplastic polymer, so a variant of the process is to incorporate a plasticizer into the polymer at the impregnation stage. This is then removed at a later point in the prepreg manufacturing process, leaving a stiff boardy prepreg. It has been amply demonstrated[1] that this method can produce fully wetted fibres, consequently the final product requires no further impregnation or chemistry to take place at the moment of shaping. Its performance is enhanced over alternative material forms.

Shaping

Flat panels can be manufactured from thermoplastic composites with little more equipment than a heated press – all that is required is for the material to be melted and welded, then cooled. Processes used for shaping thermoplastic composites are drawn for a variety of sources, including the thermoset and metals processing industries. A more complete discussion of the subject can be found in Section 3.4 on fabrication in thermoplastic composites.

Mechanical performance

Although the behaviour of advanced thermoplastic composites can be discussed on its own, it is most useful to compare it with more conventional epoxy based composites used as the baseline. Table 16 shows a summary of the behaviour of the best researched thermoplastic composite, carbon fibre–PEEK APC-2/AS4. Discussion of thermoplastic mechanical performance is largely confined to this material as the paradigm of thermoplastic composites – one may expect that the behaviour of other thermoplastics will lie somewhere between APC-2/AS4 and 'conventional' epoxy composites. Comparison with other data in this handbook will show that the properties of the material differ little to those of epoxy composites, with one or two notable exceptions. The most outstanding of these is the toughness of the material, measured in terms of critical strain energy release rate in mode I failure, G_{Ic}. Although some of the new generation of multiphase thermoset composites may begin to approach this value, typical tough thermoset composites have values of G_{Ic} smaller by factors of 5 or 10. Interlaminar shear strength is

Table 16. Typical properties of carbon fibre/PEEK APC-2/AS4 at room temperature[3]

Property	Value
0° Tension	
Strength (MPa)	2068
Modulus (GPa)	138
Failure strain (%)	1.45
Poisson's ratio	0.30
0° Compression	
Strength (IITRI) (MPa)	1208
Modulus (IITRI) (GPa)	123
90° Tension	
Strength (MPa)	86
Modulus (GPa)	10.2
Failure strain (%)	0.88
Short beam shear strength (MPa)	105–127
G_{Ic} (kJ m^{-2})	2.3
G_{IIc} (kJ m^{-2})	3.0
Open hole compression strength (MPa)	327
Open hole tension strength (MPa)	387

also high in comparison to thermoset materials, and these two properties combine to provide excellent resistance to delamination after impact.[2] The high toughness of PEEK also confers an unusually high resistance to microcracking in the composite.[3] This leads to a major difference in the behaviour of PEEK and thermoset composites during fatigue loading; the thermoplastic has a reduced capability to microcrack to relieve stresses and has the advantage that its overall properties, particularly stiffness, are less affected up to the fatigue limit. But there is little difference in the ultimate fatigue life of tough thermoplastic and thermoset composites.[4]

Somewhat surprising is the comparison between creep behaviour of thermoplastic and thermoset composites. As one might expect, parallel to the fibre direction in fibre dominated lay-ups, both types of material show little time dependent deformation. Away from fibre dominated loading directions, thermoplastic composites do show significant, highly predictable creep, and in short-term testing crosslinked thermoset materials can show improved behaviour. However, it is noteworthy that thermoset systems are highly susceptible in long-term testing, most likely due to microcracking. Since microcracking leads to progressive failure, thermoset systems tend to experience acceleration of the creep process with time. In the longer term, microcrack resistant thermoplastics appear to confer some advantage.[5] Finally, the high toughness of some of the thermoplastics systems also confers excellent abrasion resistance.

Conclusions

In some respects, thermoplastic composites can offer significant performance advantages over thermoset materials, though the performance margin has been degraded since the introduction of the first high performance materials in the early 1980s[6]. A wide portfolio of matrices is available, often in different final product forms from different manufacturers. The principal benefit of thermoplastic composite technology appears to lie in the economics of fabrication, where, in many circumstances, costs can be demonstrably much lower than those for competitors, if the composite is provided in the form of a fully wetted prepreg.

References

1. F N Cogswell, *Thermoplastic Aromatic Polymer Composites*, Butterworth-Heinemann, Oxford, 1992.
2. M Akay, *Composites Sci. Tech.*, **33**(1), 1–18 (1988).
3. S Liu, J Nairn, *J. Reinf. Plast. Comp.*, **11**, 158 (1992).
4. S Kellas, J Morton, P T Curtis, *Fatigue Damage Tolerance of Advanced Composite Materials Systems*, Virginia Polytechnic Institute, 1990.
5. A Horoschenkoff, J Brandt, J Warnecke, O S Bruller, in *New Generation Materials and Processes, Proceedings of the European SAMPE Conference*, Milan, 1988, pp. 339–349.
6. *ICI Thermoplastic Composites Materials Handbook*, 1991.

2.2 Conventional thermoplastics

R S BAILEY

Definition

Composite materials that comprise discontinuous reinforcing fibres in thermoplastic matrices are extremely high volume materials and represent one of the largest families of commerical composite systems. In recognizing this factor, it is clear that these materials need to be produced and subsequently manufactured by high throughput processes, such as extrusion and injection moulding, in order to be economic. The

Table 17. Properties of conventional thermoplastics, PET is polyethylene terephthalate

Property	Polypropylene	Nylon 6.6	Polycarbonate	PET
Density ($Mg\,m^{-3}$)	0.9	1.14	1.06–1.2	1.21
Young's Modulus ($GN\,m^{-2}$)	1.0–1.4	1.4–2.8	2.2–2.4	2.7
Poisson's ratio	0.3	0.3	0.3	0.3
Tensile yield strength ($MN\,m^{-2}$)	25–38	60–75	45–70	62
Thermal conductivity ($W\,m^{-1}\,°C$)	0.2	0.2	0.2	0.2
Coefficient of thermal expansion ($10^{-6}\,°C^{-1}$)	110	90	70	65
Melting point (°C)	175	264	–	254–259

most commonly used matrix polymers are shown with some of their properties in the Table 17. In spite of the long established market for these materials, there is considerable growth and opportunity for new markets by using new processing techniques, reinforcements, interfaces and polymer matrices. Properties are continually being extended upwards and encroach into the diecast metal casting market. Two recent developments, either using liquid crystal polymer matrices or longer aspect ratio fibre composites from a pultrusion compounding technique offer some further possibilities.[1]

Thermoplastics and engineering compounds

Thermoplastic polymers differ from thermosetting resins. They are not crosslinked and derive their strength and stiffness from the spatial arrangement of the monomer units that form high molecular weight chains. In general, fibre reinforcement is used in discontinuous or 'short fibre' form to extend the mechanical property range available in bulk unreinforced thermoplastics. The range of conventional thermoplastic composites and their resultant property advantages depend on the average length of the fibres on moulding. While the property gain from adding fibre reinforcement may seem modest in comparison with continuous fibre systems, the improvements in either Young's modulus or toughness justify their use for many applications.

In amorphous materials, heating leads to polymer chain disentanglement and a gradual change from a rigid to a viscous liquid. In crystalline materials, heating results in the sharp melting of the crystalline phase to give an amorphous liquid. Both amorphous and crystalline polymers may have anisotropic properties resulting from molecular orientation induced during processing and solidification. In composite systems, however, the properties are largely dictated by the character of the fibre reinforcement.

These polymer composite systems are often termed 'engineering compounds' because their mechanical properties extend beyond the commodity levels expected from bulk polymers and they are used in multicomponent compounds for specific applications. Compounds may include colorants, viscosity modifiers, stabilizers, matrix modifiers, interfacial bond enhancers or secondary fillers.

These commodity polymers are typically used in conjunction with short fibre reinforcement; mineral fillers are also commonly used in lower performance applications. E-glass fibres are used in the vast majority of reinforced injection moulding compounds on cost grounds, but carbon fibres (usually high strength type) and Kevlar type fibres may be compounded for specialist applications. Conductive fibres such as steel are also used to impart dielectric properties.

Compounding

These materials are compounded using a variety of processing routes, although plasticizing extrusion products account for the bulk of the market. Extrusion compounds are produced in a granular feedstock form, easily transported and handled in commercial injection moulding operations. Recently, alternative compounding routes have emerged. Pultrusion compounding[2] offers scope for longer fibres to be retained on moulded parts by allowing production of feedstock granules that contain significantly longer fibres (determined by the cutting operation of the

(a)

(b)

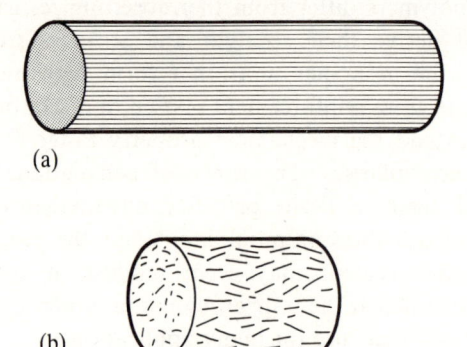

Figure 28 The two basic types of moulding pellet used for injection moulding of fibre reinforced thermoplastics. The illustration (a) is the 'long' fibre type with fibres extending the full length of the pellet (typically 10 mm). The more common type is (b), where the fibres are fully dispersed by extrusion compounding. In this case the fibres are much shorter than the pellet (typically 0.2 mm)

pultrudate precursor). Figure 28 illustrates the granules from these compounding routes.

Alternative compounding routes such as paper spraying, pultruded prepregs and tapes offer specific property opportunities using random-in-plane or continuous fibres for niche applications. These materials are used in laminates, panels by compression moulding or in filament winding.

Processing and microstructure

The processing of discontinuous fibre filled thermoplastics owes much to the rheology of these systems. Shear thinning and adiabatic heating, associated with the broad molecular weight polymers in these compounds, offer attractive processing conditions for engineering applications.[3]

Since this family of composite materials is used almost excusively in injection moulding, it is worth considering the process and its effect on the microstructure of the produced components. Unlike filament winding or hand lay-up techniques employed in continuous fibre systems, the fibres are positioned by the management of the flow into the mould, either by process control or mould design. Although the morphology of the polymer matrix is of some significance, the fibre microstructure dominates the anisotropy in moulded parts.

Fibre length

The fibre length distribution is determined by three stages in melt processing: (i) the plasticization stages of extrusion compounding; (ii) remelting during injection moulding; and (iii) the high shear during injection and mould filling. It is the plasticization stages that dominate the fibre attrition process.[3] This leads to a broad fibre length distribution ranging from fragments up to the chopped glass fibre strand length.

Fibre orientation

The flow into the mould cavity during injection dictates the fibre orientation in the final moulded part and hence the properties of the component. The addition of fibres makes most polymer systems more shear thinning and thermally conductive. These factors influence the velocity and shear profiles of the melt front as it passes through the narrow mould channels; they also influence the rate of solidification. The

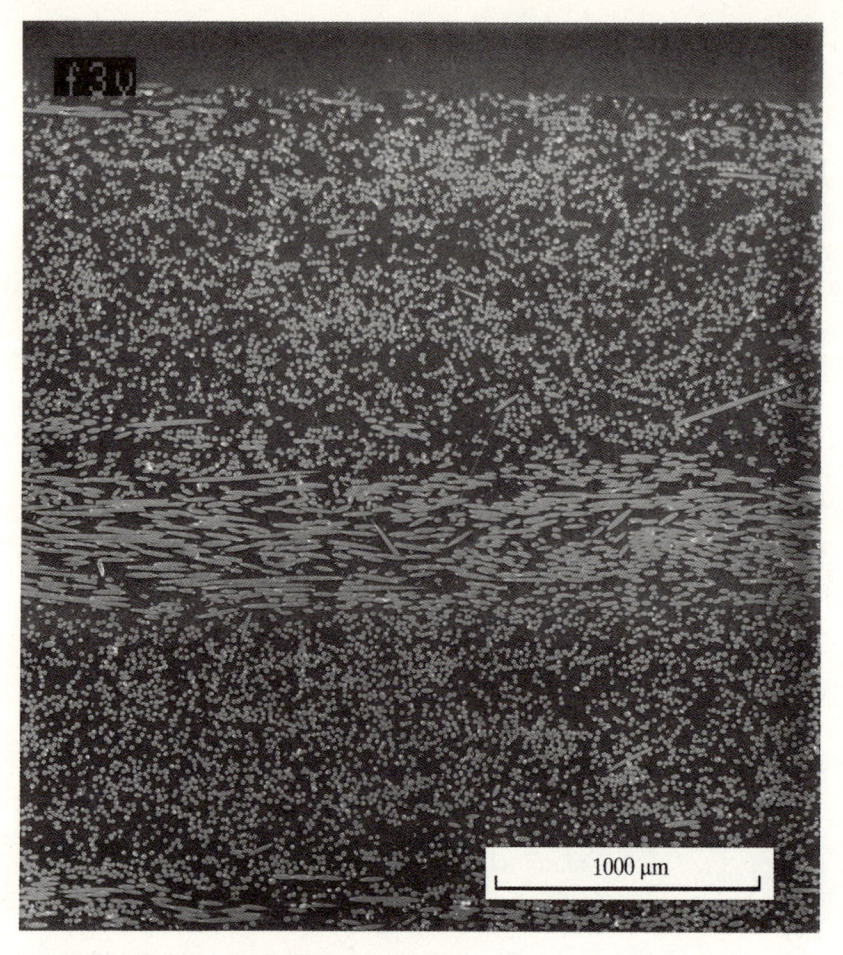

Figure 29 A typical cross-section through and injection moulded part showing a sandwich structure. The flow direction is into the page resulting in fibres aligned in the same directions (skin) and a random-in-plane core.

melt front proceeds by a 'fountain flow' mechanism, which advances leaving behind oriented fibres frozen in situ at the cold mould walls. The central region is last to solidify and is free to relax into random-in-plane orientations constrained by the decreasing channel thickness. A so-called skin–core structure as shown in Fig. 29 arises. To some extent, it can be controlled in critical mould sections.[4] Novel processes to manage the microstructure are becoming available,[5] see Section 3.10. Pulsing flows, oscillating packing pressure, gas injection and moulding over low melting temperature alloy preforms; each technique assists the designer to meet a specific mould design or microstructural goal.

Recycling

One advantage of thermoplastic systems lies in repeated melt processability. This is possible with high polymers unlike crosslinked thermosets. The property cascade associated with either degradation of the polymer molecular weight or fibre length is inevitable in most cases, but segregation of specific polymer types, to avoid contamination, is a more critical logistical issue.[5] See also Section 3.13.

References

1. F N Cogswell, in *Proceedings of the 2nd International Conference on Short Fibre Reinforced Thermoplastics*, Solihull, Oct. 1988, Plastics and Rubber Institute, London, 1988.
2. C R Gore, G Cuff, D Ciannelli, *Mater. Engng*, **103** (3), 47–50 (1986).
3. J M Lunt, J B Shorthall, *Plastics Rubber: Proc.*, **Sept**, 108–14 (1979).
4. R Bailey, B Rzepka, *Intl. Polym. Proc.*, **VI**, 35 (1991).
5. A Weber, in *Developments in Science and Technology of Composite Materials* (ECCM 4), eds J Füller et al. Elsevier Applied Science, New York, 1990, p. 15.

2.3 Conventional thermoplastics – pultruded composite tapes

G CUFF

Introduction

During the 1980s, many companies sought to develop advanced thermoplastic composites by pultrusion driven processes (see Section 2.1). In general, they have been motivated by the initial promise of defence and aerospace applications. It is not surprising they have neglected more conventional thermoplastics, because their composite production processes and means of fabrication into components are slow and costly.

Those companies which have looked at the logic of rapid impregnation have found that the lead times to fully understand their products, develop fabrication techniques and educate end-users have not formed a happy marriage with the latest strategic plans of their management. Some companies have withdrawn from the arena, others are seeking to license their technology. This is unfortunate. Many of the problems of producing low cost composites and demonstrating realistic economic routes to fabrication have been solved. There is little doubt that pultruded composites offer an important advance in upgrading conventional thermoplastic composites and making them suitable for use in competition with more established metals or thermoset products in many large volume markets.

Definition

Pultruded conventional thermoplastic composites are produced by pulling collimated continuous fibres through a heated zone, where they come into contact with a thermoplastic resin which is induced to flow around the surface of all the fibres to achieve a very high level of impregnation. These products are produced in the form of unidirectional fibre reinforced foils (prepregs) or tapes. They comprise three vital components:

(i) a thermoplastic resin,
(ii) a continuous reinforcement fibre,
(iii) an interface or interphase region.

In theory, a wide range of amorphous and semicrystalline thermoplastic polymers can be combined with any continuous reinforcement fibre, e.g. glass, carbon, aramid or steel to give a range of fibre volume contents from 15–60%. To minimize costs and make materials as attractive as possible to large potential customers, such as the automotive industry, these composites will be dominated by E-glass fibre and affordable resins that flow easily around fibre surfaces. Examples of resins are polypropylene, thermoplastic polyurethanes, thermoplastic polyesters and polyamides. And to achieve a compromise between significant property enhancement and cost effective prepreg production, fibre volume contents are likely to be initially in the order of 25–50%.

Production routes

Many routes have been investigated. One of the most encouraging has been melt impregnation. This approach was developed by ICI and by Polymer Composites, from whom it was transferred to Celanese and Hoechst. Tradenames which relate to melt impregnated pultrusions include 'Fiberod', from PCI and 'Plytron' from ICI. Bayer also carried out development work with this method before seeking to license their technology in 1991. Statoil purchased an exclusive licence from ICI for polyolefin-based 'Plytron's for Europe' in October 1992.

One of the earliest routes consisted of pulling fibres sequentially past a crosshead extruder to achieve a resin melt coating and then through a shaping die to produce a lace or rod (Wilson Fiberfil). This technology was developed further by Akzo, who discontinued work in 1991.

Several companies (including ICI, ATO, Spieflex-Batignolles and Flexline) have examined powder coating techniques. Methods have varied from fluidized bed to 'electrostatically' aided processes with powder particle sizes ranging from a few micrometres up to several hundreds of micrometres cross-section. There is no doubt that very good quality products can be achieved by powder impregnation, but many resins have to be ground from pellets to produce powder and the rate of melting before powder can coat fibres effectively is diffusion limited. This restricts the choice of resins on cost grounds, and attempts to produce composite tape products quickly may result in poor or variable interface properties. By 1992, Spieflex had sold their technology to Baycomp and ATO and Flexline have been active in seeking licensees. ICI stopped their work on powder impregnation in favour of the melt impregnation route many years ago.

Similar problems apply to 'hybrid fibre' impregnation; it costs money to convert resin into fibres, there is a problem of dispersing reinforcement and resin fibres uniformly and there is a diffusion limitation on the rate at which resin fibres will melt and coat the surfaces of reinforcement fibres. Some work has been carried out in this area for aerospace materials by BASF.

Solvent assisted impregnation is not recommended. In theory, solvents can be used to reduce the viscosity of thermoplastic matrices so that good impregnation may result. Unfortunately, most solvents that will dissolve thermoplastic resins provide potential health hazards. Furthermore, the added cost of solvent recovery and its elimination from saleable products is prohibitive for low cost materials aimed at bulk markets.

Properties of pultruded composites

High collimation of fibres provides the maximum possible reinforcing effect along the axis of the fibres. Transverse to that direction or through the thickness of composite layers there is little effective reinforcement. This anisotropy gives the designer freedom to arrange reinforcement in the materials according to load paths in the structure.

The matrix resin transfers stress between the fibres and plays a vital role in determining the service temperatures, environmental suitability and the processability of the material.

The interface between resins and fibres can vary significantly from one combination of resin and fibre to another, e.g. polyamides and glass fibre form a very good interface with a number of proprietary glass size treatments, but polypropylene is naturally unreactive and may need extra chemical modification together with optimized size formulations to form a reasonable interface with glass fibre. This ability to control different levels of interface, or interphase, can be used to design components with some degree of confidence to meet specific impact threats ranging from low speed collision to high speed ballistic.

Table 18. Selected properties of unidirectional fibre reinforced thermoplastic composites

Resin	Polypropylene	Nylon 6:6/6 copolymer	48D TPU	70D TPU
Glass content (vol%)	35	42	40	39
Density (g cm^{-3})	1.48	1.74	1.76	1.75
Tensile strength (MPa)	720	800		
Tensile modulus (GPa)	28	32	26	
Flexural strength (MPa)	436	915	105	192
Flexural modulus (GPa)	21	29	6	19
Shear stress (MPa)	19	59		
Compression strength (MPa)	366	955		
HDT at 1.81 MPa (°C)	156	245		

TPU = thermoplastic polyurethane.
Source: ICI 'Plytron' data.

Table 18 lists a number of properties for different glass fibre reinforced resin products. By careful selection of a suitable resin, flexural, shear, compression and other properties can be varied considerably.

Table 19 compares strength properties of a glass/polypropylene composite with literature values for a variety of traditional engineering materials.

One of the bonuses of pultruded collimated fibre composites is that many of their properties can be predicted from classical laminate theory or the 'Rule of Mixtures' before products are even made. If the end use requirements are known, the correct product can be made first time. Tensile moduli are directly proportional to the fibre volume fraction. Using raw material values, coefficients of thermal expansion can be readily computed for any orientation of fibres in the plane of a laminate.

Table 19. Strength properties of 35 vol% glass reinforced polypropylene compared with various traditional engineering materials

Material	Average density (g cm^{-3})	Tensile strength (MPa)	Specific tensile strength (MPa cm^3 g^{-1})	Normalized tensile strength
Glass/PP [0]$_8$ laminate	1.48	720	486	1.00
Glass/PP [0/90]$_S$ laminate	1.48	360	243	0.50
Stainless steel	7.8	286–500	36–64	0.07–0.13
Mild steel	7.8	220	28	0.06
Copper alloys	8.3	60–960	7–116	0.01–0.24
Aluminium alloys	2.8	100–627	36–224	0.07–0.46
Aluminium	2.6	40	15	0.03
Magnesium alloys	1.8	80–110	44–61	0.09–0.13

Source: ICI 'Plytron' data.

The shear modulus of any laminate can be calculated. Other properties such as flexural modulus and flexural strength have straight line relationships with the fibre volume up to 55–60% fibre content. Much of the characterization of such materials can be effected by measurement of the basic elastic constants of a resin/fibre combination. Table 20 shows a series of constants for ICI's 35 vol% glass/polypropylene 'Plytron'. This lists previously unpublished work in which measurements have been carried out both mechanically and ultrasonically. Differences in the absolute values result from the speed of generation of test data by the two techniques. The important feature is that they show there is a slight, but significant, difference in properties of laminates through their ply thicknesses. This does actually happen and can give problems in thermoforming composites to specific bend angles or degrees of curvature.

As mentioned, the impact properties of pultruded thermoplastic composite laminates varies with the nature of the resin/fibre interface. It also varies considerably with laminate construction and size, fibre volume content and the speed and sharpness of the impactor. As an example, a well impregnated glass/polyamide crossply laminate is much better than glass/polypropylene in stopping sharp knife penetration, but glass/polypropylene is much more effective in stopping many ballistic threats, because the energy absorbed in interply delamination mechanisms can arrest the passage of a bullet and ensure that no harmful fragments spall off the tensile surface of the laminate.

Table 20. Elastic constants for 35 vol% glass fibre reinforced polypropylene 'Plytron'

Constant	Mechanical test measurements	Ultrasonic measurements
E_{11}	28.7	26.2
E_{22}	4.07	8.22
E_{33}	–	7.15
ν_{12}	0.4	n.m.
ν_{13}	0.39	0.31
ν_{21}	0.04	n.m.
ν_{23}	0.62	0.41
G_{12}	1.35	n.m.
G_{23}	1.26	2.18
G_{13}	–	2.78

E, tensile modulus; ν, Poisson's ratio; G, shear modulus.
Source: ICI 'Plytron' data.

Another interesting aspect of this technology is the change in the high temperature safety of these composites, e.g. a red hot element can be held in contact with a 35 vol% glass/polypropylene for 30 s, according to Glow Wire Test IEC 695-2-1, at a temperature of 850°C without risk of fire. (The base polypropylene without reinforcement has an autoignition temperature just above 340°C, as measured by ASTM 1929.)

Design and fabrication

Typical design assemblies are shown in Fig. 30, to indicate some of the ways in which plies of unidirectionally reinforced composites can be assembled to give reinforcement in specific directions. Other assemblies can be produced from piles formed by weaving tapes on slightly modified conventional looms. Fig. 31 shows the variation in flexural modulus that can be achieved by differences in laminate construction.

Products based on glass/polypropylene have a number of very important advantages in developing end use markets at this stage.

(i) Prepregs have low cost potential.
(ii) They have longer shelf life, require shorter fabrication cycles and give rise to less concerns over health and safety than some thermoset materials.
(iii) Products can be thermoformed at temperatures below 190°C, so can be evaluated on a range of existing capital equipment, e.g. sheet stamping, vacuum forming and multi-daylight presses.

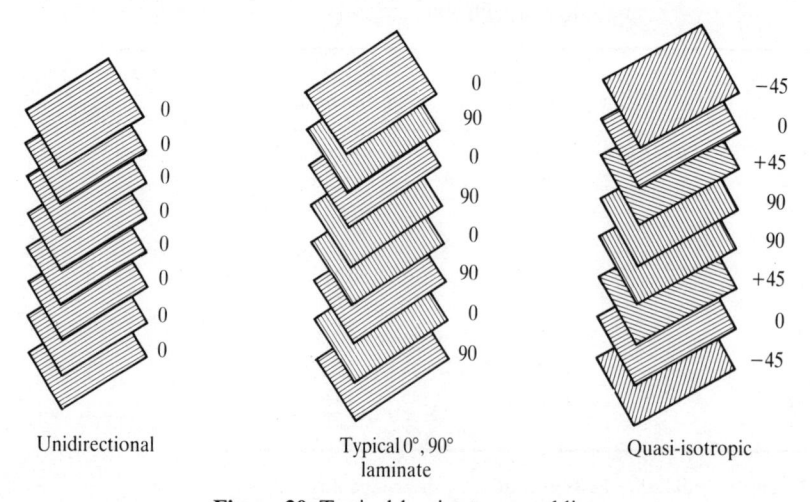

Figure 30 Typical laminate assemblies

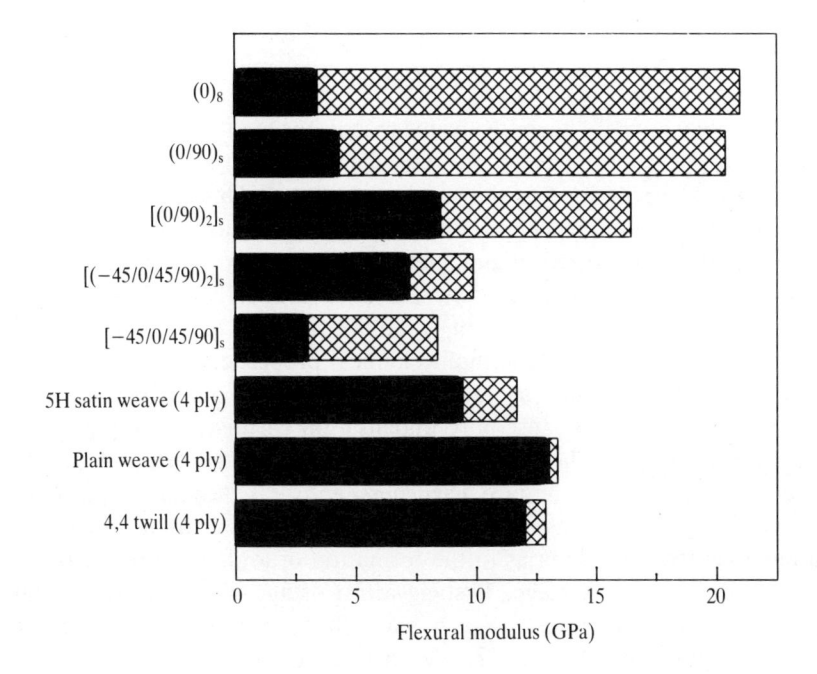

Figure 31 Flexural modulus of Plytron GR/PP laminates. Effect on ply lay-up on Anisotropy. ■ y direction ▨ x direction

(iv) Products have already been used in a number of tape winding or tape placement processes (including braiding), as well as in developments of rubber block stamping, matched metal moulding and section pultrusion.
(v) All scrap materials can be recycled, of particular interest to any environmentally conscious society.

References

1. F N Cogswell, *Thermoplastic Aromatic Polymer Composites*, Butterworth-Heinemann, Oxford, 1992.
2. G Cuff, *Long Fibre Reinforced Thermoplastics – Extending the Technology*, BPF Congress, Blackpool, Nov. 1988, BPF Publication No. 293/4, pp. 15–17.

2.4 Epoxy resins

F R JONES

Introduction

Cured epoxy resins[1-4] form the prime matrix for high performance glass, aramid and carbon fibre composites, whereas the unsaturated polyester resins are primarily used in industrial applications. This is because epoxy resins combine good mechanical performance with 'easy' processing that satisfies the generalized requirements of a matrix polymer: (i) low viscosity for fibre impregnation, (ii) high reactivity on curing, (iii) chemical control of cure without volatile formation, (iv) low shrinkage, (v) good mechanical and thermomechanical properties.

True thermosets such as the phenolics and the polyimides do not satisfy criterion (iii) because polycondensation chemistry is employed in the curing stage. The epoxies are more correctly referred to as chemosets since they require hardeners and/or catalysts to promote crosslinking. As a consequence the curing process cannot simply be interrupted by quenching as in the formation of a phenolic resit from a resole. The terms, A stage, B stage and C stage were coined for the gradual increase in viscosity associated with an increase in the extent of polymerization of a phenolic resole on heating, i.e. resole, resitol, resit, respectively, but they are occasionally used in epoxy resin chemistry. In epoxy resin matrices for prepregs, the term B-staging is used analogously to describe an increase in viscosity, after fibre impregnation, which occurs on partial reaction of the epoxy with the hardener. For high performance composites, prepreg with the required 'tack' can be achieved in this way. However, the lifetime of the resulting prepreg may be limited, and other latent hardeners are often preferred, but viscosity modifiers are also needed.

Types of epoxy resin

The difunctional diglycidyl ether of bisphenol A (e.g. Ciba Geigy MY750) and related oligomers of degrees of polymerization up to four form the basis of the general purpose resins for both adhesives and composites. These are formed from the reaction of (i) epichlorohydrin (ECH) with (ii) bisphenol A (BA) in the presence of sodium hydroxide. The molar ratio of these two monomers dictates the degree of polymerization and the molar mass of the final epoxy resin (iii).

With $n = 0$ in (iii), the resin is referred to as diglycidyl ether of bisphenol A (DGEBA) and is commonly used as a component in an epoxy resin matrix. The higher polymers ($n > 0$) have higher viscosities and are less useful in high performance composite applications. Further, the hydroxyl content makes them of more interest as adhesives (through their hydrogen bonding capability) but makes them more polar and water sensitive and therefore of less interest in composite materials. In addition, a larger value of n dictates a more flexible molecular structure to the cured resin with consequently lower T_g and modulus but increased ductility. It also follows that n is fractional, which illustrates the fact that the resins are a mixture of oligomers of integral degree of polymerization. The high molar mass polymers are known as phenoxies. The properties of these resins are given in Table 21.

The glycidyl amines have a higher functionality and potentially higher crosslink densities and glass transition temperatures and form the basis of the resins for aerospace composites. The latter are used as prepreg resins and are often mixtures based on the tetrafunctional tetraglycidyl amine, TGMDA, e.g. Ciba Geigy MY720. Typical resins for laminates are given in Table 22. Judicial choice of resin blend and hardener is required to provide controlled cure and viscosity profiles during care. Furthermore, thermoplastic modifiers such as polyethersulphone (see Section 2.1) and phenoxy resins are also used to control process variables such as viscosity and tack.

The Ciba Geigy 913/914 systems are based on this technology, as are some Fiberite systems. Both the molecular weight of the thermoplastic and the end-groups contribute to processing and performance. For example amine end-group terminated polyethersulphones ensure chemical bonding with the epoxy resin.

Moisture absorption leads to a significant reduction in the glass transition temperature of the cured epoxy resins and seriously limits the

Table 21. Different epoxy resins from bisphenol A, BA, and epichlorohydrin, ECH (diglycidyl ethers of bisphenol A)

n (structure iii)	ECH:BA (mole ratio)	Ideal M (g mol^{-1})	Observed M (g mol^{-1})	Epoxy groups (mol^{-1})	*EPEW (g mol^{-1})	†S pt (°C)	Viscosity (cps)	‡Epikote designation
0	4	340	380	2	170	<RT	4000–6000	825
0.07	–	360	–	2	180	<RT	7000–9000	826
0.14	–	380	–	2	190	<RT	10,000–16,000	828
1	2	624	451	1.39	314	43	–	–
2	1.4	908	791	1.34	592	84	–	–
2.3	1.33	993	802	1.10	730	90	–	–
3	–	1192	–	1.99	500	65–75	–	1001
4	1.25	1476	1132	1.32	862	100	–	–
34	1.20	9996	1420	1.21	1176	112	–	–
	–		–	1.99	5000	155–165	–	1010

*Epoxy equivalent weight or epoxy molar mass.
†Softening point.
‡Shell Company Epikote or Epon resin.

Table 22. Typical epoxy resin matrices for composites

Difunctional
DGEBA, $n = 0$, e.g. MY750

Trifunctional
For example, MY0510

THPM-1, e.g. XD 7342.00L

Tetrafunctional
TGMDA, e.g. MY 720

PGA-X

THPM-2, e.g. XD 9053.00L

maximum service temperature. To mitigate this effect, highly crosslinked resins are preferred. However, to ensure that microcrack resistance of the composite is maintained, toughening modifiers are employed. The thermoplastic modifiers described above can also act in this way.

Other important systems employ thermosetting modifiers such as bismaleimide or cyanurate resins to achieve high temperature performance and durability.

Curing of epoxy resins

The epoxide or epoxy is a highly reactive functional group. It can either be homopolymerized to a polyether with a catalyst or copolymerized with a hardener. Hardeners are more important and generally have active 'hydrogen' atoms, which add to the epoxy group

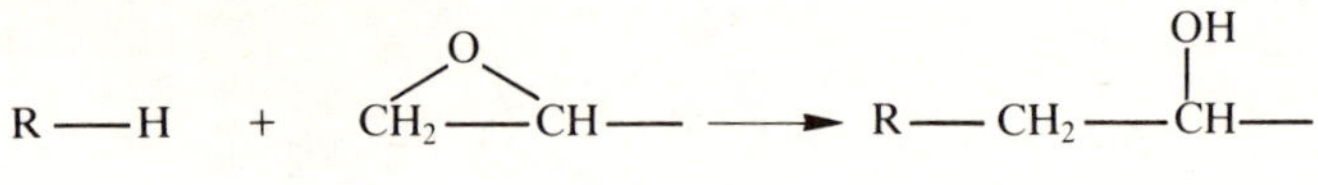

$$R-H + CH_2-CH(O) \longrightarrow R-CH_2-CH(OH)-$$

(iv)

As a consequence, the hardener is added at a stoichiometric concentration dependent upon the epoxy equivalent weight (molar mass per functional group or weight per epoxy group) and the molar mass of the hardener. The useful concentration can therefore vary from a few parts per hundred of resin (phr) to 100 phr. This demonstrates that the cured resin has a molecular structure and properties that are strongly dependent upon the nature of the curing system. There are two main types of hardener: the acid anhydrides and the amines and polyamides. These react with epoxies to form β hydroxyl amines (v). At first sight, this might be expected to promote adhesion to the fibres through the hydroxyl groups but the polarity promotes moisture sensitivity. Hardening or catalytic systems that provide a less polar molecular structure are preferred.

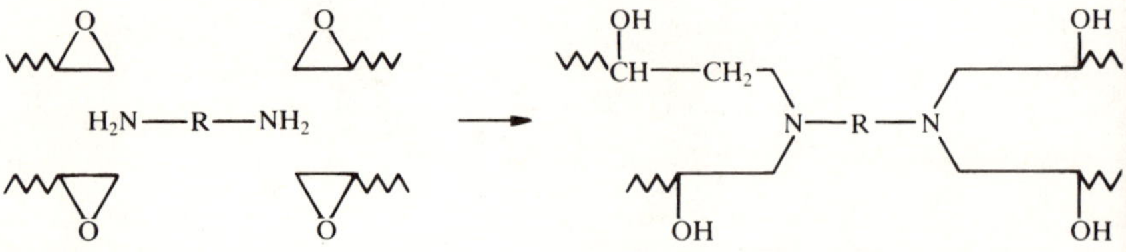

(v)

With anhydride (and acid) curing agents, several reactions are possible giving a cured resin structure that consists of ester linkages and pendant hydroxyl groups at irregular intervals. The final structure and polarity is dependent on the kinetics of the various individual reactions and the presence of adventitious moisture. The rate of cure is too low for composite fabrication and catalytic quantities of 0.5–2 phr of a tertiary amine give more control and initiate an alternating copolymerization of the epoxy and anhydride.

Nadicmethylenetetrahydrophthalic anhydride (NMA) is commonly employed with the diglycidyl ether (MY750) and the tetrafunctional glycidyl amine (MY720) with benzyldimethylaniline (BDMA) or the similar Epikure K61b. This system provides a matrix with lower water absorption, which can be attributed to the polyester copolymer molecular structure (vi), and the fact that hydroxyl groups will only form in side reactions.

(vi)

The catalytic curing agents such as tertiary amines and boron trifluoride complexes (at 2–6 phr) initiate an anionic or cationic addition homopolymerization of the epoxy groups. A polyether structure (vii) originates from the use of these catalytic systems of which the cationic boron trifluoride amine complex is favoured.

$$BF_3:NH_2Et \longrightarrow H^+ + [BF_3NR_2]^-$$

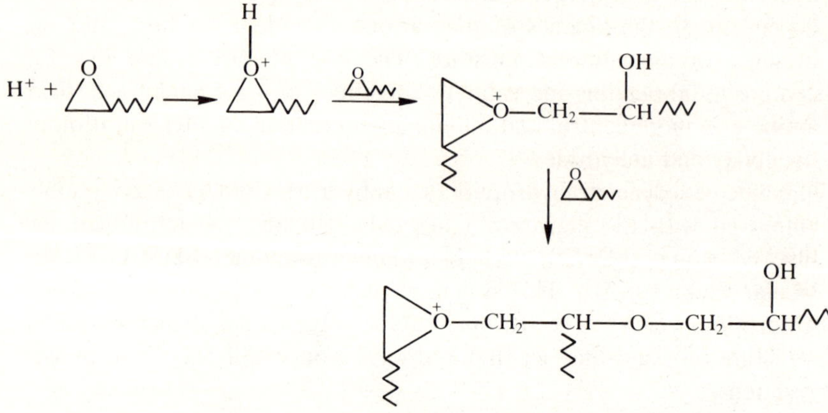

(vii)

For advanced composites, BF_3 complexes, often in combination with diamino diphenylsulphone (DDS) (viii) act as latent hardeners becoming active at $\simeq 160°C$. The viscosity of the resin and hence the tack of the prepreg can be adjusted by advancing DDS/epoxy reaction. The resin viscosity profile with temperature is also optimized for the consolidation of the prepreg stack and removal of entrapped air during lamination by vacuum bag/autoclave techniques before the onset of cure.

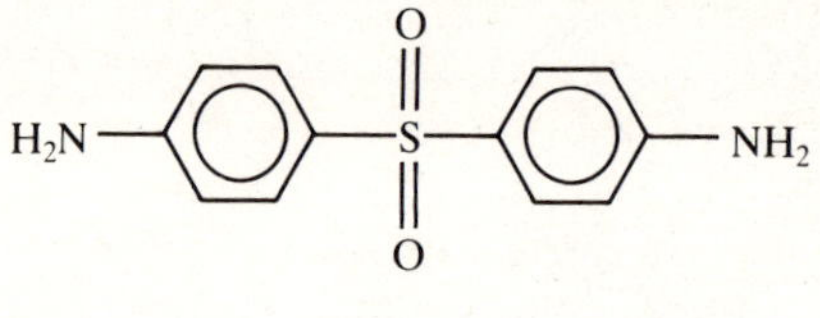

(viii)

Micronized dicyandiamide (DICY) (ix) also provides latent curing ability to a prepreg with improved shelf life. In these systems prepreg resin viscosity and tack is adjusted by incorporating a thermoplastic modifier such as polyethersulphone or phenoxy and employing mixed epoxies. Ciba Geigy refers to these as flow controlled prepreg. Both high and low temperature curing systems are available. Without the presence of 'Diuron' or 'Monuron' amine accelerators, the rate of curing is determined by the melting of DICY at 165°C at which temperature it dissolves into the epoxy resin. With 'Monuron' the curing temperature can be reduced to 120°C. While low temperature systems have clear advantages, not least the control of the exotherm during cure, careful control of the curing schedule is required for the production of good quality void-free laminates. Thus a dwell time at 'minimum' viscosity has to be incorporated, usually $\simeq 90°C$ for these systems. The other disadvantage of the low temperature curing systems based on DICY is the increased probability of residual DICY hardener in the resin. Another observation is the inconsistent appearance of the microstructure in these cured systems. Much of the variability can be reduced by ensuring that the hardener is micronized to a uniform fine particle size before prepreg manufacture. However, microstructural variations can still occasionally appear but these could also be attributed to phase separation of the thermoplastic modifier. DICY is used in catalytic quantities, 4–6 phr, but the curing chemistry is complex. It produces an approximate structure[1] expected of a polyamine, as indicated below.

$$H_2N-\underset{\underset{NH_2}{|}}{C}=N-C\equiv N \rightleftharpoons NH_2-\underset{\underset{NH}{\|}}{C}-NHCN$$

(ix)

$$3 \;\triangle\!\!\!\!\!\overset{O}{} \;+\; H_2N-\underset{\underset{NH}{\|}}{C}-NHCN$$

$$\downarrow$$

$$\begin{array}{c} \sim\!\!\!\!\sim CH-CH_2 \\ | \\ OH \\ \sim\!\!\!\!\sim CH-CH_2 \\ | \\ OH \end{array} \!\!\!\! N-\underset{\underset{NH}{\|}}{C}-N-CN \\ | \\ CH_2 \\ | \\ CH-OH \\ | \\ \sim\!\!\!\!\sim$$

$$\sim\!\!\!\!\sim C\equiv N \;\bigg\downarrow$$

$$\begin{array}{c} \sim\!\!\!\!\sim CH-CH_2 \\ | \\ OH \\ \sim\!\!\!\!\sim CH-CH_2 \\ | \\ OH \end{array} \!\!\!\! N-CH-N-CN \\ | | \\ NH CH_2 \\ | | \\ \sim\!\!\!\!\sim C=NH CH-OH \\ | \\ \sim\!\!\!\!\sim$$

(x)

It should be pointed out that the complete reaction of the epoxy groups is not normally achieved except after long times at high temperature, so higher heat distortion temperatures and chemical resistance can only be achieved by appropriate post-curing. The time temperature transformation curing diagrams of Gilham (see Section 2.9) are particularly valuable in this context. The effect of post-curing on mechanical properties can be generalized by appreciating that the heat distortion temperature and Young's modulus increase with crosslink density and the concentration of aromatic groups within the network. But there are exceptions to this where a higher glass transition can arise alongside a lower modulus, since the glass state is frozen in at higher molecular free volume.

Mechanical and thermomechanical properties of cast epoxies

The various curing mechanisms cause the molecular structure of the crosslinked epoxy resins to differ, consequently the mechanical performance and useful temperature range can vary across a wide range. This is illustrated in Table 23, where the 'hardener' dependent properties are summarized. These should be considered as a guide to epoxy resin performance. The effect of curing conditions also has a profound effect on properties. A further aspect of resin selection is

Table 23. Effect of curing agent on the properties of an epoxy resin, diglycidyl ether of bisphenol A (DGEBA)

Type	Agent	Base resin structure	Level (phr)	HDT max (°C)	σ_{um} (MPa)	E_m (GPa)	ϵ_{um} (%)
Anhydride hardener	NMA	Polyester (vi)	80	218	90	3.3	2.7
	THPA		60	120	81	2.7	4.5
	PA		50	110	82	3.0	4.1
Amine	DDS	Polyhydroxyamine (v)	33	193	78	3.1	6.0
	DTA		8	95	70	3.4	5.3
	TETA		13	96	62	3.6	2.6
Catalyst	BF_3.MEA	Polyether (vii)	4	176	58	–	1.8
	DICY	Polyhydroxyamine (x)	4	125	–	–	–

DDS = diamino diphenyl sulphone.
DICY = dicyandiamide.
DTA = diethylene triamine.
HDT = heat distortion temperature (maximum achievable value given).
NMA = nadicmethylenetetrahydrophthalic anhydride.
PA = phthalic anhydride.
TETA = triethylene tetramine.
THPA = tetrahydrophthalic anhydride.

moisture absorption during service. This is particularly important for high humidities. These aspects are discussed further in Chapter 6 but of crucial importance is the average reduction in T_g caused by water plasticization. For each per cent of moisture T_g is reduced by 20 K. Since the maximum moisture concentration is generally determined by the polarity of the cured resin structure, the maximum useful temperature will be up to 100°C below the dry T_g, depending on the resin/hardener combination. Thus a 150–200°C composite resin may only be used where service temperatures are limited to 100–150°C. This has led to much interest in higher functional epoxies; bismaleimide (BMI) and bismaleimide triazine (BT) modified epoxies, which can have a wet T_g of 190°C; and alternative BMI, BT, polyimide matrices (see Sections 2.5 and 2.6). For example, the tri(hydroxyphenyl) methane triglycidyl ether (THPM-1) resins in comparison to the tetraglycidyl amine (TGMDA) resins have improved dry and wet mechanical performance (see Table 22).

Rapid curing resins for structural RIM

Much progress has been made recently in the development of rapid curing epoxy systems for alternative composites processing (see Section 3.19). For example, it is reported[5] that gel times of <25 s at 40°C can be achieved with BF_3 diol complexes (xi) of dianol 240 and 320 to cure DGEBA.

$$BF_3 : HO{-}[{-}CH_2CHO{-}]_x {-}\phi{-}C(CH_3)_2{-}\phi{-}[{-}OCHCH_2{-}]_y{-}OH$$

(xi)

where with R = H (xi) the average value of x and y = 2.15. With R = CH_3, $x = y = 1$.

To provide control over resin viscosity for RIM processing, comonomers for the cationic polymerization of the epoxy groups, tetrahydrofuran yielded a resin of low T_g but with butyrolactone (xii) a higher T_g of 103°C could be achieved. Gel times of 64 s and 80 s were observed.

(xii)   HO—CH_2CH_2O—⌬—CH_2CH_2—OH

(xiii)

The shortest gel time of 35s was observed with the BF_3 complex with diol (xiii). The maximum cure temperature was 70°C.

Conclusions

Epoxy resins is a generic term that covers a range of chemoset or crosslinked plastics, many of which are used as matrices for fibre and particulate reinforced composites. The nature of the hardener and epoxy components governs the final cured properties but those for composites generally provide the highest temperature performance with convenient processing and environmental durability.

References

1. B Ellis, ed., *Epoxy Resins*, Blackie, Glasgow, 1992.
2. C A May, *Epoxy Resins, Chemistry and Technology*, Marcel Dekker, New York, 1988.
3. R G Weatherhead, *Fibre Reinforced Resin Systems*, FRP Technology, Applied Science, London, 1980.
4. G E Green in *Composite Materials in Aircraft Structures*, ed. D H Middleton, Longman, Harlow, UK, 1990, Ch. 4.
5. S Mortimer, A J Ryan, J L Stanford in *Proceedings of the 5th International Conference FRC 1992*, Newcastle upon Tyne, Plastics and Rubber Institute, London, 1992, Paper 2.

2.5 High temperature resins – thermosetting polyimides

J N HAY

Polyimides[1–3] represent the largest class of high temperature polymers in use in composites today. A recent survey of high temperature polymers by Kline and Co. showed that only aramids and fluoropolymers exceeded the sales of polyimides in 1988. Sales of polyimides are forecast to rise by over 12% per annum. They are used in a wide variety of applications, from microelectronics to structural composites. Polyimides are available as linear or thermoplastic polymers (Ultem; Kapton) or as thermosets (so-called addition polyimides). Thermoset polyimides are considered in more detail below: the two main ones are the PMR types and the bis-imides, such as BMI.

PMR polyimides

PMR polyimides are made by the polymerization of monomeric reactants. This refers to the fact that the materials are polymerized in

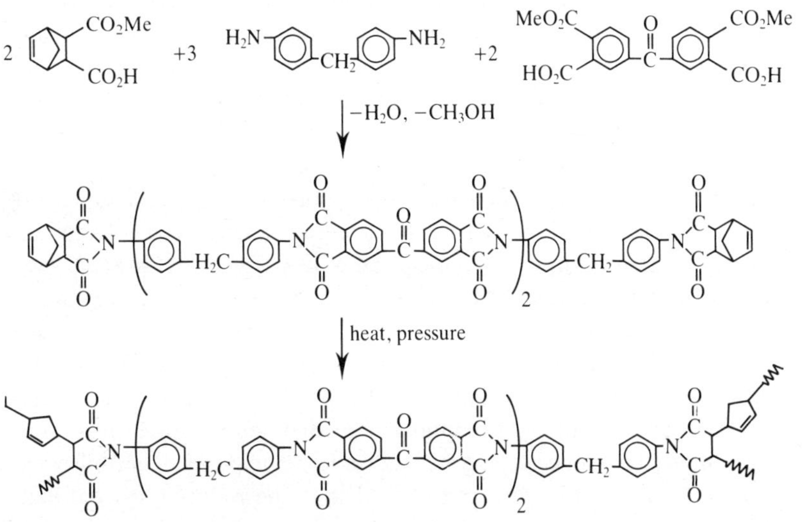

Figure 32 PMR approach (PMR-15)

situ during processing of a component. The materials are usually supplied as prepregs, made by impregnating glass or carbon fibres with a solution of the monomers in methanol or by a pseudo hot melt route. The chemistry of the PMR polyimides is illustrated in Fig. 32 for the most common type, PMR-15, which was developed by Serafini at NASA. A PMR-15 prepreg normally contains the monomers MDA (4,4'-diaminodiphenylmethane), BTDE (benzophenone tetracarboxylic acid dimethyl ester) and NE (5-norbornene-2,3-dicarboxylic acid monomethyl ester), often with residual solvent to provide tack and drape. Components are made by laying up prepreg plies in the desired configuration followed by processing using compression moulding or an autoclave. The first stage ('staging') leads to an imidized prepolymer or oligomer, of molecular weight 1500 in the case of PMR-15. Volatile by-products need to be removed, either in the autoclave or in a vacuum oven, before compression moulding. In the second stage, the imidized prepolymer is crosslinked via reaction of the norbornene (nadic) end cap. High temperatures and pressures (e.g. 316°C and 200 psi) are required to provide good consolidation and limit void formation.

PMR-15 composites have good thermo-oxidative stability and mechanical properties at elevated temperatures up to around 300°C. Room temperature properties of representative polyimide resins and carbon fibre composites are given in Tables 24 and 25. The high stiffness/weight and strength/weight ratio of PMR-15 composite has led to applications in aero-engines, for example in bypass ducts. Despite its generally good properties, PMR-15 does suffer from a number of drawbacks, which have hindered its more widespread use. These include the toxicity of the constituent MDA, microcracking of fibre composites on repeated thermal cycling between temperature extremes, high temperature processing and problems due to irreproducibility in processing. Much effort has been expended in trying to solve these problems. BP has approached the toxicity problem by replacing MDA with a less toxic monomer in its B1 prepreg system, while Rohr and American Cyanamid have introduced an MDA-free prepreg, believed to be based on pre-imidized PMR-15. SP Systems markets an apparently similar PMR-15 prepreg. Microcracking eventually reduces mechanical properties such as interlaminar shear strength, but joint BP/Rolls-Royce work has shown this can be improved by reducing the cure temperature to 288°C, while retaining acceptable elevated temperature properties. There is limited evidence to suggest that microcracking susceptibility can also arise from volatile entrapment and that the reduced susceptibility could arise from polycyclopentadiene crosslinks of differing degrees of polymerization.[4] NASA has reduced microcracking by forming a semi-interpenetrating network of PMR-15 with the tough thermoplastic NR-150B2 polyimide in their LaRC-RP40 system.

Other crosslinkable end caps can be used. Acetylene end caps form the basis of National Starch and Chemical Co.'s Thermid 600 series of resins. These suffer from a narrow processing window due to the relatively low cure temperature of the acetylene groups. This problem has been partially alleviated by using the isoimide, which has better flow characteristics but rearranges to the imide during processing.

Resins that can operate at higher temperatures than PMR-15 have been developed. Operating temperatures of 371°C (700°F) are claimed for these composites. They are based on Hoechst's fluorinated monomer, 6F, and are more expensive than PMR-15. NASA developed a nadic end capped analogue called PMR-II-50 which has a prepolymer molecular weight of 5000. A more thermo-oxidatively stable resin, AFR700B, has been developed recently by Serafini at TRW under a US Air Force contract.

Bis-imides; BMI

Bis-imides are materials derived from monomers containing two preformed imide groups. By far the most common members of this family are the bismaleimides, BMIs. Allylnadic imides are a recent addition to the family. Bis(benzocyclobutene imides), BCBs, are considered in Section 2.6.

BMIs occupy the temperature regime below PMR resins and have a continuous use temperature in the range 200–230°C. BMI monomers are made by the reaction of diamines (normally aromatic) with maleic anhydride. A wide range of BMI resins is commercially available. They are often supplied as eutectic mixtures to lower the melting point. Suppliers of base resins and formulated BMI resins include Shell-Technochemie, Ciba Geigy and DSM. BMI prepregs can be made by impregnating a fibre reinforcement from the melt or from a solution of the monomers in dichloromethane or methylethyl ketone (MEK). Although BMI resins can be cured simply by thermal radical polymerization, the product formed is very brittle, so comonomers and reactive diluents are most commonly used to improve the toughness of the final product. The modifiers also serve to reduce the melting point and viscosity of the resin. The most commonly used comonomers are aromatic diamines, allylphenyl and propenyl compounds. The diamines react initially by a Michael addition to form an aspartimide. Allylphenyl modifiers form the basis of the Ciba Geigy 5292 system and the Technochemie TM120 and TM121 comonomers. They react with BMI resins via an ene reaction, followed by a Diels-Alder addition. Propenyl compounds coreact initially via a Diels-Alder reaction. Examples of propenyl modifiers include Technochemie TM122 and TM123.

BMI resins can be processed by a wider range of techniques than PMR resins, largely because no volatiles are evolved during cure. Autoclave processing of BMI composites is typically carried out under 100 psi pressure at a final cure temperature of 204°C (400°F), followed by a post-cure. Compression moulding, filament winding and resin transfer moulding (RTM) can also be used with BMI resins. The low viscosity resins required for filament winding and RTM can be obtained by using the modifiers described above, although cure or post-cure times can be rather long (several hours). Resin viscosities of a few hundred centipoise and an acceptable pot life can be realized at temperatures below 100°C. DSM Desbimid resin is modified with reactive diluents such as styrene or 2-hydroxyethylmethacrylate to produce low viscosity grades. In these cases, cure is effected by using a peroxide as a free radical initiator.

Typical BMI resin properties are shown in Table 24 for the Compimide and Matrimid systems, in comparison with the higher temperature polyimides and a tetrafunctional epoxy. Composite properties of two formulated BMI resins are given in Table 25. As mentioned previously, unmodified BMI resins are very brittle. Addition of comonomers increases the fracture toughness, as can be seen from the G_{Ic} values. This normally occurs at the expense of both the resin modulus and the glass transition temperature. Further toughening of the BMI resins can be achieved by blending in thermoplastics such as polyimide (Matrimid

Table 24. Typical properties of polyimide resins

Resin	T_g (°C)	Flexural modulus (GPa)	Flexural strength (MPa)	Elongation (%)	G_{Ic} (J m^{-2})
PMR-15	340		176	1.5	87
Thermid MC-600	320	4.5	145	1.5	
Avimid NR150B2	340			6.0	
Compimide 796	>300	4.6	76	1.7	63
Compimide 796-TM123 (60:40)	261	3.7	132	3.7	439
Matrimid 5292	273	4.0			170
TGMDA – DDS	250	3.6	90	1.8	53

5218), polyetherimide (Ultem) or polyarylene ether. Rubbers may also be used but will tend to reduce the thermal properties of the BMI. Thermoplastic toughening is common in commercial BMI prepregs. Despite this, microcracking of BMI laminates can still be a problem. A significant advantage of the BMI resins is their improved hot-wet properties compared to epoxy resins. A point to note is that the properties of BMI composites can be affected significantly by the choice of fibre and fibre sizing resin.

The good hot-wet properties of BMI resins have allowed them to find aerospace applications, for example in wing skins, nacelles and missiles. The improvements in toughness and microcrack resistance achieved by formulators might be expected to lead to more widespread use. Suppliers of BMI prepregs include BP-Hitco, BASF-Narmco, American Cyanamid and Hysol. BMI resins are also widely used in printed circuit board

Table 25. Typical properties of polyimide/carbon fibre laminates

Resin	Fibre	0° Flexural modulus (GPa)	0° Flexural strength (GPa)	0° Short beam shear strength (MPa)	G_{Ic} (J m^{-2})	T_g (°C)
PMR-15	Celion 3k 8HS fabric	66	845	69		345
PMR-15	60% V_f UD tape	134	1645			
PMR-II-50	Celion 6000 UD		1840	112		370
Thermid 600	HTS UD	104	1346	86		354
Compimide 796-TM123 (65:35)	HTA-7	126	1884	92	399	
USP V391	Hitex 46-8B UD	169	1765	124		263

laminates; some show excellent adhesion to copper, in particular Ciba Geigy Kerimid 601.

Allylnadic imides are a relatively recent class of thermoset polyimides developed by Ciba Geigy for use as composite matrices and adhesives. These are preformed bis-imides containing both allyl groups and norbornene double bonds. Thermal polymerization is slow, but sulphonic acid catalysts catalyse curing at 250°C for 2 h. Some of the monomers exhibit low viscosities at temperatures < 100°C. Glass transition temperatures above 300°C have been reported for the homopolymers. The monomers have also been cured in blends with BMI resins and cyanate ester resins.

Conclusions

The main types of thermosetting polyimides commercially available are PMR and BMI resins. The generally good thermal and mechanical properties of their fibre composites suits them to applications in areas as diverse as aerospace and electronics.

References

1. B Sillion in *Comprehensive Polymer Science*, Vol. 5, eds G C Eastmond, A Ledwith, S Russo, P Sigwalt, Pergamon Press, Oxford, 1989.
2. D Wilson, H D Stenzenberger, P M Hergenrother (eds), *Polyimides*, Blackie, Glasgow, 1990.
3. J W Verbicky in *Encyclopedia of Polymer Science and Engineering*, 2nd edn, Vol. 12, eds H F Mark, N M Bikales, C G Overberger, G Menges, John Wiley & Sons, New York, 1988.
4. M Simpson, P M Jacobs, F R Jones, *Composites*, **22**, 89, 99, 105 (1991).

2.6 *High temperature resins – other thermosets*

J N HAY

Although the current market for high temperature thermosetting composite matrices is dominated by polyimides, a number of alternative resin systems have been developed either to fill the gap in the service temperature range between epoxies and polyimides or to overcome performance shortcomings of the polyimides.

Cyanate ester resins

Cyanate esters[1] are probably the most well developed group of high temperature thermosets after the polyimides. In terms of temperature capability, simple cyanate esters occupy a position just below the BMI resins. Homopolymer glass transition temperatures, T_g, normally range from below 200°C to nearly 300°C (phenolic triazine resins are an exception, see below).

Cyanate esters are made by cyanation of phenols, based on technology originally developed by Bayer. They cure via a trimerization reaction to give a thermoset network containing highly thermally stable triazine rings. Cyanate ester resins are now manufactured by Ciba Geigy (previously Hi-Tek Polymers), Dow and Allied Signal. Ciba Geigy AroCy resins are esters of bisphenols. They are supplied as solid or semisolid monomers or prepolymers, or more recently as liquid monomers. AroCy L-10 has a viscosity of only 140 cP at 25°C. The low viscosities attainable make cyanates amenable to processing by a variety of techniques, including prepreg routes, resin transfer moulding (RTM) and filament winding. The monomers can be homopolymerized or copolymerized using a trimerization catalyst such as cobalt or copper acetylacetonates, or zinc naphthenate. Latent cure can be achieved by using trivalent cobalt catalysts. A few phr of nonylphenol is normally added as a cocatalyst. Cure cycles are typically 1–2 h at 177°C, followed by a post-cure at up to 250°C.

Dow's cyanate ester is based on dicyclopentadiene phenol novolak. In this case, the cured resin exhibits low moisture uptake due to the hydrophobic nature of the polymer. Allied Signal's cyanates are structurally related to Dow's cyanates. They are derived from cyanation of phenolic novolaks. Allied Signal calls its resins PT (phenolic triazine) resins. A range of resins is commercially available under the tradename Primaset, with glass transition temperatures up to around 400°C after post-cure. Cure can be effected uncatalysed, or by using a catalyst to lower the cure temperature. Like conventional cyanate esters, PT resins can be processed by a wide variety of processing techniques including liquid moulding methods.

Cyanates exhibit a number of desirable properties, apart from their rheology. In general, they give cured thermosets that are tough, exhibit low water absorption and have low dielectric constant. Monomers and precursors also tend to have low toxicity. Drawbacks include their relatively high cost and the sensitivity of the cyanate group to hydrolysis. Cyanates can be toughened further by blending with thermoplastics, as in ICI Fiberite prepregs. PT resins have thermal performance in excess of conventional cyanate esters, with thermo-oxidative stability comparable

Table 26. Typical properties of cured resins

Resin	Cure Temperature (°C)	T_g (°C)	Flexural modulus (GPa)	Flexural strength (MPa)	Elongation (%)	ϵ_x* (1 MHz)
AroCy L-10	250	259	2.76	162	7.7	3.0
RTX 366	204	175	2.8	121	5.1	2.6
PT Resin		400	4.7	97	2.5	2.97 (12 GHz)

* Dielectric constant

to BMI resins and PMR polyimides. A particular advantage of PT resins is their excellent fire properties. Selected properties of some typical cyanate resins are given in Table 26. The properties of cyanate esters make them suitable for use in structural aerospace components and electronic applications such as printed circuit boards.

Cyanates can be copolymerized with epoxy resins, giving products with increased T_g and improved electrical properties compared to the epoxy itself. Cyanates are also available as blends with BMI resins, as in Mitsubishi Gas Chemical BT resins, although it is now accepted that the two resins do not coreact.

Bisoxazoline–phenolic resins

Bisoxazoline–phenolics are a relatively new class of thermosetting resins developed by Ashland Chemical Company for composite applications. The resins are thermosetting poly(amide ether)s prepared by reaction of phenolic resins with bisoxazolines ('PBOX' monomers). Use of catalysts permits curing at 175°C, with post-curing at 225°C. Addition of unspecified reactive diluents can be used to lower the resin viscosity to around 200 cP at 100°C, suitable for RTM. Using a catalyst, gelation occurs in minutes at 175°C. Glass transition temperatures can be varied to in excess of 200°C, depending on the ratio of monomers. The flammability characteristics of the bisoxazoline–phenolic resins are excellent. Composites exhibit generally good mechanical properties, including hot-wet properties. Selected composite properties are given in Table 27 along with those of typical epoxy composites. Other advantages claimed for these resins include good electrical properties, low shrinkage on cure and low thermal expansion coefficient. The cost of these materials is not known. The properties of the resins and composites make them candidates for use in structural applications, including aircraft interiors, and in electrical laminates.

Table 27. Bisoxazoline phenolics – comparison of carbon fibre laminate properties with other resins (AS4 UD CF or equivalent) reprinted by permission of the Society for the Advancement of Material and Process Engineering from 20th Int. SAMPE Technical Conference, 27–29 September 1988

Resin	0° Flexural modulus (GPa)	0° Flexural strength (MPa)	0° Short beam shear strength (MPa)	Compression strength (MPa)
PBOX-phenolic	113	1593	108	1524
TGMDA-DDS	102	1572	73	
Hercules 8551-7 (tough epoxy)	128	1834	111	1682

BCB resins

Benzocyclobutene (BCB) resins are relatively new. They have been under development in recent years by Dow Chemical. The US Air Force Materials Laboratory has also been active in this area. A family of BCB monomers has been developed by Dow. Other reports have described imide oligomers terminated by benzocyclobutene groups. The BCB resins exhibit a number of attractive characteristics that should lead to significant commercial use in the future. The monomers cure by ring opening of the four-membered benzocyclobutene rings to give *o*-quinodimethane structures, which subsequently polymerize without evolution of volatiles. Bis-BCB monomers prepared by Dow include a divinylsiloxane monomer and a diketone product. Use of partially polymerized prepolymers improves stability and reduces the polymerization exotherm.

The prepolymers can exhibit viscosities of a few hundred centipoise when melted above 100°C, with excellent pot life, making them amenable to liquid moulding techniques such as RTM. Cure can be effected at 220°C, followed by a post-cure. T_g values greater than 300°C can be attained. BCB/carbon fibre composites appear to have better thermal performance and hot-wet properties than toughened BMI resins, while compression after impact (CAI) values are comparable (about 32 ksi).

A recent development by Dow has been the synthesis of a range of AB monomers containing both a BCB and a maleimide end group. These groups can copolymerize giving products with a high degree of linearity. This results in high values of resin fracture toughness, G_{Ic}, and carbon fibre composites with better values of CAI, up to $2\,kJ\,m^{-2}$ and 42 ksi respectively. Excellent thermo-oxidative stability is claimed for the cured resins.

In addition to RTM, the BCB resins can be used in prepregs and adhesives. NASA work has suggested these resins are strong candidates for high speed civil transport applications. The BCBs are also recommended for use in electrical and electronic applications where their low dielectric constant (<3 is possible) is an advantage. The expected high cost of these materials may limit their use to specialized high value applications.

Phthalonitrile resins

The phthalonitrile resins have been under extensive development in recent years by Keller at the US Naval Research Laboratory. The phthalonitrile monomers are solids containing four nitrile groups that can be polymerized to give crosslinked polymers, originally thought to be phthalocyanines but now in doubt. The cure reaction is catalysed by acids (organic, inorganic or Lewis), amines, metals or metal salts. A typical cure cycle is 16 h at 315°C. The T_g of around 300°C can be increased by a higher temperature postcure. Typical properties of the cured resin include tensile strength of 94 MPa, G_{Ic} of 0.12 kJ m^{-2} and K_{Ic} of 0.45 MN m$^{-3/2}$. Advantages of the phthalonitrile resins include indefinite shelf life, easy melt processing (although long cure cycles) and extremely good fire properties. The phthalonitriles may be competitive with high temperature polyimides such as PMR-15. They have potential as matrices for composites, adhesives and in electronics applications. Phthalonitrile monomers are expected to be available commercially in the United States in the near future.

Other resins

Other thermoset resins with potential applications in composites have been introduced recently or are under development.

Bayer has introduced a series of so-called EPIC resins (tradename Blendur) based on the catalysed reaction of epoxy resins with isocyanates. A combination of trimerization and cycloaddition leads to highly crosslinked thermosets. Uncured resins can have low viscosities and latent catalysts are available. Curing can be effected at 160°C followed by post cure at 250°C. T_g values well in excess of 200°C can be achieved. The resins are proposed for use in electrical and electronic applications.

Polystyrylpyridine (PSP) resins were developed by SNPE in France. The resins can show low viscosities (about 100 cP) at elevated temperatures (120°C), allowing processing by RTM. The resins can be cured at 200°C, but an extended post cure may be necessary at 250°C. Properties are good up to 200°C. Drawbacks are volatile evolution during cure and brittleness of the cured resins.

Sumikin Chemical in Japan has begun commercializing a resin based on condensation polynuclear aromatic (COPNA) resin. The resin has good heat resistance (up to 260°C), good abrasion resistance and excellent lubricating properties. Processing can be carried out by injection moulding and compression moulding. The resin has potential applications in carbon and glass fibre composites.

Dow Chemical is evaluating acetylene chromene terminated (ACT) resins for high temperature use. The uncured resin has a low viscosity (about 200 cP) at reasonable temperatures (70°C) and cures to give a product with a T_g of 350°C, high modulus and low water absorption. Dow was last known to be working to improve the toughness of the polymer.

Reference

1. R B Graver in *International Encyclopedia of Composites*, Vol. 1, ed. S M Lee, VCH Publishers, New York, 1990.

2.7 Phenolic resins – laminates

P A SHEARD

Resoles and novolaks are the two main categories of phenolic matrix commonly used for the manufacture of reinforced laminates. Both types are normally supplied as a fluid suspension with solid contents in the region 70–90%.

Resoles are formed by a condensation reaction between phenol and an excess of formaldehyde in the presence of either ammonia or caustic soda. These additions result in a spirit (ethanol) soluble or a water soluble resin respectively. Novolak resins are prepared from a small excess of phenol and formaldehyde reacted in the presence of an acid catalyst such as oxalic acid.[1]

During the preparation of the laminate, the resin solution impregnates and wets out the fibre reinforcement and is hardened (cured) to form a solid thermosetting phenolic matrix. Resoles are hardened by the application of heat and/or an acid catalyst, whereas novolaks are usually cured by adding hexamine, which is a source of the required additional formaldehyde.

Novolak resins have traditionally been used for general purpose moulding compounds and the water soluble resoles for paper making and decorative laminates. Hybrids or blends of the two are gaining in importance and are now used for the production of high quality structural composite components.

Phenolic resins are particularly attractive for the reinforced plastics industry because they have high service temperatures, good electrical properties and excellent chemical resistance. Though they are generally more brittle than their polyester or epoxy counterparts, their inherent fire resistance and ultralow smoke and toxic fume emission during combustion make them suitable for applications in high fire risk areas.[2]

Phenolic laminates have high glass transition temperatures (> 150°C) and retain their structural integrity during heating (Fig. 33). This makes them superior to steel structures, which can collapse during a fire.[3] Table 28 presents property data for a typical pultruded component.

The final properties of phenolic laminates are very sensitive to the fibres employed, the processing route and the curing schedule.[4] There is not yet a general phenolic compatible glass fibre sizing, therefore the interfacial performance is specific to individual combinations of phenolic resin systems and fibres. During cure, phenolics undergo a condensation reaction that can generate up to 25% by volume of water vapour together with the suspension medium. If not accounted for, this can result in very high void contents and entrapped moisture levels as well as damage to processing equipment due to pressure build-up. Typically, voidage of 10% by volume results from standard processing routes, although with

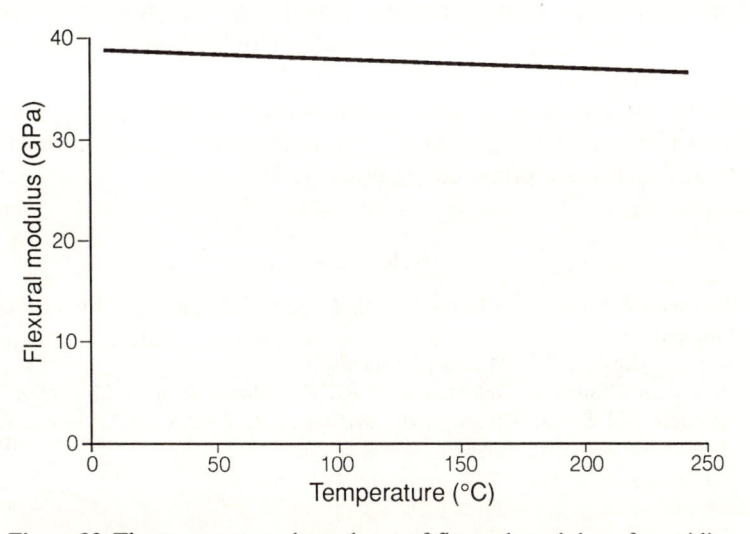

Figure 33 The temperature dependence of flexural modulus of a unidirectional glass fibre phenolics laminate (V_f 58%)

Table 28. Typical properties of a pultruded unidirectional, $V_f = 58\%$, glass phenolic composite

Physical properties	
Density (g cm^{-3})	1.9
Mechanical properties	
Tensile modulus (GPa)	45
Tensile strength (MPa)	550
Flexural strength (MPa)	800
Flexural modulus (GPa)	42
Temperature performance	
Coefficient of thermal expansion (°C^{-1})	25×10^{-6}
Thermal conductivity (W m^{-1} K^{-1})	0.35
Fire performance	
Temp. at 50% of ambient flexural strength (°C)	200
Time to this temp., NPD furnace/hydrocarbon test – 20 mm thick (min)	4
Flammability (UL-94)	V-O

modified autoclave and pultrusion techniques less than 3% has been achieved. (Autoclave techniques are discussed in Section 3.2 and pultrusion in Section 3.12.) A high residual moisture content in the laminates can lead to violent delamination during fire testing. This can be easily prevented by careful processing and permanently remedied by thermal or environmental ageing treatments.

Phenolic laminates are commonly available as compression moulded panels and increasingly as resin transfer moulded, filament wound, pultruded and autoclaved products. Structural applications for these items are emerging and phenolic laminates are becoming more attractive to the offshore, civil engineering, marine and mass transit industries, where fire safety is a prime design criterion.[5]

References

1. *Engineered Materials Handbook*, Vol. 1, ASM International, Metal Parks, Ohio, 1987.
2. K Forsdyke, *I.R.P.I.*, Mar./Apr. 1989, p. 6.
3. G Bishop, P Sheard, *Construction & Building Materials*, **6**(1), 31 (1992).
4. M Chen-Chi, S Wen Chang in *Proceedings of the 33rd SAMPE Symposium, Mar. 1988*, SAMPE, Anaheim, 1988, p. 767.
5. P Edwards, *Modern Plastics Intl*, Jun. 1989, p. 30.

2.8 Phenolic resins – moulding compounds

A SMITH

Introduction

Moulding compounds based on phenolic resins are well recognized and offer an attractive combination of important properties at a competitive price. Thus, the benefits include:

(i) high temperature stability,
(ii) retention of modulus over a wide temperature range,
(iii) excellent compressive creep resistance,
(iv) excellent chemical resistance,
(v) good flame retardance,
(vi) low smoke generation in a fire situation,
(viii) good electrical properties.

Whilst traditional wood-flour filled moulding compounds are still extensively used for products such as pan handles and electric meter cases, advances in fibre and resin technology have resulted in the development of high mechanical performance compounds. Among their automotive applications are products such as water pumps, inlet manifolds and brake components; their performance under stressed conditions at under bonnet temperatures is particularly valuable.

Raw materials

The main constituents of a phenolic moulding compound are resin, curing agent, fillers and/or reinforcing fibres, pigments and lubricants.

Resins[1]

Novolaks or two-stage resins are used for the majority of products. They are prepared from phenol (P) and formaldehyde (F) using a P:F molar ratio of about $1:0.8 \pm 0.05$. Acid catalysts are generally used; oxalic acid is the most popular. Under these conditions, phenol reacts with formaldehyde to produce a mixture of o and p-methylol phenols.

These then react quickly with more phenol to produce dihydroxy-diphenyl methanes, (DMP), i.e.

$$HO-\text{C}_6\text{H}_4-CH_2-\text{C}_6\text{H}_4-OH$$

4,4'-DMP

2,4'-DMP   2,2'-DMP

The 4,4'- and 2,4'-isomers are present in greatest quantity. These react further with formaldehyde and phenol; the degree of condensation is controlled by the excess of phenol. A typical novolak molecule contains 5–7 benzene rings and will not react further since it contains no unreacted methylol groups. After the compounding stage the final reaction is accomplished using hexamine (HMT) as curing agent.

This acts as a formaldehyde donor on heating and enables an irreversible crosslinked network structure to develop through methylene bridge formation.

Resoles, or single stage resins, are produced under alkaline conditions with formaldehyde in excess. Reactive methylol groups are thus present and the action of heat is sufficient to crosslink a resole resin.

Compounds based on resole resins are used, for instance, where contact with brass or copper parts will occur. Here the ammonia produced as a by-product of HMT cured novolaks would cause corrosion.

The resin content of a moulding compound is dependent on the ultimate property requirements and on the method of processing. Compounds intended for injection moulding have a higher resin content (about 40–50%) than those intended for compression moulding.

Fillers and reinforcing fibres[2]

A wide range of fillers have traditionally been used in phenolic moulding compounds. The most common of these is wood-flour, a low cost filler that provides strength and helps to reduce shrinkage during the curing process in the mould.

Ground mineral fillers, such as calcium carbonate, china clay and wollastonite are particularly useful in improving thermal and electrical properties together with improved compressive strength.

Cotton flock is used to improve the impact strength of moulding compounds but an accompanying increase in water absorption is usually noted.

Most work on fibrous reinforcement has been carried out with glass fibre because of price/performance considerations.

Good results have been obtained using chopped strand, specially surface modified to provide chemical groups that are active in the resin condensation reaction and develop a chemical bond between resin and fibre. The stronger this interfacial bond, the more readily will load be transferred from the continuous resin phase to the discrete fibre phase, and the greater will be the mechanical strength of the composite product.

In addition to surface treatment, the fibre must be free flowing with no tendency to lump. This ensures it can be fed to processing equipment in a controlled way by volumetric or preferably gravimetric feeders.

Resin selection is also very important to the achievement of maximum mechanical strength in the compound. Thus the viscosity of the resin under compounding conditions must be low enough to allow maximum wet-out of the fibre surface and provide an opportunity to develop interfacial bonding. A low viscosity is also important in minimizing fibre damage and facilitating a uniform fibre distribution.

Lubricants and pigments

Metal stearates (such as zinc stearate), stearic acid and a variety of waxes are used for lubrication purposes. Lubricants serve as processing aids during compound manufacture and provide easy tool release when moulding.

Nigrosine type dyes, rather than carbon black, have traditionally been used for pigmentation, to preserve the electrical insulation ability of phenolic compounds.

Manufacturing process

The traditional process for the preparation of phenolic moulding compounds involves the preparation of a dry blend of all compounding ingredients. These are then fed to a heated two-roll mill, where the resin melts and the rest of the ingredients are compounded into it to form a hide around the front roll. Some advancement of resin cure occurs during this compounding stage and the degree to which this occurs is controlled to produce the flow characteristics required in the final compound. The hide is taken from the roll in the form of a strip and partially cooled before feeding to a granulator. The granulated material is sieved and blended before packaging.

The milling process has some drawbacks when producing glass fibre filled materials. Thus the hide is tough and difficult to control on the mill. Substantial fibre damage can also occur as fibres pass through the nip, and there may be some separation of fibre from compound during granulation.

Extrusion technology provides a better means of compounding, allowing fibres to be added when resin viscosity is at a minimum. Less fibre damage occurs and optimum fibre dispersion can be readily achieved. This leads to the realization of excellent mechanical properties in the resulting compound.

Much effort is being expended to optimize both the compounding process and the interfacial bond between glass fibre and resin. Further improvements in mechanical performance are therefore anticipated.

Moulding process[3]

Compounds can be processed by the well-known methods of compression, transfer and injection moulding, but special care must be taken when processing glass fibre filled grades, otherwise the built-in beneficial properties may be destroyed.

With compression moulding, best results in terms of mechanical properties, surface finish and cycle times are normally achieved when the material has first been preheated, using either a screw preplastifier or an HF preheater. Flow distances are relatively short with compression moulding so fibre damage is minimized and there is little tendency for fibre orientation to occur.

In the case of transfer moulding, a preheated slug of material is pushed through a small orifice (gate) into a closed mould. The compound is more easily degassed compared with compression moulding and there is less danger of gas trapping in mouldings having thick walls or variable wall thickness.

Injection moulding (see Sections 3.9 and 3.10) is the widely used processing technique for phenolic moulding compounds, and it offers potentially fast cycle times. With glass fibre reinforced materials, good tool design and the achievement of optimum process conditions are essential if the full performance potential of the material is to be realized. Thus the barrel temperature profile must be selected to avoid too much friction and consequent fibre attrition. Mould gates, for the same reason, must not be too small. The method and position of gating will affect the way material flows in the tool. This is very important, since orientation effects are significant in injection moulded components. Orientation effects together with shrinkage can be reduced using the injection/compression moulding technique. Material is first injected into a partially open mould. The mould is then closed and the mould clamp force is used to complete the mould fill under compression. The lower injection pressures cause less fibre damage and the open tool allows effective degassing and reduced shrinkage.

Properties[4]

Phenolics, particularly the fibre reinforced types, show their superiority over those engineering thermoplastics with which they would normally compete on price in applications that require performance under load at elevated temperatures. Figure 34 illustrates their performance and Table 29 compares typical property levels for different filler types.

In addition to their mechanical and thermal performance, phenolics have excellent chemical resistance. They are not attacked by glycols, alcohols and esters, nor by gasoline, vegetable oils, mineral oils and lubricants. They are also resistant to attack by inorganic acids, except for the concentrated forms.

Phenolics are inherently flame retardant and can be formulated to give a UL 94 V-O rating without the need for the addition of halogenated flame retardants. The toxic combustion products normally associated with these additives are thus avoided. In fact, the combustion products of phenolics are relatively non-toxic, and combined with low smoke

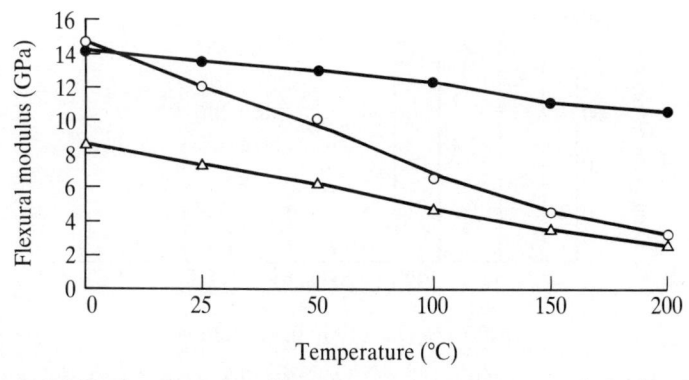

Figure 34 Engineering phenolics: comparative temperature dependence of flexural modulus of 40 wt% short glass fibre reinforced mouldings: ● = phenolic resin; ○ = polyethylene terephthalate (PET); △ = nylon.

Table 29. Typical property values for different fillers, obtained on compression moulded test specimens

Property	Wood-flour filled	Mineral filled	Cotton filled	Glass fibre filled
Specific gravity	1.4	1.5–1.6	1.35–1.39	1.6
Flexural strength (MPa)	75	65	75	120
Flexural modulus (GPa)	7	9	7	12
Tensile strength (MPa)	55	45	55	75
Heat distortion (°C)	155	175	165	185
Blister temperature (°C)	180	220	175	240
Compression strength (MPa)	200	230	210	320
Impact strength, notched (kJ m^{-2})	1.5–2.0	1.5–2.0	3.0–3.5	3.5–4.0
Water absorption (%)	0.2–0.5	0.1–0.3	0.5–0.9	0.1–0.2

generation in a fire, this makes them advantageous when human safety is at a premium, e.g. where exits are restricted.

Applications

The glass fibre filled materials in particular are used extensively in the automotive industry, particularly for under bonnet components. Water pumps, inlet manifolds, thermostat housings, cam covers and pulleys are all being produced from these materials. They meet the mechanical and thermal performance requirements and their excellent compressive creep resistance (Fig. 35) is of particular value for bolt-on applications where retention of initial bolt torque over a wide temperature range is critical. Reinforced phenolics can be used without the need for expensive moulded-in metal inserts.

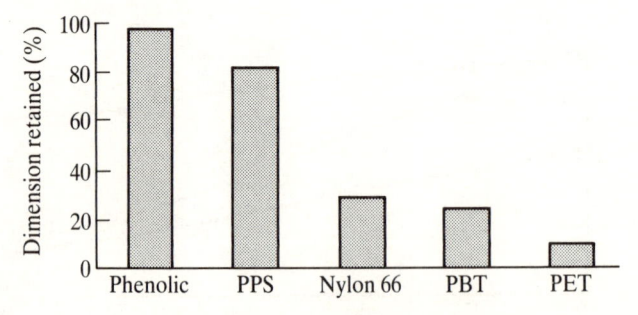

Figure 35 Engineering phenolics: comparative compressive creep at 150°C and 70 MPa of 40 wt% glass fibre filled materials. PPS is polyphenylene sulphide; PBT is polybutylene terephthalate; PET is polyethylene terephthalate.

Chemical resistance, particularly to glycol/water and fuel again demonstrates their suitability for under bonnet components.

Other application areas include commutators, valves and seal housings with some outlets in white goods and military products.

Conclusions

Phenolics, in particular fibre reinforced phenolics, have a unique combination of properties. This makes them the natural choice for a number of important applications. One prominent outlet is the automotive industry.

Advances in fibre technology and in manufacturing process technology are expected to result in further performance improvements and could open up new product opportunities.

References

1. A Knop, and L A Pilato, *Phenolic Resins*, Springer-Verlag, Berlin, 1985.
2. H S Katz, J V Milewski (eds), *Handbook of Fillers and Reinforcements for Plastics*, Van Nostrand Reinhold, New York, 1978.
3. J F Monk (ed.), *Thermosetting Plastics (Practical Moulding Technology)* George Godwin, London, 1981.
4. C A Harper, *Handbook of Plastics and Elastomers*, McGraw-Hill, New York, 1975.

2.9 Time–temperature–transformation diagrams for thermosets

B ELLIS

The concept of time–temperature–transformation or TTT diagrams, Fig. 36, was introduced by Professor John Gillham to provide a basis for the specification of the cure of thermosetting resins. During cure, relatively low average molecular weight monomers or oligomers are linked together by reaction with hardeners or curing agents resulting in an increase in molecular weight and molecular complexity. Thus, there is a sequence of 'events' intrinsic to the cure process. The major events are gelation, vitrification (glass formation), phase separation, devitrification and degradation. Cure temperature, T_c and cure time, t_c, have to be specified to optimize the properties of the cured resin for a specific application. Control of T_c and t_c is obviously desirable but not always readily achieved because the cure reactions are exothermic and the thermal history of thick sectioned mouldings or castings will be a

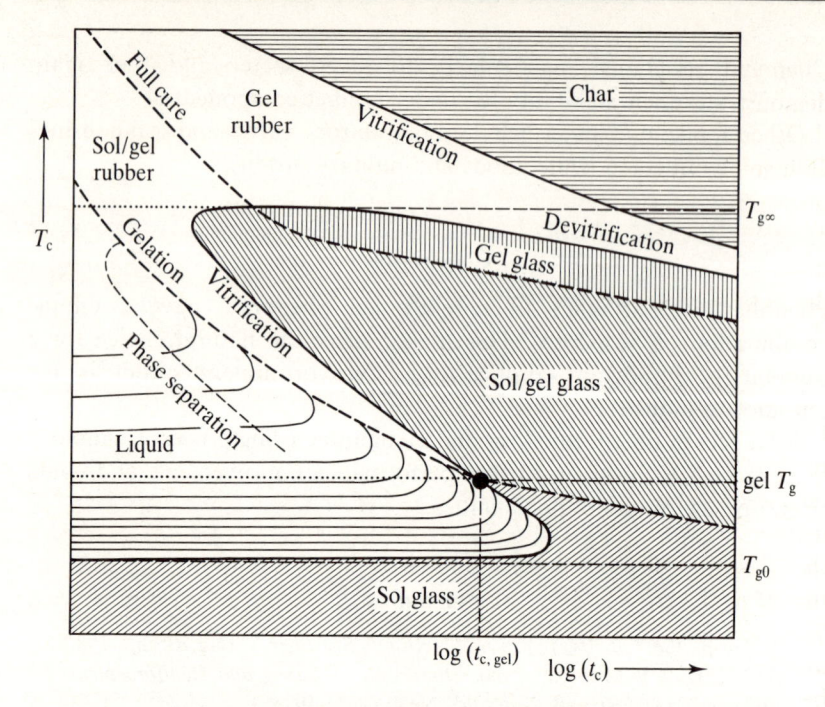

Figure 36 Time–Temperature–Transformation (TTT) Diagram (adapted from Aronhime and Gillham[3]). t_c, T_c are the cure time and temperature; T_g, T_{g0}, $T_{g\infty}$ are the glass transition temperatures at time = t_c, $t = 0$ (prepolymer/hardener mixture) and for the fully cured resin, respectively. gel T_g is the T_g at $t_{c,gel} = t_{c,vit}$ the cure times for gelation and vitrification respectively (gelation, vitrification lines).
Sol glass is soluble, whereas the gel glass will only swell in solvents. In the liquid region isoviscous contours, differing by 10, for an homogenous system are shown.

function of position within the product. To avoid an excessive exothermic temperature rise, the initial cure temperature is often relatively 'low', then a post-cure at a 'high' elevated temperature ensures that a 'fully cured' state is attained. The relative terms low and high cure temperature depend on the reactivity of the resin and hardener. It is essential to avoid devitrification and degradation caused by either high cure temperatures and/or long cure times, conditions represented by the top right-hand corner of the TTT diagram, Fig. 36. Systems such as rubber modified epoxies show phase separation, essential to attain high fracture toughness. Gillham[1] also considers the application of TTT diagrams for the polymerization of difunctional liquid monomers, which form non-crosslinked, non-network polymers, but these systems will not be discussed further here.

An outline of the concepts involved in the construction of a TTT diagram has been given by Gillham[2] and a detailed exposition of the theory and application of TTT diagrams is that of Aronhime and Gillham.[3] TTT diagrams have been treated in more detail for epoxy than other thermosetting resins although Prime[4] has used TTT diagrams as the basis for a comprehensive review of the cure of phenolic, amino, allyl and unsaturated polyester resins as well as epoxies. The latter have also been discussed in detail by Ellis.[5]

Cure kinetics

For the cure of thermosetting resins, the initial rates of reaction between resin and hardener are determined by the chemical kinetics of the specific reactions involved; these are often complicated.[4,5] However, with the increase in molecular complexity as cure progresses, there is an increase in the glass transition temperature, T_g (see below), of the reacting mixture, that is $T_g = T_g(t_c)$, where t_c is the cure time at constant cure temperature T_c. When the difference between T_c and T_g becomes small, molecular mobility is greatly reduced and the effective rates of the cure reactions become slower and slower. This is because mutual diffusion of the reactive species is slow. The rates of reaction are now under 'diffusion control', in contrast to 'kinetic control', for the initial stages of cure. The chemical reactions are eventually quenched when $\Delta T = T_g(t_c) - T_c$ becomes large enough to inhibit molecular motion, so that not all of the reactive groups are consumed. As mentioned previously, a common industrial practice, especially for composites, is to post-cure the resin to ensure more complete reaction of the functional groups, and attain the limiting glass transition temperature, $T_{g\infty}$. The upper use temperature of composites is often determined by the glass transition temperature of the resin matrix; this is clearly a function of the cure treatment, T_c and t_c. It is important to note that the glass transition temperature will also be affected by degradative reactions when the composite is in use for prolonged periods at elevated temperatures, see the TTT diagram Fig. 36. Hygrothermal stability (see Chapter 6) is also dependent on cure treatment.

Monitoring cure

Experimental methods for monitoring cure have been discussed by Prime[4] and Ellis.[5] These include chemical assay of the concentrations of reactive groups, differential scanning calorimetry (DSC), nuclear magnetic resonance (NMR) and infrared spectroscopy. Gillham[3] has developed torsional braid analysis, TBA, to characterize the extent of cure. In this technique a braid is impregnated with the liquid resin precursor/hardener mixture then mounted as the active element in a torsion pendulum so the whole cure process from the fluid to a viscoelastic solid can be studied. From the oscillation period and logarithmic decrement, Δ, of the torsion pendulum the relative change in the components of the complex modulus, $G^* = G' + iG''$, can be calculated where G' and G'' are the real (or storage) and imaginary (or loss) moduli respectively. With progressive cure, G' increases more rapidly than G'' and there are maxima in $\tan \delta = G''/G' \simeq \Delta/\pi$, which have been discussed in detail by Aronhime and Gillham.[3] When gelation occurs, there is a change in slope of log G'' versus t_c, i.e. d(log G'')/dt_c decreases at the gel point.[5]

Structural changes during cure

Initially, that is before cure has started, thermosetting resins are viscous liquids at processing temperatures so that reinforcements can be impregnated and the resin composites moulded, both of these processes involve flow of the liquid resin. With increase in mass average molecular weight ($\bar{M}_W$) during cure, there is an increase in viscosity, which would prevent efficient impregnation of fibrous reinforcements or flow during moulding. The viscosity is a function of the molecular weight of the curing resin, $\eta = f(\bar{M}_W)$, where the exact function depends on the magnitude of $\bar{M}_W$. Initially, the molecular weight increases slowly as cure progresses, and the average molecular weight is still relatively low.[5] However, a precipitate rise in viscosity is observed, either due to gelation and/or the onset of vitrification when $T_g(t_c)$ approaches the cure temperature, T_c. When both these processes coincide at the same time of cure $t_{c,gel} = t_{c,vit}$, the temperature is specified by $_{gel}T_g$ in the TTT diagram.

With increased cure times, the resin molecules become larger and more highly branched so that eventually an incipient three-dimensional network, or gel is formed.[3,5] This is the gel fraction, w_g which swells when 'solvents' are imbibed, but is not soluble. For $t_c < t_{c,gel}$, only sol, that is soluble molecules, are present and the sol fraction, $w_s = 1$. For $t_c > t_{c,gel}$, the gel fraction, w_g, increases and the sol fraction decreases ($w_s + w_g = 1$). Branching theory[5] can be used to predict the critical conditions necessary for gelation. Aronhime and Gillham[3] have discussed the application of Flory's gelation theory and also the more recent methodology introduced by Macosko and Miller for specification of the conditions for gelation, which are also reviewed by Ellis.[5]

The glass transition temperature of the curing resin is a very important parameter, since $T_g = T_g(t_c)$ at constant T_c. There is no agreed operational definition of the glass transition temperature; different methods of determination yield values of T_g that may differ by maybe 30°C or so.[5] However, a TTT diagram can be self-consistent provided the method of estimating T_g is defined unambiguously. With a temperature difference of the order of 30°C, the degree of molecular motion and mobility at the lower temperature will be considerably less than at the upper estimate. The precise position of the vitrification curve on a TTT diagram will depend on the definition of T_g, but it is general experience that the cure reactions are not completely inhibited until T_g exceeds T_c. With lower cure temperatures, not all of the reactive groups are consumed. To attain the limiting 'fully cured' glass transition temperature, $T_{g\infty}$, high cure temperatures are required. With high cure temperatures or exposure of a cured resin to high operational temperatures for long times, degradation can occur. Such reactions lead to bond scission and an increase in free volume so that there is a decrease in T_g. When conditions are extreme, the processes are represented in the top right-hand corner of the TTT diagram.

Cure–property diagrams

Wang and Gillham[6] recently introduced a T_g temperature–property (T_gTP) diagram to represent the structural changes that occur during the cure of a thermosetting resin. The structure was related to the glass transition temperature of the cured resin, $T_g(t_c)$, which is obviously a function of the extent of cure. A slightly modified version is the cure temperature–property or CTP diagram, Fig. 37, in which the structure is related to a cure parameter, C, defined as

$$C = \frac{T_g - T_{g0}}{T_{g\infty} - T_{g0}}$$

where T_{g0} is the glass transition temperature of the initial uncured system and $T_{g\infty}$ is the glass transition temperature of a fully cured resin. Wang and Gillham[6] point out that the relationship between the glass transition temperature of a partially cured resin, T_g, is not a linear function of the extent of reaction. Thus, representation of the structural changes in relation to the introduction of the T_gTP diagram avoids this difficulty.

The glass transition temperature (or the cure parameter, C) linearizes the relationship, cTg vs C, as can be seen from Fig. 37. Not only can the change in glass transition temperature with cure be represented but also the β relaxation, which has important implications for the energy loss mechanisms and hence the application of thermosetting resins.

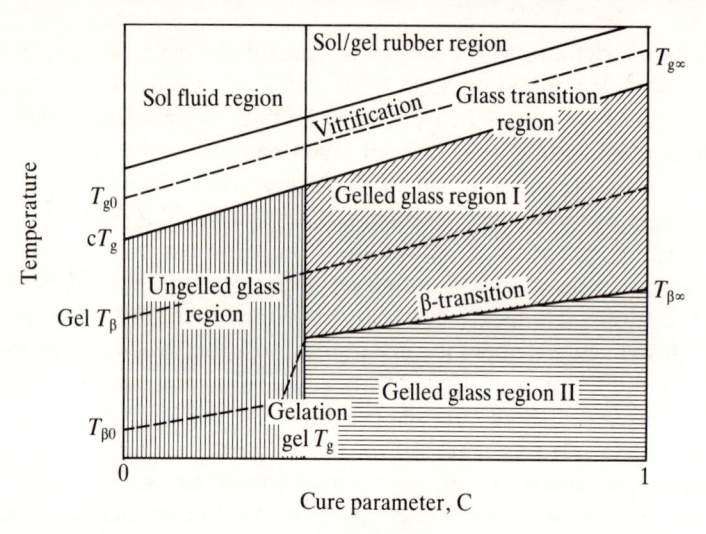

Figure 37 Cure-temperature property (CTP) diagram after Wang and Gilham[6] and Ellis[5]. $C = (T_g - T_{g0})/(T_{g\infty} - T_{g0})$; cT_g is the end of glass transition. See Fig. 36 for other definitions

Conclusions

For the cure of thermosetting resins there are three critical temperatures T_{g0}, $_{gel}T_g$ and $T_{g\infty}$, which are represented on the TTT diagram, Fig. 36. For $T_c < T_{g0}$, cure is inhibited. For $T_g(t_c) < T < {}_{gel}T_g$, it is possible to impregnate fibrous reinforcements and to process reinforced resins; it is essential that such processes are virtually complete before either gelation or vitrification of the resin. For a cure time $t_{c,gel}$, the glass transition temperature is $_{gel}T_g$ with the formation of an incipient network which will prevent flow of the resin. However, for $T_{g0} < T_c < {}_{gel}T_g$, the curing resin will vitrify before gelation; the cure reactions are quenched so that gelation is precluded. A 'full cure' may be defined by a resin system which has the limiting glass transition temperature, $T_{g\infty}$.

Acknowledgement

Stimulating discussions with Professor John Gillham on the specification of the cure of thermosetting resins are gratefully acknowledged.

References

1. J K Gillham, *Polym. Eng. Sci.* **26**, 1429 (1986).
2. J K Gillham, in *Encyclopedia of Polymer Science and Engineering*, Vol. 4, 2nd edn, John Wiley & Sons, New York, 1986, p. 519.
3. M T Aronhime, J K Gillham, *Adv. Polym. Sci.* **78**, 83 (1986).

4. R B Prime in *Thermal Characterization of Polymeric Materials*, ed. E A Turi, Academic Press, New York, 1981, Ch. 5.
5. B Ellis (ed.) in *The Chemistry and Technology of Epoxy Resins*, Blackie, Glasgow, 1993, Ch. 3. See also pp. 4–7.
6. X Wang, J K Gillham, *J. Coatings Tech.*, **64**, 37 (1992).

2.10 Unsaturated polyester resins

R A PANTHER

The term 'polyester resin' is applied to the condensation reaction products of diacids and diols (glycols). They are, therefore, strictly alternating polymers of the type shown below.

$$—A—B—A—B—A—B—$$

A good example is polyethylene terephthalate, PET, made from ethylene glycol and terephthalic acid and used in the form of a fibre, Terylene, for clothing or moulded into bottles for carbonated drinks. Polymers of this type are termed 'saturated polyesters'.

Unsaturated polyesters[1-4] are derived in a similar manner but at least one of the raw materials used is an ethylenically unsaturated compound. These polymers are therefore capable of further reaction.

In their natural state, unsaturated polyesters are usually hard glassy solids and as such are difficult to process further into useful materials. To overcome this difficulty, they are normally supplied as solutions in an ethylenically unsaturated monomer. This not only facilitates easier handling but also is capable of reacting with the polymer chain. The monomer of choice is usually styrene, although others can be used. The styrene performs the vital function of enabling the resin solution to cure from a liquid state into a solid by the formation of crosslinks between the polyester chains.

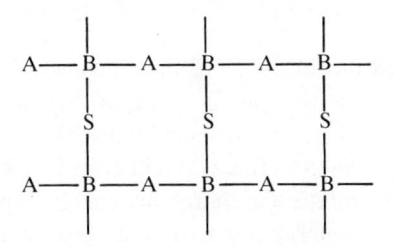

```
      |       |       |
  A—B—A—B—A—B—
      |       |       |
      S       S       S
      |       |       |
  A—B—A—B—A—B—
      |       |       |
```

The curing reaction is an example of vinyl copolymerization that proceeds without the evolution of any by-products to produce an extended three-dimensional network. The statistical nature of the copolymerization results in a heterogeneity of the crosslink density and leads to the formation of a microstructure on the 0.01–0.05 μm scale within the cured resin (see Section 6.3)[5]. Curing is accomplished by the addition of suitable initiators, capable of producing free radicals under the right conditions. These are usually organic peroxides or hydroperoxides, although other materials such as azonitriles or photo-initiators can be used. The simplest curing process is to heat the resin with an organic peroxide. Thermal decomposition of the peroxide produces free radicals and these initiate the copolymerization of styrene with the unsaturated polyester. Curing at room temperature is accomplished using an organic peroxide and a suitable reducing agent, termed an accelerator. Suitable accelerators include tertiary aromatic amines and transition metal soaps such as cobalt or vanadium octoate or naphthenate. Curing agents are usually employed in small concentrations, of the order of 1–2% peroxide, 100 ppm cobalt metal. The curing reaction is non-reversible and the cured resin is usually a hard strong material with varying degrees of chemical and heat resistance.

The choice of raw material available to formulate polyester resins is vast and it determines the final cured properties of the material. By careful selection of both the raw material type and its molar concentration in the polymer, polyester resins can be tailored to suit a very wide variety of structural applications.

The key raw material is, of course, the one that provides the unsaturation. In the vast majority of polyesters, this is fumaric acid or maleic acid used in the form of its anhydride.

fumaric acid   maleic anhydride

Other unsaturated diacids, such as itaconic acid, can be used but maleic anhydride is by far the most common for a number of reasons. Esterification of each mole of maleic anhydride requires the liberation of 1 mole of water, whereas fumaric acid gives 2 moles. Removal of water from the reaction mixture requires an energy input, therefore maleic based resins are less energy intensive and more economical. In addition, over 90% of the maleate esters produced during the esterification process are isomerized to fumarate esters. Although there seems little to justify the use of fumaric acid, it is needed to obtain maximum benefit from the crosslink density of the cured product and is commonly used in high performance resins. The reason lies in the particularly favourable reactivity ratios of fumarate unsaturation with styrene unsaturation; this represents another good reason for choosing maleic anhydride or fumaric acid as a prime raw material. On an industrial scale, maleic anhydride has the important advantage of being a relatively low melting point solid, so it can be used in its molten state to facilitate easier bulk storage and transfer.

Some resins are produced commercially using maleic anhydride or fumaric acid as the only acid constituent of the polymer. Such resins have a very high crosslink density and as a result have generally good high temperature resistance and chemical resistance. Usually, however, they are also very brittle with low elongation at break, low impact resistance and a general lack of overall toughness. They are, therefore, only of interest for certain special applications.

To produce a resin with more useful properties, it is necessary to reduce the crosslink density; this is achieved by using saturated acids as well as maleic anhydride. For most applications, resins with molar ratios from 2:1 to 1:2 saturated:unsaturated acid are quite satisfactory.

The most commonly used saturated acids are the isomers of phthalic acid with the 1,2-isomer being used in the form of the anhydride.

phthalic anhydride   isophthalic acid   terephthalic acid

Of these, phthalic anhydride and isophthalic acid are the most common. Terephthalic acid has the disadvantage of requiring higher temperatures to effect esterification and the properties of resins derived from it are very similar to those derived from isophthalic acid. Isophthalic resins generally have better properties than those derived from phthalic anhydride, particularly in terms of water, chemical and weathering resistance. They are generally more durable and also tougher but are more expensive to process for the same reasons given for the choice between maleic anhydride and fumaric acid.

The last major components of polyester resins are the glycols. A wide choice is available. The three most commonly used are given below.

HO—CH$_2$CH$_2$—OH

CH$_3$—C(H)(OH)—CH$_2$OH

ethylene glycol propylene glycol

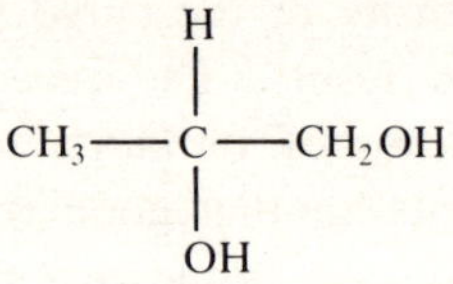

HO—CH$_2$CH$_2$OCH$_2$CH$_2$—OH

diethylene glycol

Mixtures of glycols are often used to give the right cured properties. Propylene glycol is the most commonly used and gives resins with generally good cured properties. Ethylene glycol increases chain stiffness but at high molar concentrations can impart crystallinity and make the polymer insoluble in styrene. Diethylene glycol imparts flexibility and toughness to the polyester but the presence of the ether group increases water sensitivity and reduces durability.

As previously mentioned, many other raw materials can be used, usually to impart specific properties. Some of the more important include halogenated materials for fire retardancy, substituted glycols and bisphenol derivatives for chemical resistance and durability.

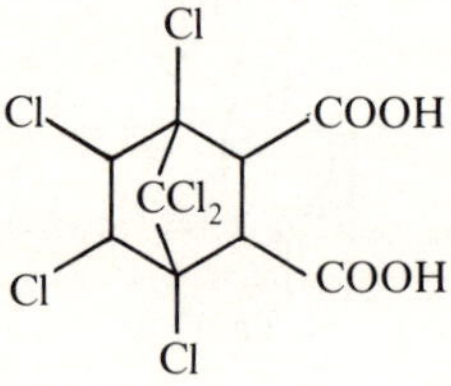

HET acid 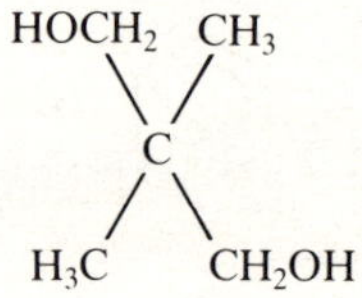   neopentyl glycol 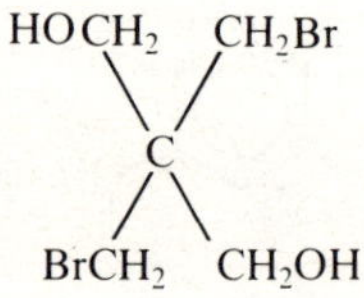   dibromoneopentyl glycol

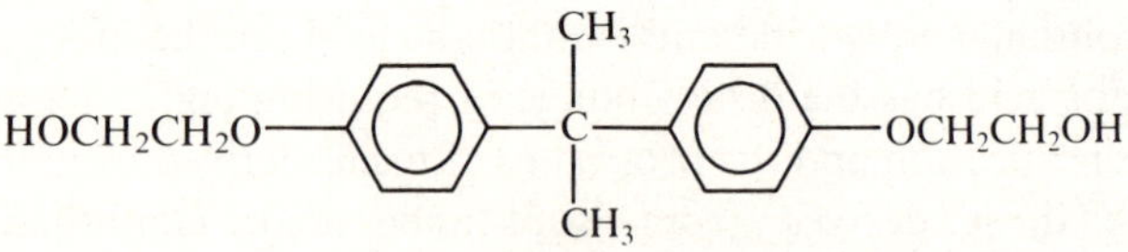

ethoxylated bisphenol A

Equimolar mixtures of glycols and diacids can theoretically grow to infinite molecular weight. It is therefore necessary to limit the molecular weight of the resin and this is achieved by using an excess of one of the reactants. Using an excess of acid or glycol will successfully limit the molecular weight but, in practice, better cured properties are obtained if the glycol is used in excess. Typical glycol excess ranges from 2 mol% to about 12 mol%.

The manufacture of unsaturated polyester resins is quite straightforward and involves charging the raw materials to a stirred vessel fitted with a fractionating column, condenser, receiver, and thermometer. The reaction mixture is blanketed with an inert gas, such as nitrogen, and heated. Distillation usually commences at around 150°C but, as the distillation rate falls, the temperature is allowed to rise to around 200°C. Most glycols are volatile in steam, are separated in the fractionating column and returned to the reaction mixture. The water vapour is removed at the top of the column, condensed and collected in the receiver. The stillhead temperature is maintained at 100°C to minimize glycol losses. The progress of the reaction is monitored by reference to the amount of distillate collected and by titration of the residual acid. The reaction is stopped when the residual acid falls to about 20–30 mg KOH/g of resin. The resin is cooled and dissolved in styrene.

At this stage, a free radical scavenger, usually a phenolic compound such as hydroquinone, is also added to the resin. This material not only provides the resin with a useful shelf life but also a controlled induction period or 'working time' after the addition of catalysts and accelerators. A small quantity (50–100 ppm) of paraffin wax is usually added to overcome air inhibition of the resin on exposed surfaces during cure. To achieve this, the solubility of the wax is selected so it remains soluble in the liquid resin but becomes insoluble as the resin cures. It is then capable of forming an air-excluding barrier layer on the surface of the curing resin.

At higher concentrations (500–1000 ppm) paraffin wax can also be used to formulate low styrene emission resins. At these concentrations, the solubility of the wax at the surface of the resin is exceeded after the evaporation of a small amount of styrene. The wax is therefore precipitated on to the surface of the resin and prevents further evaporation of styrene. The effectiveness of this technique can be readily appreciated in that the styrene emission of a normal resin may exceed $100 \, g^{-2} h^{-1}$, whereas a low styrene emission resin may be below $10 \, g^{-2} h^{-1}$. Obviously, this is very important from health and safety and environmental viewpoints.

Thermoplastic polymers are another type of additive worthy of note; they control the shrinkage of a resin on cure. Molecules present in a resin are spaced at distances equivalent to van der Waals' radii, during cure they move to covalent distances which causes the resin to shrink. Typical shrinkages for unsaturated polyesters are 8–12% by volume. Shrinkage is a very important consideration where highly dimensionally accurate mouldings are required and where high quality surfaces are desired. The shrinkage can be controlled by the addition of a thermoplastic polymer such as polyvinyl acetate or polystyrene. Such systems are termed 'low profile' and are usually limited to hot curing applications. There are many theories to explain why such a system works and the precise mechanism is currently the subject of much debate. What is known is that in order to be effective, the thermoplastic has to become incompatible with the resin during cure and separate out to form a dispersed phase throughout the polyester matrix. One possible explanation of the shrinkage control is that the thermal expansion of the discrete thermoplastic particles is sufficient to overcome the shrinkage of the resin as it cures. By use of suitable materials, it is possible to formulate a low profile system that actually expands slightly on cure; this is the key to high quality surface finishes.

In conclusion, it can be seen that a very wide range of polymer types can be prepared with a very wide range of useful properties, Tables 30 and 31. The usefulness of these materials is clearly demonstrated by the growing use of unsaturated polyester resin composites (GRP) in many domestic and industrial areas.

Table 30. Typical properties of cured polyester resin

Property	Value
Specific gravity	1.28
Hardness, Rockwell M scale	110
Tensile strength (MPa)	70
Compressive strength (MPa)	140
Tensile modulus (GPa)	3.5
Elongation at break (%)	2.5

Table 31. Comparative properties of glass reinforced polyester resin laminates with other materials

Material	Tensile strength (MPa)	Tensile modulus (GPa)	Specific strength (10^2m)
Polyester/glass rovings	800	30	400
Polyester/glass mat	100	7	70
Mild steel (structural)	310	200	40
Duralumin	450	70	150
Douglas fir	75	13	150
Hickory	150	15	200
Portland cement	10	17	5

References

1. Bjorkstein Research Labs, *Polyesters and Their Applications*, Reinhold, New York, 1959.
2. D Othmer (ed.) *Encyclopaedia of Chemical Technology*, 3rd edn, Vol. 18, Interscience, 1982.
3. A V Boenig, *Unsaturated Polyesters*, Elsevier, Amsterdam, 1964.
4. R G Weatherhead, FRP Technology, *Fibre Reinforced Resin Systems*, Applied Science, London, 1980.
5. W Funke, *Kolloid Z.Z. Polym.* **197**, 71 (1964).

2.11 Urethane methacrylates

F R JONES

Introduction

These resins, which are manufactured by ICI Chemicals and Polymers Ltd as Modar, have recently begun to find application in composite materials, especially where rapid processing is required. They are similar to the vinyl ester resins (c.f. Section 2.12) in that they are based on a urethane skeleton with terminal methacrylate groups and dissolved in a reactive diluent. The solvent can be styrene, but is usually methyl methacrylate or a mixture of the two. They are cured by free radical chemistry using peroxides. In this way the curing process is analogous to the vinyl ester and unsaturated polyester resins, in that it involves copolymerization of the terminal unsaturations with methyl methylacrylate and/or styrene. Methyl methylacryate polymerization occurs more rapidly than styrene copolymerization with an unsaturated polyester or vinyl ester, which makes these resins unsuitable for rapid fabrication processes. Furthermore the reactivity ratios for the copolymerization of methyl methylacrylate with the terminal methacrylate groups of the urethane acrylate are probably similar, thereby reducing the likelihood of a heterogeneous crosslink density in the cured resin casting. This contrasts with the nature of the curing of an unsaturated polyester resin (see section 2.10).

Mechanical properties

As with many polymeric resins the mechanical properties can be modified by altering the chemical structure of the base resin and/or the nature and concentration of any comonomer. In these resins the base urethane methacrylate structure can be modified to affect both the mechanical

Table 32. Typical properties of urethane methacrylate resins

Resin (Modar designation)	835	535S	836	865	855
Reactive diluent	M	S/M	S/M	M	M
Cast resins					
Tensile failure strain (%)	7–8.5	6.5	4.2	6.2	4.6
Tensile strength (MPa)	56*	62*	75	80*	88
Tensile modulus (GPa)	2.4	2.4	2.9	2.8	3.1
Flexural modulus (GPa)	2.3	2.9	3.6	3.4	3.7
Glass transition temperature (°C)	148	–	–	–	–
Heat distortion temperature (°C)	–	83	98	107	117
Fracture toughness (K_{IC})(MPa m$^{1/2}$)	1.2	–	–	–	–
Uncured resin					
Viscosity at 20°C (mPa s)	–	60	–	150	750

M = methylmethacrylate; S/M = styrene and methylmethacrylate; * = yielding

properties and the viscosity of the undiluted and diluted resin. Typical mechanical properties are given in Table 32. Of particular note is the range of achievable failure strains. This is seen as an advantage when they are used as matrices for continuous and long-fibre composites in that matrix cracking occurs at higher levels of applied stress approaching that for failure.[1]

Processing of urethane methacrylate resin composites

The reactivity (i.e. cure times of less than one minute) of these resins means that they find applications in a wide range of processes such as resin transfer moulding (RTM), SRIM, pultrusion, infiltration through cold pressing and dough moulding (see Chapter 3). Grades of resin (see Table 32) with differing viscosities enable the resin to be optimized for the appropriate process. Resins with viscosities ranging from 50 to 2000 mPa s at 20°C are available. It is argued that the equivalent hand-layed composite properties can be achieved automatically by RIM, with properties close to the predictions of the law of mixtures (see Section 4.2). Futhermore it is claimed that this results from controlled and rapid wet-out of the glass fibres. The effect of fibre volume fraction on the flexural strength and modulus of differing laminates is given in detail elsewhere.[1]

Reference

1. M L Orton, I M Fraser, S H Rogers, *Eng. Plastics*, **2**, 274 (1989).

2.12 Vinyl ester resins

F R JONES

Vinyl ester resins were developed to incorporate the generally superior cast resin properties of epoxies with the ease of fibre impregnation during the more conventional fabrication routes and hence provide glass fibre reinforced plastics with superior corrosion resistance. They have a backbone resin structure akin to epoxies with terminal unsaturated vinyl groups which can be cured by copolymerization with the reactive styrene diluent analogously to the unsaturated polyester resins. Thus peroxide catalysts and similar accelerators are used to initiate a free radical polymerization. It is, however, usual to promote the curing reaction with dimethylaniline accelerator in addition to the cobalt soap, to ensure cold curing in a reasonable timescale. A typical vinyl ester resin (i) contains terminal vinyl ester groups, obtained by reacting DGEBA (ii), the diglycidyl ether of bisphenol A ($n = 1$), or higher epoxies, with acrylic acid, R = H, (iii) or methacrylic acid, R = CH_3, (iv).

Derakane 411–45 is considered to be based on (i) with $n = 1$ and R = CH_3 in a solution containing 45% styrene.

The reduced concentration of ester groups in the backbone polymer gives improved hydrolytic stability and hence chemical resistance to the cured resin. The terminal vinyl groups have differing probabilities of copolymerization with styrene compared to the fumarate groups of a conventional unsaturated polyester. Whereas there is a tendency for alternate copolymerization in the initial stages of cure of the unsaturated polyester, for the vinyl esters there is a bigger probability of the formation of longer polystyrene blocks, which act as crosslinks between the vinyl ester chains. As a consequence, the combination of the longer backbone and crosslink network chains leads to a higher extensibility compared to the equivalent polyesters. Furthermore, a resin microstructure, with extremes of crosslink density known to form in the unsaturated polyesters probably does not occur. As a result, these cured resins, in comparison with the unsaturated polyesters, have a higher resilience and fracture toughness. These differences appear to be less pronounced in the presence of absorbed moisture.

Styrene emission and air inhibition at free surfaces are similar to the unsaturated polyesters. This can be controlled by small additions of paraffin wax, which migrate to the free surface, reduce the partial pressure of styrene and act as an oxygen barrier.

The major use of vinyl ester resins is in chemical plant (tanks, pipes, ducts) where corrosion resistance is required.

Specialist vinyl ester resins

For fire retardant applications the tetrabrominated bisphenol A (v) replaces the equivalent in (i) and (ii).

(v)

For high temperature performance, an epoxy novolak (vi) replaces DGEBA (ii) in the formation of the methacrylate based vinyl ester, $R = CH_3$ in (i).

(vi)

Related corrosion resistant resins

For chemical resistance, the equivalent unsaturated polyesters are also based on bisphenol A (vii), see Section 2.10.

The standard bisphenol fumarate resin has the average structure given in (viii).

However, the Crystic 600 and Atlac 382 systems incorporate a chemical structure based on a more flexible backbone (ix).

The premium chemically resistant styrenated resin for anticorrosion applications of GRP has a related structure and is analogous to the vinyl esters but based on urethane chemistry. The reported structure of Atlac 580 polyester base is given below (x). The resin is generally available as a 50% solution in styrene (Atlac 580-05). Typical properties of these cast resins are given in Tables 32 and 33. Structures vii–x are shown overleaf.

Table 33. Typical properties of cast vinyl ester resins

Property	Value
Flexural strength (MPa)	130–140
Tensile strength (MPa)	70–80
Tensile modulus (GPa)	3.3
Failure strain (%)	5–6
Heat distortion temperature (°C)	100–150

Table 34. Heat distortion temperatures (HDT) of chemically resistant resins

Resin	HDT (°C)	Type
Derakane 411–45	100	Vinyl ester
Derakane 470–45	150	High temperature vinyl ester
Derakane 510-A-40	110	Brominated fire resistant vinyl ester
Atlac 580–05	118	Chemically resistant vinyl urethane
Atlac 382–05	140	Chemically resistant bisphenol A – fumarate
Crystic 600 (47% styrene)	120	Chemically resistant bisphenol A – fumarate
Crystic 272 (41% styrene)	75	High performance isophthalic

Derakane is the tradename of Dow Chemical.
Atlac is the tradename of ICI.
Crystic is the tradename of Scott-Bader.
For a complete compilation of resin properties, see Weatherhead.[1]

Reference

1. R G Weatherhead, FRP Technology, *Fibre Reinforced Resin Systems*, Applied Science, London, 1980, Chs 9, 13.

132 Vinyl ester resins

CHAPTER 3

Fabrication of Polymer Composites

3.1 A guide to selection

F R JONES

The choice of manufacturing technology for the fabrication of fibre reinforced plastics or composite materials is intimately related to the performance, economics and application of the material. The range of performance achievable with glass fibres is illustrated in Fig. 38, where the effect of fibre volume fraction on mechanical properties is described.

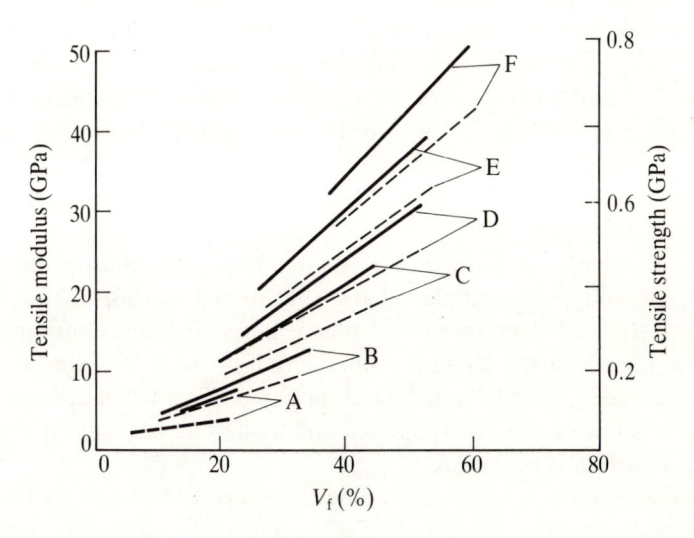

Figure 38 Stiffness (——) and tensile strength (- - -) of polyester-glass laminates, showing effect of different forms of reinforcement over practical range of glass content. A, random short fibre plastic; B, chopped strand mat; C, balance woven rovings; D, cross-ply laminate or woven rovings; E, unidirectional woven rovings; F, continuous aligned rovings

Two main factors contribute to the wide range in performance, namely fibre length and fibre orientation. As discussed individually in Chapter 4, a critical length, l_c, exists for a particular fibre matrix combination. For reinforcing efficiency equivalent to a continuous fibre, a discontinuous fibre needs to be at least 10 times that length. Since the best properties are obtained with aligned fibres, a random-in-plane arrangement will also lead to a reduced reinforcing effect. Most discontinuous fibre composites are manufactured in moulding processes that lead to randomization of the fibres, so a non-linear relationship exists between fibre volume fraction, composite modulus and strength.

The other major aspect is the range of fabrication processes that exists. For high performance, continuous aligned fibres are generally required; this can usually be achieved only by prepreg technology (Section 3.2) or pultrusion (Section 3.12). Prepreg technology leads to the highest performance but in a process relatively slow, highly skilled and labour intensive. Pultrusion is a rapid fabrication route limited to long lengths of uniform cross-section. It therefore, requires post-fabrication assembly into useful objects. Attempts have been made to incorporate additional shaping operations. For example, haul-off onto a large 'wheel' is used to put a curvature into the length of the profile, in the fabrication of automotive springs.

The use of woven cloth (Section 1.10), or unidirectional cloth, i.e. unidirectional fibres with transverse 'stringers', improves the complexity of the shapes possible with *autoclave* or *prepreg* routes or by hand lay-up (Sections 3.7 and 3.18) but these processes are still labour intensive. There is, however, a performance penalty for the convenience of fabrication. Related techniques involve resin injection or transfer, RTM (Sections 3.15 and 3.16) or structural RIM (Section 3.19) which are primarily used for larger structures.

Injection moulding (Sections 3.8, 3.9 and 3.10) offers the advantage of rapid processing into complex shapes but at the expense of the fibre length retention. The more complex shapes require thermoplastic matrices for which the minimum fibre length is lower than that for the thermoset based, dough moulding compounds. This technique requires high capital investment for equipment and tooling so that only large-scale mass production is economic.

Reinforced resin injection moulding (Sections 3.14 and 3.19) has the benefit of lower capital and tooling costs; it can be used for less complex but often larger shapes, such as automotive body panels but at a penalty of reduced fibre volume fraction, fibre reinforcing efficiency and rate of manufacture.

Compression moulding or stamping of glass mat reinforced thermoplastic (**GMT**), sheet moulding compound (**SMC**) are other fabrication processes for artifacts of intermediate complexity but with potential for rapid production (Section 3.17). This brief discourse illustrates the trade-off between mechanical performance and ease of fabrication. As a general rule mass production processes give rise to lower performance material whereas those for high performance composites are less readily automated. Intermediate rate processes, which utilize longer fibres or continuous fibre preforms, woven cloth or random mats (Section 1.10), give rise to average properties because lower achievable fibre volume fractions add to the fibre orientation effect. Carbon (Sections 1.6, 1.7 and 1.8) and aramid fibres (Section 1.3) are expensive reinforcements and are preferably used in continuous fibre mouldings.

Economic aspects of material/process selection[1]

The specific moduli and strengths of continuous fibre composites are generally higher than metals. Glass fibre epoxy has a similar stiffness to aluminium alloy with three times the specific strength; this makes it highly competitive. The high strength and high stiffness to weight ratios of carbon fibres make CFRP the superior material in high performance structures.

The specific strengths, but not the specific moduli, of conventional plastics are higher than those of more conventional metals. Discontinuous fibre reinforcement enhances this difference. However, as discussed above, the application can demand the use of a short fibre reinforced moulding material. The economics of the selection of a moulding material are given in Fig. 39, where the relative costs of materials to give similar rigidity and strength to steel are compared[1]. It is immediately clear that the composites solution is not the most economic from a materials cost point of view. However, the mouldability of plastics based materials generally means that economies can be made in fabrication when compared to the additional costs incurred through waste and post-fabrication of conventional materials. For high performance carbon fibre composites used in airframe structures, the fuel efficiency of the design and lower running costs can also help to offset the high materials cost. In the automotive field, the composites solution (usually from long discontinuous fibres) is overall more cost efficient but replacement has been delayed by the development of appropriately rapid mass production moulding techniques. However, composites technology does provide materials design capability by which the impact performance can be tailored. Vertical body panels can be manufactured from low speed impact resistant material, whereas for horizontal panels, higher stiffness (but less impact resistance) is required. Furthermore, composite members

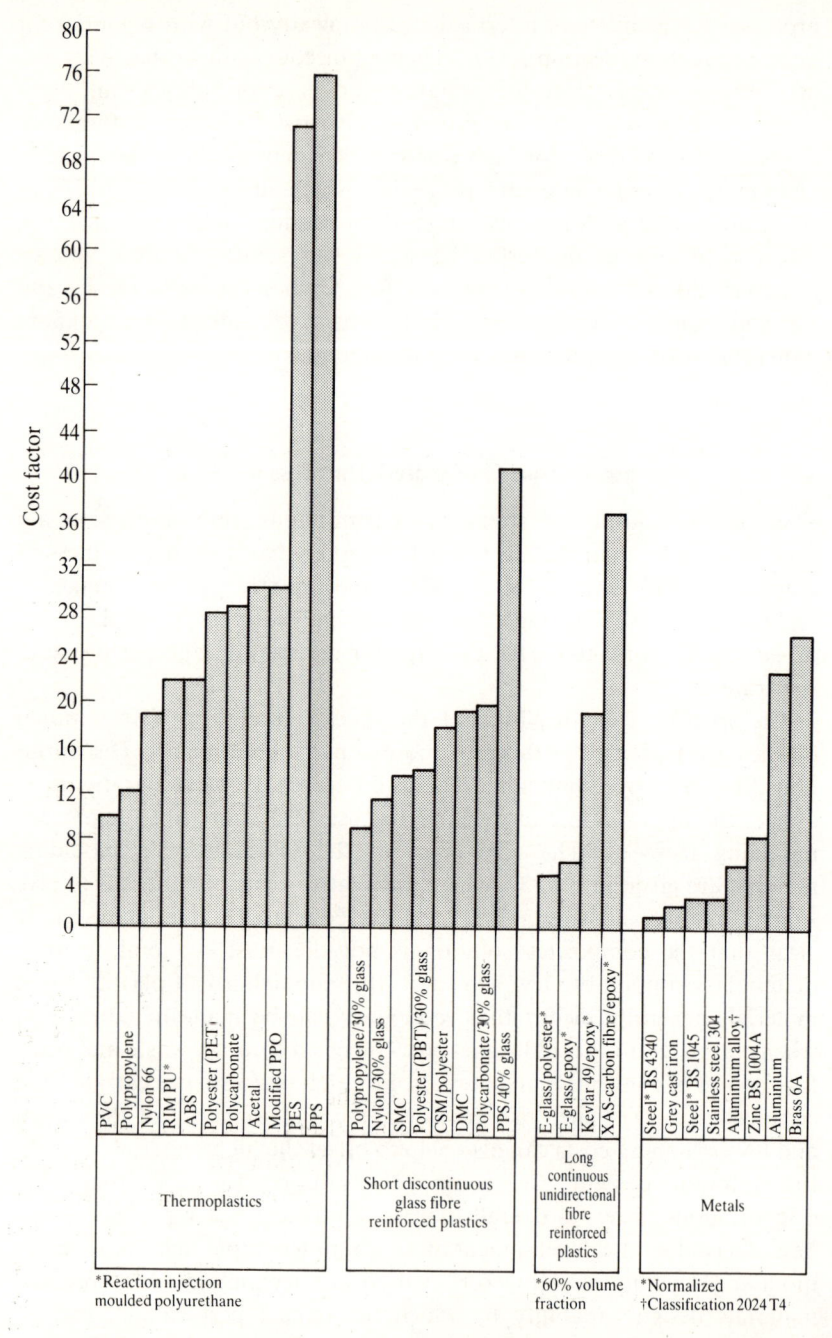

Figure 39a Relative costs of materials of thicknesses to give similar strengths to that of steel (Reproduced from reference 1 with permission)

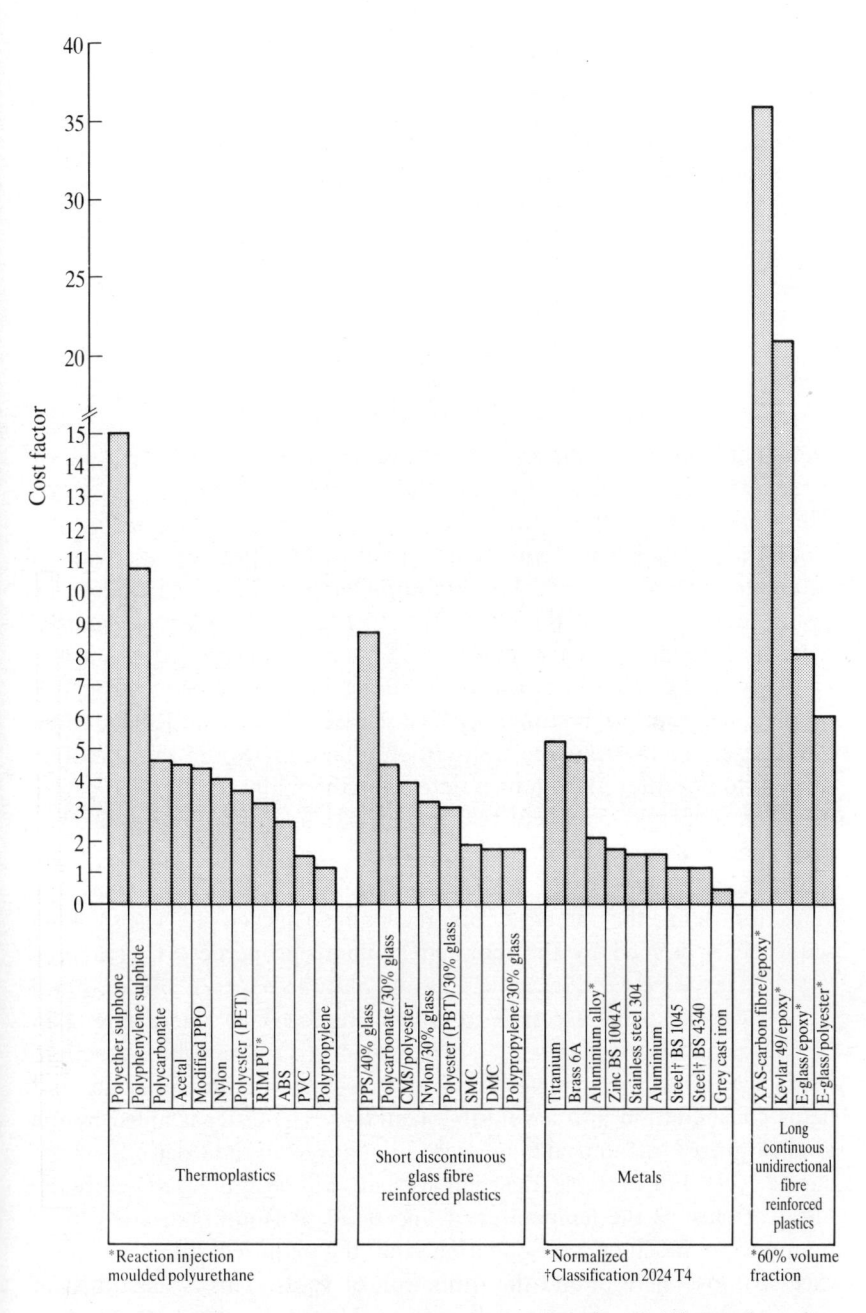

Figure 39b Relative costs of materials to give a similar rigidity to that of steel to BS4340 (Reproduced from reference 1 with permission)

that collapse in a controlled manner in a high speed crash, to provide energy absorption analogous to metal buckling, can also be designed[2].

Reference

1. *Replacement of Metals with Plastics*, National Economic Development Office (Plastics Processing EDC), London, 1985.
2. D Hull, *Composites Sci. Tech.* **40**, 377 (1991).

3.2 Autoclave moulding

F R JONES

High performance composite laminates are mainly manufactured by consolidating prepreg material in an autoclave or a laboratory pressclave. This is because the fibres can be arranged at predetermined angles in a laminate of high and consistent quality. The prepreg consists of collimated fibres or woven cloth pre-impregnated with the thermosetting resins (e.g. epoxy) or thermoplastic (PEEK). The latter is 'boardy' whereas the former has a designed level of tack that allows the laminae to be stacked together. For ease of handling and cleanliness, the epoxy prepreg, for example, is stored between release films in a sealed container in a freezer. To prevent condensation on to and ingress into the prepreg, it needs to be removed from the freezer, and thermally equilibrated before use. The prepreg sheet is cut to appropriate dimensions and stacked at predetermined angles to produce a balanced laminate. For unidirectional material, a typical lay-up would be $0_2^\circ/\pm45^\circ/90_2^\circ/\pm45^\circ/0_2^\circ$ where the subscripts define the number of prepreg laminae in each ply. The stacking sequence is defined by the required laminate properties. The prepreg 'stack' is layed between release cloths and blotting paper or similar 'bleed pack' to soak up the excess resin. Some prepreg does not require resin bleed for consolidation into a laminate with correct fibre volume fraction. Resin bleed aids the removal of air and residual volatile impurities and helps consolidation into a void-free laminate. The latter is aided by the application of vacuum and pressure at the appropriate periods in the curing cycle. For example, the resin viscosity will be minimized before the onset of cure as the temperature is increased. An optimum viscosity is required to facilitate consolidation and the required 'bleed' without excessive loss of resin and the formation of voids. This is illustrated in Fig. 40, where a typical cure cycle is given. The appropriate cure schedule for each resin system will need to be identified independently.

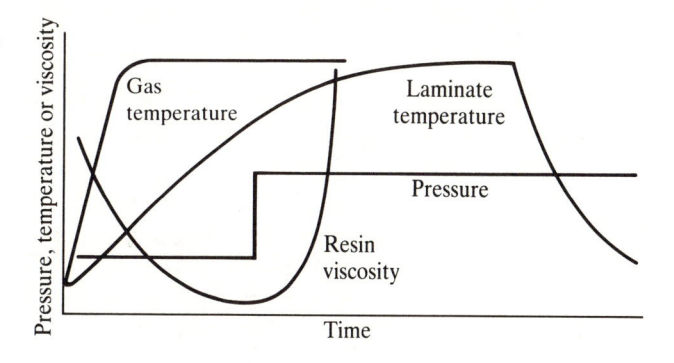

Figure 40 Schematic of an autoclave cure cycle

Moulding procedure

A typical lay-up in the autoclave is given in Fig. 41. The 'stack' of prepreg is placed on the tool surface and covered by a porous release layer. Any excess resin exudes through this into the overlying absorber. The prepreg stack is surrounded by a close-fitting dam. The top caul plate is optional but will provide flat laminates with a more even finish. The 'stack' is surrounded by a membrane or 'bag' which separates it from the gas used to apply the pressure and is sealed at its periphery with an appropriate grease or sealant. In order to evacuate the 'bag' evenly, in a technique analogous to vacuum packaging, a porous membrane is incorporated between the membrane and the stack. The evacuation can be accomplished by incorporating an appropriate port into the bag to which the vacuum line can be attached, or by venting it directly through the tool. The whole process of cutting, stacking and bagging of the 'stack' must be done under clean conditions, since contamination of the prepreg surfaces with the release agents, grease, hand barrier cream or other will lead to a poor quality product. Whereas large areas of delamination or poor wet-out can be observed by C-scan (Section 5.16), areas with impaired adhesion cannot be observed.

The prepreg

A number of variables can affect the quality of the laminates manufactured. These are fibre distribution, resin content and viscosity profile. The prepreg is impregnated uniformly with an exact proportion of resin to give zero bleed or, with a slight excess, 5% bleed. Removal of large volumes of resin can disturb the fibre alignment. The flow of the resin is clearly an important property of a prepreg and one that

140 Autoclave moulding

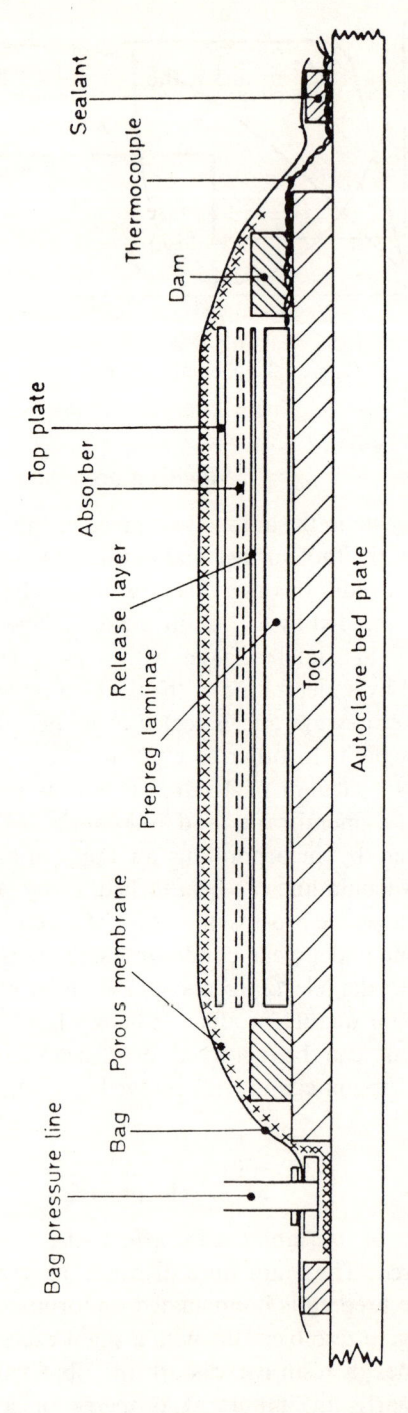

Figure 41 Autoclave moulding assembly (reproduced with permission of Butterworth Heinemann Ltd © from Purslow and Childs)[1]

determines the ease of fabrication. The temperature dependence of viscosity can be controlled by advancing the cure of the epoxy/hardener mixture or by incorporating a thermoplastic modifier to produce a flow controlled prepreg.

It is also possible to advance the cure of the prepreg during the cure cycle by choosing an appropriate temperature schedule, probably involving a 'dwell' in the increase in temperature before pressurization. This may be a valuable technique for ensuring viscosity consistency between batches of prepreg manufactured at different times.

The above discussion clearly indicates that complex shapes cannot be accomplished readily using this autoclave moulding. For moulds with excessive curvature, the drape of the prepreg can be improved by using a woven cloth reinforcement, otherwise it can be cut into appropriate small shapes to be layed-up on the mould surface in a curved fashion without large overlaps.

The role of the absorber

Even with zero bleed systems, some surplus resin will be exuded from the stack and collected in the absorber. The absorbency has been shown to be inversely proportional to its thickness and the applied pressure.[1] The rate of transport of the resin through the absorbent is clearly an important factor. The absorbency of the chosen material needs to be experimentally determined so the correct quantity can be selected, otherwise excess resin absorption can lead to void formation. Care must also be taken to ensure the porous membrane does not act as an additional absorber.

The cure schedule

This is chosen to ensure the laminate heats up at $\approx 2\,\text{K}\,\text{min}^{-1}$ and this may be achieved by heating the autoclave at a differing rate, sometimes as high as $10\,\text{K}\,\text{min}^{-1}$. This is illustrated in Fig. 40. A dwell in the heating cycle may be incorporated to enable thermal equilibration of the large thermal mass of the 'stack' and tooling. Even though this is used at low temperatures relative to that for cure, a 'dwell' can alter the gelation point in the cycle, as a result of partial curing. It should be considered that the state of cure is increasing with time, independent of dwells in temperature.

Since convection heating is used in the autoclave, the gas pressure will also influence the rate of temperature increase and hence the gel time. The cure temperature is determined by the resin system; it can be up to 200°C for epoxy but may be in excess of 300°C for other thermosets and some thermoplastics. The pressure cycle involves applying vacuum to the bag to bring the pressure to below 1 atm (101 kPa) and enhance the removal of air and volatile residues. With the chamber vented to the atmosphere, the 'stack' is under a compaction pressure of 1 atm. The evacuation stage is crucial to the production of good quality laminates, where volatiles are present. For the polyimides, condensation volatiles need to be removed at appropriate temperatures.

The pressure can be applied to the bag either with the vacuum maintained or the bag vented to the atmosphere, as shown in Fig. 40; the pressure would generally be applied at the temperature at which the resin viscosity is minimized. This minimum viscosity is not the lowest achievable since it may have proved necessary to increase it by partial curing. For example, a flow controlled prepreg matrix has a minimum viscosity of 45 P, whereas a conventional TGDMA/DDS resin has a minimum viscosity of 1–2 P.[2] (In SI, $1P = 10^{-1}$ kg m^{-1} s^{-1}). An optimum pressure of 150 kPa is reported[1] to avoid overcompaction, since higher pressures will serve only to distort. However, with systems where volatiles are present, these may be kept in solution and prevented from causing voids by applying higher pressures. For PMR-15 lamination, a pressure as high as 1.3 MPa is routinely employed. Typically pressures of ≈ 600 kPa are employed for epoxy composites.

Voidage and defects

The main origin of voidage is entrapped air and other volatiles, which can be largely removed in the vacuum cycle. However, voids also arise from the relationship between the interfibre spacing and the eventual resin content. The inherent fibre packing fraction is largely determined by the nature of the prepreg but a degree of control can be exercised by careful choice of temperature at which the pressure is applied, so that resin is not removed excessively from the fibres by force to create voids. Low compaction pressures will therefore lead to resin-rich areas between the plies, whereas excessive autoclave pressure will lead to fibrous areas devoid of resin. Overcompaction can also lead to misaligned fibres in unidirectional laminates. Thus void formation and final fibre volume fraction are intimately related, as shown in Fig. 42.

Advances

Most improvements in the autoclave moulding process have arisen from the need to automate the cutting of the prepreg into patterns for lay-up onto a complex mould surface. A reciprocating ultrasonic knife can be used in combination with a large vacuum bed on which the prepreg is

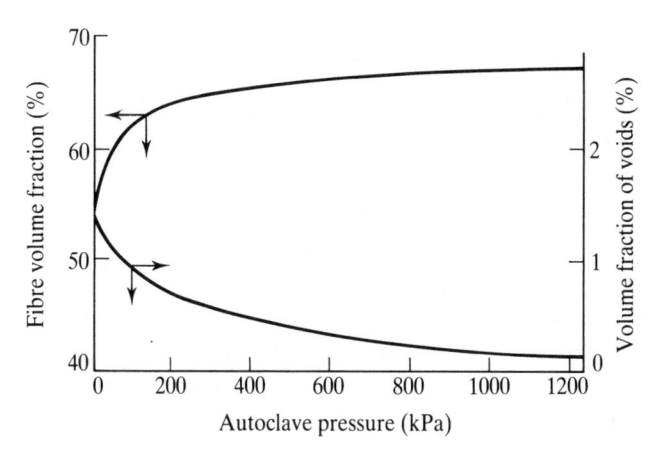

Figure 42 Effect of lamination pressure on fibre volume fraction and void content

placed to cut accurate profiles at high speed, leaving the backing paper intact. Hand lay-up of the profiles is still required.

Tape laying provides for automatic lay-up of prepreg strips at appropriate positions on the mould surface; the positioning and cutting of the individual pieces can be fully computer controlled. Both techniques are used in the fabrication of helicopter rotor blades and have been described elsewhere.[2]

Tooling

There are large differences between the thermal expansion coefficients of conventional metals and composite materials, especially CFRP; this presents difficulties in dimensional control. But tooling from composite materials can be manufactured with matched thermal expansion characteristics. Prepreg materials have been developed specifically for tooling fabrication. These combine modest temperature curing, e.g. 95°C with free-standing post-curing at 175°C, with adequate thermal stability for the temperatures employed in the moulding process. The use of this technology also has benefits of good finish on complex mould surfaces at economic costs. Further details are given in Middleton.[2]

In many applications, for example a helicopter rotor blade, it is impractical or impossible to remove the tooling; it has to be left in situ. In these cases, high temperature resistant, lightweight cellular material can be used as the mandrel onto which the prepreg is layed, and left in situ after curing. The main requirement is for adequate thermal dimensional stability at the curing temperatures and pressures employed. Apart from the honeycombs based on aluminium or Nomex/phenolic resin paper, a polymer foam can be employed. Most conventional polymeric foams have insufficient thermal stability and polymethacrylimide (Rohacell) is preferred; this has compressive strengths up to 1.5 MPa, depending on density, and withstands epoxy curing temperatures.[2]

Conclusions

Autoclave moulding is a developing technology specifically for high performance applications such as aerospace and racing car bodies. Its use is expanding from military to civil aircraft with the need for automation. Autoclave moulding enables consistent high performance laminates to be manufactured.

References

1. D Purslow, R Childs, *Composites*, **17**, 127 (1986).
2. D H Middleton (ed.), *Composite Materials in Aircraft Structures*, Longman, Harlow, UK, 1990, Chs 4, 10, 14.

3.3 Centrifugal casting

F R JONES

This technique is analogous to the process used for 'spun concrete pipe' and consists of a rotating mould into which resin and reinforcement are introduced. It generally involves spinning the mould in one plane with a longitudinally movable lance to deliver the resin and chopped fibres. In principle, a fixed quantity of resin (and reinforcement) can be placed into a closed mould for multiaxial rotation. In the plastics industry, this latter process is called rotational casting but is not normally considered to be a fabrication process for composites and is restricted to thermoplastics, both filled and non-filled.

Centrifugal casting is used for manufacturing pipes employing cold curable resins such as the unsaturated polyesters and vinyl esters. After demoulding, the pipe can be post-cured in an oven or by suspended heaters. It has the advantage that the rotational speed can be adjusted to increase or decrease the centrifugal and hence compaction forces that operate. Furthermore, by varying the resin type and/or reinforcement, a complex through thickness structure can be designed to withstand the environmental and stress state requirement of the pipe in service. For example, under flexural stresses, the central position of the pipe wall thickness will not need to withstand a significant tensile stress so the fibre reinforcement can be replaced by a filler, the equivalent to an I-beam in engineering. The fibre and filler volume fractions within the strata can be varied across a wide range by varying the compaction centrifugal forces during the introduction of resin/reinforcement at appropriate concentrations.

In contrast to 'lamination' techniques, interfaces are not generated between the individual layers. For example, in pipes intended for corrosive environments, a corrosion resistant resin inner surface can be graded into the structural resin alongside a graded increase in the fibre volume fraction away from the service surface. Under aqueous acidic conditions, environmental stress corrosion cracking of GRP can occur, as discussed in Section 6.5. The centrifugally cast pipe can be designed to resist this type of failure by incorporating a barrier resin of high fracture toughness which grades into a structural resin as the fibre volume fraction increases gradually in a controlled manner.[1-3] Furthermore, the simple mechanical operations involved can be readily controlled by microprocessor.

In contrast to filament wound pipes where socket and spigot ends can often be incorporated directly into the moulding, for centrifugally cast pipe, these need to be bonded on in a post-fabrication process. For tapered pipes, glass fibre preforms in combination with a tilted mould can be employed. Curved pipes can be formed using flexible moulds.[4]

References

1. M A Kanona, R D Currie in *Composite Structures 4*, ed. I H Marshall, vol. 1, ch. 16, pp. 1.223–1.234
2. F R Jones in *Proceedings of a Symposium on Plastics for Pipeline Renovation and Corrosion Protection*, Plastics and Rubber Institute, London, 1985, paper 13.
3. Johnston Pipes Armaflow Literature, Telford, UK, 1984.
4. R G Weatherhead, FRP Technology, *Fibre Reinforced Resin Systems*, Applied Science, London, 1980, pp. 73–76.

3.4 Continuous fibre reinforced thermoplastic composites – shaping

J A BARNES

Introduction

Since the early 1980s, considerable effort has been expended on the development of continuous fibre reinforced thermoplastic composites based on aromatic polymers such as poly ether ether ketone (PEEK), polyether sulphone (PES), and polyether imide (PEI). The principal difference between such materials and the more conventional high performance composites based on thermoset polymer technology lies in the behaviour of the polymer matrix during processing. Thermoplastic polymers are those which possess a capacity for melt processing and undergo no chemical change during their processing cycle. Hence the need for curing through crosslinking is removed and the forming process involves only three stages: melting, flow and solidification.

Amorphous and semicrystalline polymers

Differences also emerge between the two types of thermoplastic polymer during cooling. We can conveniently divide thermoplastic polymers into two families, amorphous (such as PES and PEI), in which the individual molecules are randomly arranged and possess no short-range order, and semicrystalline (such as PEEK), in which a proportion of the molecules adopt a regular arrangement with respect to each other. In amorphous polymers, no significant physical changes occur during the cooling cycle other than freezing. But in the case of semicrystalline polymers, some care must be taken to control the cooling rate in order to reach the optimum level of crystallinity. Additionally, in semicrystalline materials it is usually necessary to ensure that all the polymer is fully melted before cooling; this is to eliminate all nuclei upon which crystalline growth may occur.

Thermoplastic composites are usually provided in one of two forms: a stiff boardy sheet with no drape or tack, or as a woven mat with some drape and no tack. This basic form is referred to as the pre-impregnated product, or prepreg.

Shaping technologies

Thermoplastic prepregs can be manufactured into finished laminates using techniques typical of thermoset materials, with a number of additional options. The primary difference is the higher processing temperatures typically required for thermoplastic composites, in general 300–400°C.

The most common route for manufacture of simple flat laminates for testing is to hot press. The prepreg is hand cut and stacked in the required lamination sequence, tack welded to aid handling and placed in a mould or frame. The assembly is then heated to some temperature above the melting point of the polymer, pressurized and cooled. Cooling can be carried out in the heating press or the mould assembly can be transferred to a cooler press to speed production.[1] The maximum temperature used is

Table 35. Processing temperatures for a number of commercially available composites (after Cogswell[2])

T_g (°C)	Approximate processing temperature (°C)	Semicrystalline matrices	Amorphous matrices	Composite suppliers
260	320–350		Avimid KIII	Dupont
260	360–390		Torlon C	Amoco
250	350–370		Radel C	Amoco
230	315–360		Victrex ITA	ICI Fiberite
230	315–360		PES	Specmat, BASF
220	315–360		PEI	American Cyanamid, Ten Cate
220	340–360		Ryton PAS-2	Schriner composites Phillips
220			Radel X	Amoco
175	390–410	Victrex ITX		ICI Fiberite
156	360–370	PEKK		Dupont
145	320–340		J Polymer	Dupont
143	380–400	PEEK		ICI Fiberite, BASF, Tech Textiles, Asahi, Nittobo, Texxes,
95	330–350		Ryton PPS	Phillips

typically $T_g + 200°C$ for semicrystalline materials and $T_g + 100°C$ for amorphous polymers, although limits on polymer stability mean temperatures greater than 400°C are rarely recommended. Recommended processing temperatures for a variety of polymer composites are shown in Table 35. Typical moulding pressures are in the region of 1.4 MPa (200 psi); in fully wetted systems, complete consolidation can be achieved using only atmospheric pressure, partially wetted materials may require higher pressures.

As an alternative to the heated press, many fabricators prefer to use existing equipment and produce panels in an autoclave; the conditions used are similar to those for the hot press, though the use of an autoclave can rarely be considered economic. To eliminate the costly process of heating the entire volume of an autoclave, work is increasingly being carried out on heated tools in pressure vessels (since the only requirement is to fully melt the matrix), with consequent savings in time and energy. Other technologies transferred from the thermoset composites industry include pultrusion and filament winding. In filament winding, layers of prepreg are welded during processing, hence non-geodesic and even re-entrant shapes can be produced. An extension of the filament winding process is tape placement, in which a moving head both melts and consolidates continuous feedstock as it moves over a surface. This is one of the processes that offer much promise as a means to supply both complete parts and sheet for subsequent processing. In tape placement and filament winding, it is usual for the completed thermoset lay-up to be removed from the manufacturing device, vacuum bagged then cured in an autoclave as a separate process; the nature of thermoplastic composites is such that all of the processing is completed in a single stage.

The metals industry has also provided manufacturing technologies appropriate for thermoplastic composites. Perhaps the most powerful is stamping using preheated blanks of composite on cold or cool tools. The required force for shaping can be provided via a rubber block (termed rubber block processing) or via a rubber membrane backed with a hydraulic reservoir (termed hydro-rubber forming). From an economic standpoint the capability to produce complex shapes at the rate of 50–100 h^{-1} from a single machine makes stamping an attractive proposition. A second metalworking process shown to be practicable for thermoplastic composites is high speed roll forming of continuous sections; other continuous 'polymer' technologies, such as double belt lamination, have also been demonstrated.

A few processes have been developed that are unique to the art of shaping continuous fibre thermoplastic composites. The most used of these is diaphragm forming, in which a preform (which may or may not be preconsolidated) is contained between two highly deformable diaphragms (either superplastic aluminium or more commonly 'Upilex' polyimide film). The assembly is evacuated, heated and used to separate two halves of a pressure vessel. By the action of differential pressure, the assembly can be made to conform to the shape of a tool. The function of the diaphragms is to allow the pressure to be applied, but also to maintain the surfaces of the laminate in tension to prevent buckling of the fibres.

A significant body of information is available describing both the thermophysical and rheological behaviour and model composites during fabrication. For a more complete exposition and a fuller description of the processes outlined above, see Cogswell.[2]

Joining

Once a shape is produced by one of the processes described above, there is usually a requirement to join it to another component. This may be accomplished by the usual means of conventional fastening, adhesive or solvent bonding (with amorphous polymers) or localized heating of the matrix polymer in areas where welding is required. In the latter case, methods more normally used to join thermoplastic polymers may be applied, including ultrasonic welding, induction heating and resistance welding. In resistance welding, a number of workers have used individual layers of prepreg to provide the interface between two components by using the carbon reinforcement to pass the required current. Thermoplastic composites have a unique capability, they provide a bonding method which relies on a technology known as amorphous interlayer bonding or 'Thermabond'.[3]

It is advantageous to provide bonding by melting of material for two reasons: it avoids sharp transitions in polymer type throughout a structure and it removes the need to provide surface treatment before application of adhesives. But complete melting of a component after shaping is obviously not sensible, and it is not always possible or practicable to locally melt parts of the structure. An ingenious solution to this problem is to incorporate a polymer in the surface layers of a semicrystalline moulding. The polymer is chosen to be miscible with the parent structure and to melt at a lower temperature; it thus provides for form stability during bonding. The most common adoption of this method uses PEI as a bonding medium for PEEK.

An alternative approach is to carry out bonding and forming simultaneously. Workers in Holland have demonstrated the power of thermoplastic composites process technology; they have developed a method for producing PEI foam filled PEI composite sandwich structures and performing limited shaping during foaming.[4] Clearly, capabilities of this kind can offer significant cost benefits over conventional manufacturing approaches.

Conclusions

A wide variety of shaping technologies are available for thermoplastic composites. The largest proportion are based on existing fabrication methods from the thermoset composites and metals industries, though some unique routes to shapes have been developed. An equally wide range of joining technologies has been developed, again based on existing ideas and new concepts in bonding.

References

1. *ICI Thermoplastic Composites Materials Handbook*, ICI plc, Wilton, 1991.
2. F N Cogswell, *Thermoplastic Aromatic Polymer Composites*, Butterworth-Heinemann, Oxford, 1992.
3. F N Cogswell, P J Meakin, A J Smiley, M T Harvey, C Booth in *Proceedings of the 34th International SAMPE Symposium*, May 1989, Anaheim, SAMPE, Covina, California, USA, 1989, pp. 2315–2325.
4. A Beukers in *Proceedings of the 12th International European Chapter of SAMPE*, May 1991, Maastricht, SAMPE, Covina, California, USA, 1991, pp. 393–405.

3.5 Bulk moulding compounds and dough moulding compounds

A G GIBSON

The terms, bulk moulding compound (BMC) and dough moulding compound (DMC) are used to describe dough-like thermosets belonging to the family of materials known as polyester moulding compounds. Like sheet moulding compounds (SMCs) (see Section 3.17) they contain unsaturated polyester resin, chopped glass fibre reinforcement and a filler. They also make use of shrinkage control additives, as described in Section 3.17, to obtain low levels of mould shrinkage and a good surface finish.

In contrast to SMCs, BMCs contain randomly oriented chopped reinforcement, with a rather lower glass content and shorter fibre length (6mm or 12mm). They generally have a lower viscosity under processing conditions and are preferred to SMC for moulding more complex parts with three-dimensional features such as ribs or deep recesses. As might be expected BMCs are cheaper than SMCs, with somewhat lower mechanical strength. They may be processed by either hot press moulding or injection moulding.

Compound manufacture

The compounds are manufactured by a low intensity mixing process, usually involving a Z-blade mixer. First the resin is compounded with all the other components except the glass. Then the glass strands are added and the process continued for the time required to wet out the reinforcement. Care must be taken to minimize the work input during this second stage, so as not to unduly degrade the fibres. The finished compound may be removed either by tipping the mixer or, as is the case with many larger mixers, by using an auger screw in the base of the compounder to extrude the material into easily handleable 'logs'. In either case the BMC is wrapped in thermoplastic film after compounding, to prevent loss of styrene monomer from the polyester resin.

Table 36 shows the composition of a typical BMC. In contrast to SMC, BMC is not usually thickened before moulding, although this can be carried out if required to render the material non-tacky and more easily handleable for hot press moulding. The effect of composition on properties is discussed by Pritchard and Gibson.[1]

Although the Z-blade mixing operation involves a relatively low work input, the fibre length that can be maintained with this method is limited. There has been development work on alternative processes which involve a more gentle compounding action and which can accommodate higher

Table 36. Composition of a typical bulk moulding compound expressed on the basis of 100 parts by weight of liquid resin

Unsaturated polyester resin	66.6
Low profile additive solution	33.3
Tertiary butyl perbenzoate initiator	1.5
Zinc stearate mould release agent	3.5
Calcium carbonate fillers	160
Chopped glass strands (6 mm, hard sized)	46

fibre loadings and fibre lengths without fibre breakage.[2] In one type of operation the fibres are wetted out by being gently squeezed between rollers. Compounds produced by processes of this type are known as thick moulding compound, kneader moulding compound and continuously impregnated compound. In contrast to Z-blade compounding, these new techniques allow chopped glass strand lengths of up to 25 mm to be incorporated. It is therefore expected that the proportion of polyester compounds manufactured by such routes will increase in the future.

Recently phenolic resin based DMC has been introduced which generally employs a mixture of a Resole and Novolak resin to provide the viscosity and cure requirements. (see Sections 2.7 and 2.8). These materials can be moulded similarly to polyester moulding compounds with due care for volatile removal, e.g. clamp breathing. These materials are competing in applications where fire resistance is paramount, such as large electrical junction boxes.

Processing

Hot press moulding, as described in the Section 3.17 for SMC, is the main processing technique for BMC. Because BMC allows a greater degree of flow than SMC it is preferred for more complex parts. The mould temperature is generally about 140°C.

Although BMC was originally developed as a hot press moulding material a substantial proportion is now processed by injection moulding. Advantages of injection moulding include shorter cycle times (in the range 20–50 s), a greater level of automation, and the fact that it is possible to produce parts with a much better surface finish.

Figure 43 shows a typical injection moulding machine for BMC. Such machines are equipped with a hydraulically assisted feed cylinder to aid the flow of material into the barrel of the injection unit. Injection units of either the screw or the plunger type may be used.

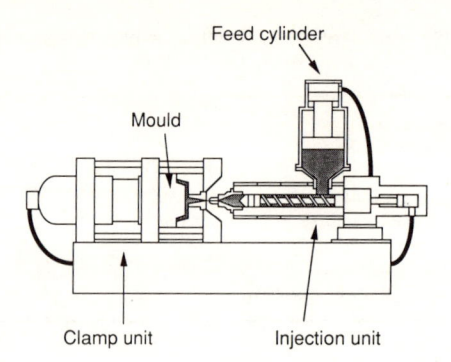

Figure 43 Injection moulding machine for processing bulk moulding compounds

Injection moulding has some potential disadvantages: the flows which take place can lead to much more pronounced fibre orientation and anisotropy of strength than with compression moulding. Moreover, unless care is taken to minimize the work done on the material during its passage through the machine there can be a significant loss of fibre length. However, improvements in the design of machinery, along with the development of special degradation-resistant hard-sized glass strands (see Chapter 1.11) have resulted in considerable strength improvements in injection moulded parts, to the extent that the property differential which used to exist between compression and injection moulding has been considerably reduced.

Properties of BMC

Typical values of mechanical properties are shown in Table 37. Such values often show considerable scatter, due to macroscopic variations in composition and fibre orientation within the material, and this variation should be borne in mind when designing parts.

Table 37. Properties of bulk moulding compounds processed by various routes

	Flexural strength (MPa)	Un-notched Charpy impact (kJm^{-2})
Hot press moulded	80	20
Injection moulded		
perpendicular to mould flow	110	23
parallel to mould flow	42	8

The anisotropy due to mould flow fibre orientation in injection moulding is significant. The fibre orientation patterns in thermoset injection moulding are different to those seen with reinforced thermoplastics. Injecting cold compound into a hot mould results in a flow profile close to plug flow, so mouldings do not have any pronounced multilayer structure. With injection moulded BMC the fibre orientation is generally in the direction perpendicular to mould flow. The transverse fibre orientation results from the divergent flow that takes place when the material enters the mould cavity from the gate.

If the fibre length is degraded in injection moulding, either by overworking the material or by using sprues and runners of too small a cross-section, then the degree of anisotropy decreases and the properties transverse to the mould flow direction fall to values similar to those shown for the mould flow direction.

Weld lines occur when two flow fronts come together during mould filling. The weld line strength of injection moulded BMC is usually low, so care should be taken in mould design to avoid having welds in critical locations in a moulding.

A key factor in many applications of BMC is the ability to achieve low or even zero levels of mould shrinkage. This permits parts to be moulded to net shape with extreme accuracy. Zero mould shrinkage also makes it possible to have parts with significant changes in section thickness and to use BMC to encapsulate components such as solenoid coils. Electrical properties, such as tracking resistance of polyester compounds, are also very good. The specific gravity of BMC is generally about 1.8.

Applications

BMC parts are used in electrical, automotive and domestic applications, where their combination of low cost, low mould shrinkage, good electrical properties and high heat deflection temperature can be used to advantage.

References

1. M Pritchard, A G Gibson, *Polymer Composites*, **9**(2), 131 (1988).
2. S P Corscadden, D J Payne, A G Gibson, *Composites Manufacturing*, **1**(3), 173 (1990).

3.6 Filament winding

V MIDDLETON

Process outline

In the wet filament winding process[1], fibre rovings are drawn from a spool through an impregnation system where resin is applied. The wetted rovings are then led through guidance eyes on to the rotating mandrel. When all layers of the winding have been applied the covered mandrel is removed from the winding machine and the composite is cured. The component is then stripped from the mandrel, which is returned for further winding.

In order to obtain consistent wet-out and accurate placement of the fibres, it is necessary to maintain them in tension throughout the process. Depending upon the type of feedstock being used (external or internal unwind) tensioning is usually done on or near the creel. Because of the tension, it is necessary to lay the fibres on a geodesic path over the surface, otherwise they will slip.

The basic stages of the process are shown in Fig. 44. The equipment comprises:

(i) the creel holder and equipment for controlling the fibre tension,
(ii) the impregnation section,
(iii) the filament guides and delivery eye which control the placement of the fibres on to the mandrel,
(iv) the winding machine on which the mandrel is mounted.

The winding machine must be capable of moving the filament payout eye accurately past the surface of the mandrel to repeatedly place the fibres so that a continuous surface is built up. The amount of material that can be laid down at any one pass depends upon the geometry of the part being produced. On large diameter tubes, many fibre tows can be laid down (i.e. up to 50), but on small precision components only a few rovings may be laid. The control data for the machine represents the path of the middle fibre. Wide fibre bands passing over small curved mandrels experience differing paths between their centres and their edges. Thus the fibres at the edge of the band are not laid geodesically and may slip.

The mechanics of winding machines vary from simple mechanically controlled systems, using gear and chain drives, to sophisticated numerically controlled machines, including robots. With advanced installations the whole process including fibre tension and resin temperature are controlled by a numerical controller. The process parameters (tension, temperature, feed rate) may be recorded together

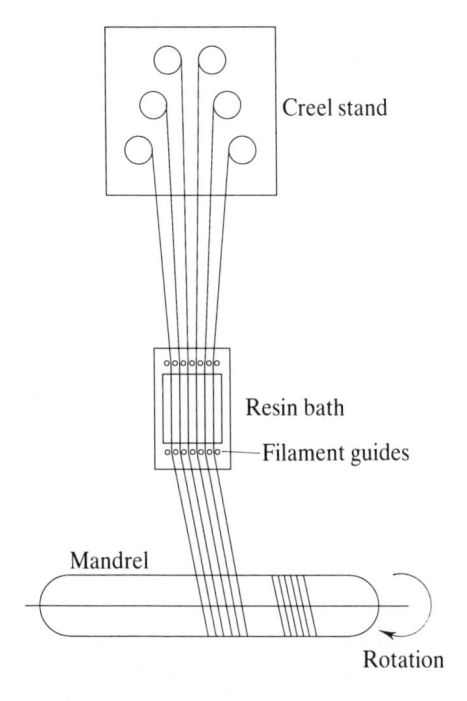

Figure 44 The filament winding process

with material usage to furnish a quality control report on completion of the winding (e.g. Pultrex Modwind System).

Fibre tension

Fibre tension has a significant effect upon the final quality and properties of the laminate. As fibre tension increases, so does the fibre content of the finished composite. Fibre volume fractions of 60% are relatively easy to achieve by this process. The production of higher volume fractions depends upon a number of factors such as component geometry, resin viscosity, payout eye design, etc. Composites with low volume fractions, e.g. 20–30%, having an even fibre distribution can only be achieved using rovings with a low tex.

Filament guides before the impregnation bath handle dry fibres and generate high friction forces, thus increasing tension; filament guides following wet-out experience less friction because of the lubrication by the resin.

High fibre tensions produce good fibre alignment, which results in a higher modulus composite and lower elongation at break in the transverse direction to the fibre. Deviations in the fibre direction are caused by waviness in the fibres and twisting within the roving, which cannot be completely eliminated. High tensions tend to damage the fibres by abrasion on the guides. It is not easy to maintain consistent low tensions, even with sophisticated equipment, because of frictional variations.

Impregnation

In the manufacture of glass fibre rovings, up to 200 individual filaments, sized to prevent surface damage, are collected together to form a strand. A number of these strands are held together in a roving by film former treatments.

In the impregnation phase, the aim is to cover all the filaments with a thin coating of resin. In order to do this, air entrapped in the roving must be expelled and replaced by the matrix material. Thus the impregnation system must break up the surface film former on the rovings and open them out into a flat ribbon, to prevent air from becoming entrapped, then apply a controlled amount of resin.

Impregnation is effected not so much by capillary attraction of the resin into the fibres but by the pressure difference that exists across the faces of the tow from the inside to the outside as the fibre passes around the roller or immersion bars.

High speed of impregnation is achieved by:

(i) high roving tension,
(ii) large contact angle of roving over the roller,
(iii) thin ribbon of glass (well opened rovings),
(iv) small impregnating roller diameter,
(v) low viscosity resin.

Increasing the roving tension too much prevents it from fanning out; compromises must be reached to provide an optimum balance between the parameters. Minimum fibre damage is a prerequisite of the fibre handling system, especially with carbon and aramid material. The most common methods of impregnation are dip impregnation and roller impregnation.

Dip impregnation

The roving is tensioned by friction bars before it is fed into the resin bath. It is held under the surface by passing under fixed bars immersed in the bath. This process applies rather too much resin for immediate winding and the excess has to be removed by passing the rovings through a system of scraper bars. The simplicity of the equipment makes this method the most commonly used for wetting out large numbers of filaments for pipe and tank winding.

Roller impregnation

Because dip impregnation applies too much resin and the bars can damage the fibres, roller impregnation is used. The amount of resin is limited by picking it up on a roller that dips into the resin bath. Excess resin is removed by a doctor blade, which limits the amount of resin passing to a thin film. The fibres pass over the top of the roller and cause it to rotate. As the resin is forced through the glass, it expels air and gives good wet-out. Further spreading of the resin may be achieved by having a squeeze roller running against the pick-up roller.

Winding

The purpose of the winding machine is to lay the fibres on the mandrel in their predetermined position. The filament guide eye and mandrel must move in a controlled geometrical relationship to one another; this requires at least two degrees of freedom – rotational and axial movements. Which function is undertaken by the mandrel and which is made by the payout eye depends upon the machine configuration; various examples are given below.

Winding machine configurations

Helical winding machines

Helical winding[2] is the most common form of winding. This mode of winding is used predominantly for pipes and other relatively long axisymmetric components. Simpler machines have mechanically actuated controls on two axes (rotation A, axial feed X) with manual adjustment for crossfeed (Y).

More advanced machines have servo control on four to six axes, usually by employing a standard NC controller. This is an advantage over specially developed (in house) controllers since maintenance service is readily available on a world-wide basis.

This type of machine can suffer from problems when winding low angle (near axial) fibres on components with small polar openings because of twist in the fibres and loss of hoop tension effect.

Lathe (bed) type machines (Fig. 45) are usually large machines for tank and pipe winding. They are made to accommodate very heavy mandrels; they have a separate bed for the pay out system; and the resin

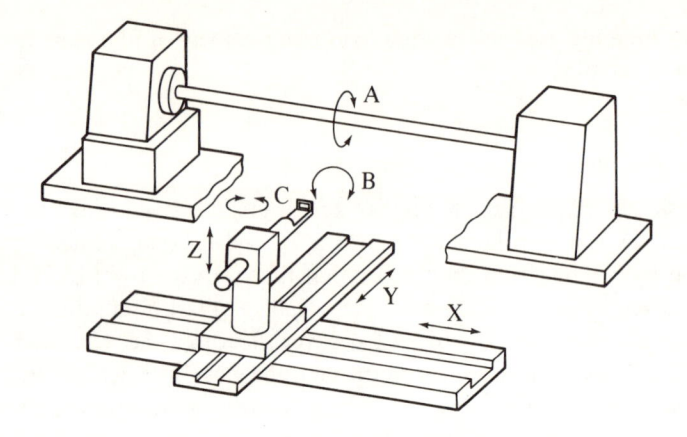

Figure 45 Lathe (bed) type winding machine

impregnation bath, and in some cases the fibre creel stand, is mounted on the longitudinal carriage (large inertia).

Gantry machines (Fig. 46) are mainly aimed at 'high tech' applications and usually have a minimum of five numerically controlled axes (AXYBZ). They are much easier to set up than bed type machines. The creel stand, fibre tensioning and resin impregnation are usually completely separate, but most designs do not achieve minimum inertia on the traversing axes. The high mounting of the carriage permits easy access

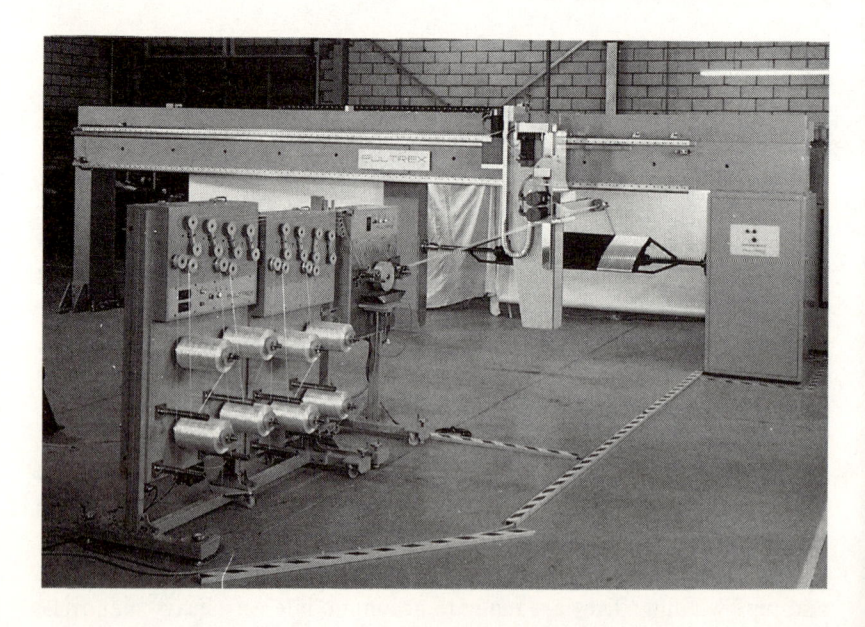

Figure 46 Gantry type filament winding machine

Filament winding

to the winding area, and the clear floor area allows disposable mats to be placed to catch resin spill. Care has to be taken to minimize the weight of the mandrel to avoid bending effects on the structure.

Planetary or whirling arm winders

Planetary or whirling arm winders[3] (Fig. 47) were developed for winding large pressure vessels (fuel tanks) for rocket applications. Helical type machines had problems supporting the mandrel and laying near axial fibres. The component is mounted on a vertical rotating axis. The material to be wound is fed along a cranked rotating arm, which circles the component from pole to pole, so only prepregged tape can be used as feed stock. They are limited in application.

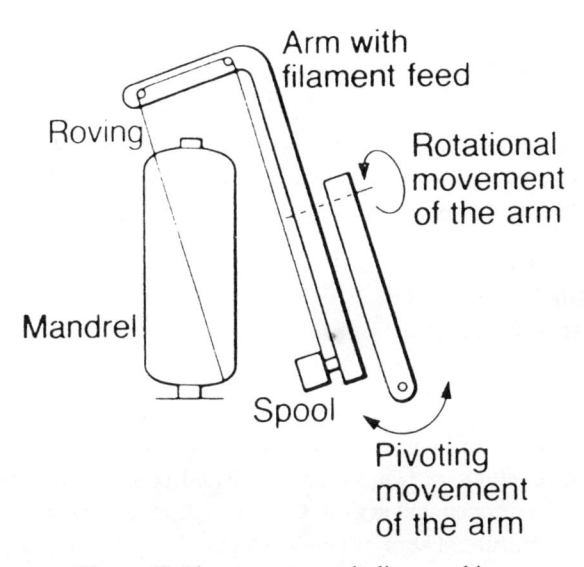

Figure 47 Planetary type winding machine

Tumble winders

Tumble winders were also developed for winding pressure vessels. The mandrel is mounted through one pole so it rotates about its central transverse axis and its polar longitudinal axis. By adjusting the tilt of the axis, different winding angles for the fibre may be generated. The fibre is led from a stationary payout eye; this permits wet material to be used. The size of the component to be wound is limited because the mandrel is cantilevered from one end. The variation in tilt is usually done manually but the drive to rotate the mandrel around its polar axis is complex.

References

1. *Filament Winding*, Shell Resins Epikote Technical Manual, 5th edn, Shell Chemicals plc, London and Amsterdam, June 1988.
2. M J Owen, V Middleton, D G Elliman, H D Rees, K L Edwards and K W Yang, *Development in Filament Winding*, Proceedings of the Reinforced Plastics Congress 86, Nottingham, 17–19 September 1986, published by British Plastics Federation, London, 1986.
3. G Lewis, *J. Composites Tech. Res.*, **9**(2), 33–9 (1987).

3.7 Hand lay

T M GOTCH

Hand lay is the mainstay of fibre reinforced plastics (FRP) processing due to its versatility in accommodating many of the difficulties traditionally associated with plastics moulding such as:

- (i) undercuts,
- (ii) parallel/vertical sides,
- (iii) sharp corners,
- (iv) incorporation of inserts and cores,
- (v) thickness variations.

The basic principle of hand lay is simplicity itself; resin and reinforcement are brought together on a moulding tool and consolidated into a form curable by heat or chemical action. Curing is generally by chemical action at room temperature. There are eight stages:

- (i) tool design and manufacture,
- (ii) tool preparation,
- (iii) gelcoat application,
- (iv) preparation of resin and reinforcement,
- (v) consolidation of the composite,
- (iv) material cure,
- (vii) item removal from the tool,
- (viii) item finishing.

Significant stages are shown in Fig. 48.[1]

Hand lay method

(a) Waxing mould

(b) Gelcoat application

(c) Resin application

(d) Reinforcement placement

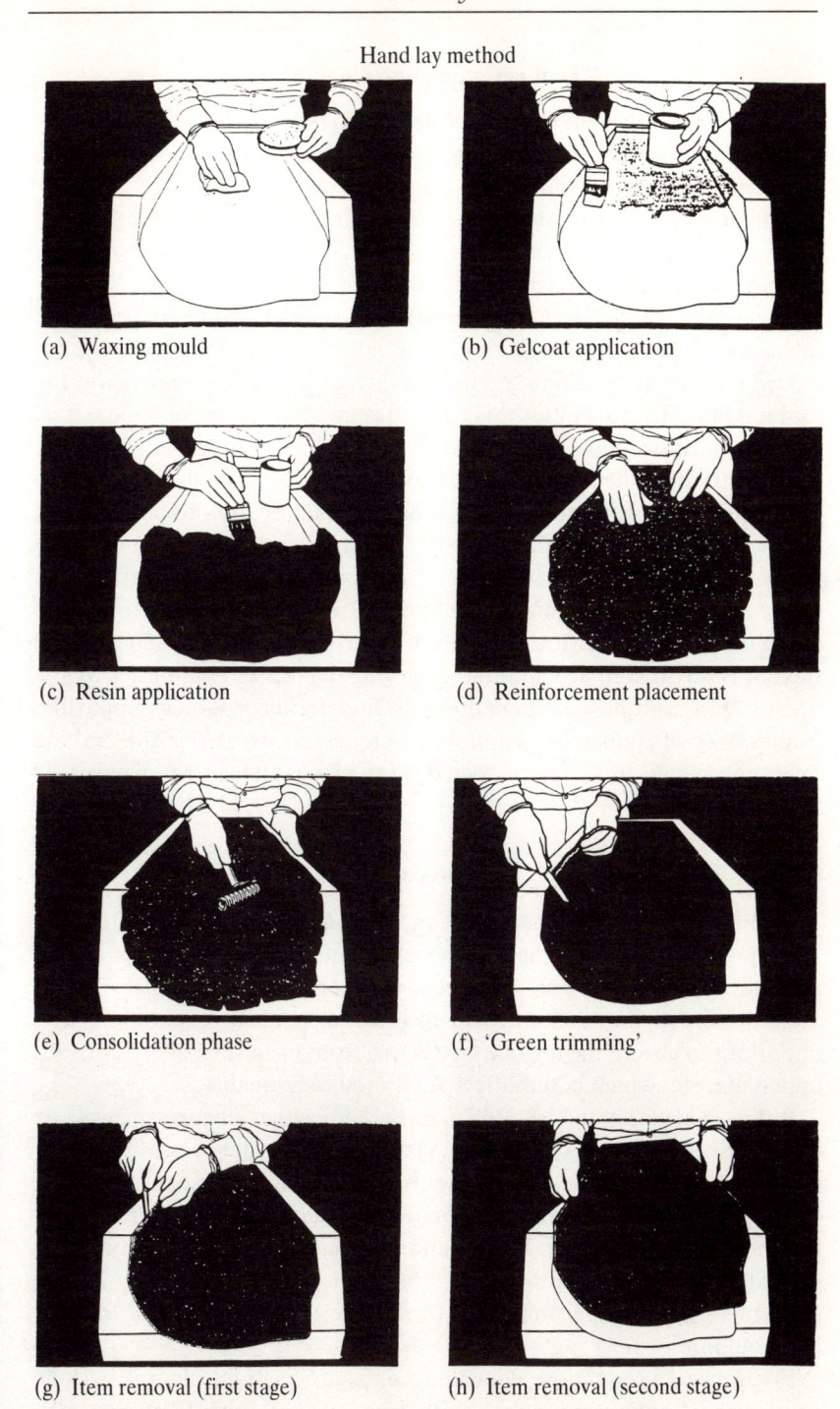

(e) Consolidation phase

(f) 'Green trimming'

(g) Item removal (first stage)

(h) Item removal (second stage)

Figure 48 Hand lay method[1] (Reproduced courtesy of BIP Chemicals Ltd)

Tool design and manufacture

This is probably the most underestimated aspect of hand lay production. Insufficient attention to detail will manifest itself later in problems during the moulding operation. Mould tools are typically manufactured in two steps: pattern manufacture and preparation followed by tool production.

Many materials can be used for pattern production but a good quality timber is typically chosen. This can be shaped into complex geometric forms and finished to a high surface quality. Patterns are usually sealed, coated with a resin and finished to a surface gloss for easy removal during toolmaking. To facilitate this, the design should accommodate as generous a draw angle as possible.

The pattern is then coated with release agents and the tool built up in FRP using the general technique for hand lay fabrication to be described below. As a general rule, the tool FRP thickness should be approximately twice the thickness of the intended component. It is imperative that particular attention is paid to removal of air during the consolidation phase otherwise premature tool failure in service could occur. During this stage it is possible to accommodate varying degrees of complexity by the use of split/multiple component tools and to incorporate supporting frameworks of timber or metal. The unique feature of this toolmaking operation is that it is only restricted by the ingenuity of the engineer.

Tool preparation

Tool preparation is often neglected and the results are often catastrophic. Tool preparation can generally be classified into three stages: initial preparation, regular mould treatment and periodic maintenance.

Before initial production, as in the case of pattern preparation, it is important to ensure the tool surface is free from blemishes, sharp corners, undercuts, etc. which could affect final moulding quality.

Release agents should be applied to the tool, either silicone-free wax or water based polyvinyl alcohol (PVA) materials (not to be confused with polyvinyl acetate, PVAc). More sophisticated chemical systems are being developed; these offer better and longer-lasting release characteristics. At this initial stage, it is desirable to build up a release layer of wax (or PVA) several layers thick with buffing between each application. As an additional safeguard, a final layer of PVA release can be added for first off mouldings.

For general production, depending on the system selected, it may be necessary to apply and polish with wax or coat with PVA for each moulding. For simple shapes or the newer release agents, it may be possible to produce a number of mouldings before recoating the tool, but it is generally a case of progressive development to define the optimum frequency.

If wax release agents are employed, there is a tendency for a wax layer to build up after a number of mouldings have been lifted. This must be removed with proprietary cutting paste on a regular maintenance basis and the tool reprepared.

Gelcoat application

A significant difference between FRP and metallic materials is that the surface finish coat (gelcoat) is the first rather than the last stage of manufacture. It is based on a specially formulated pre-accelerated resin (usually unsaturated polyester) which when catalysed will cure to a coherent film but with an air inhibited surface to promote adhesion to the back-up material.

The gelcoat is generally applied to the tool at a coating weight of 450–600 g m^{-2}. This must be allowed to cure to an intermediate or 'green' stage, so a good bond will be obtained with subsequent layers of resin/reinforcement, but the finished surface of the moulding will not be affected by the solvent in the resin or fibres being worked through the gelcoat and presenting a fibre pattern in the surface.

Preparation of resin and reinforcement

Since the cure mechanism of the resin imposes a constraint on the time available for the consolidation phase, it is essential that materials are prepared before commencement of manufacture.

In the case of reinforcement, this should be cut to templates to ensure consistency. Where any joints are involved, the use of butt and overlaps is to be avoided to prevent weak points occurring. Jointing should be achieved by 'teasing' fibres and should not be in the same position for successive layers.

Resin should be weighed out to provide the required resin/glass ratio, typically between 2.5:1 and 3:1 for chopped strand mat, higher for woven rovings and continuous filament. If this is not closely controlled, any increase will result in a reduction of physical properties, notably strength and impact properties. If the resin is not pre-accelerated, the catalysing and accelerating should be undertaken in two stages, ensuring that each component is throughly mixed into the resin before adding the second constituent. On no account should accelerator and catalyst be mixed together since they can react with explosive violence.

Consolidation

This is perhaps the most critical stage in the production of quality products and can be accomplished by the use of brush, roller or doctor blade. In each case, the essential point is to bring together the resin and reinforcement to produce a coherent material with the minimum of air inclusions. Each layer of reinforcement should be consolidated independently. Failure to consolidate adequately can result in inferior physical properties and other performance problems if operating under arduous environmental conditions.

Material cure

The time required to achieve adequate cure will generally be dependent upon the temperature of cure and/or the quantity of curing agent employed. However, to prevent distortion and shrinkage, it is preferable that the heat of exotherm is kept within manageable levels. Typically for ultra high quality surface finish requirements this could involve leaving the item on the tool for up to 24 h, but times considerably less than this could be used for less significant items by adjustment of the curing agent content or cure temperature.

During the cure mechanism, the material, as with the gelcoat, goes through a transitional 'green' stage where it exhibits a rubbery nature. This permits the material to be trimmed relatively simply with a sharp knife, hence reducing the amount of post-cure trimming.

Item removal from the tool

After full cure has been achieved, the item can be removed from the tool. Depending on geometry, this can range from simply lifting the moulding to the use of compressed air or extraction points located in the tool.

For more complex mouldings, this may also involve splitting of the tool and/or the removal of closure pieces. It may be possible by careful tool design to incorporate one or more of these features as an aid to item extraction.

Item finishing

Provided high quality tooling has been employed, it should not be necessary to undertake any further treatment to the moulding face unless damage has occurred. In this case, rectification procedures are available but with time these repairs will become evident and may be considered detrimental. Areas that could not be trimmed at the 'green' stage can be removed with the use of high speed carborundum tipped tools in a booth with suitable extraction.

From the above, it can be seen that hand lay is a relatively inexpensive manufacturing technique and the technical/economic advantages and disadvantages can be summarized as follows.

Advantages
- (i) simple technique,
- (ii) low capital investment requirement,
- (iii) inexpensive tooling,
- (iv) complex shapes can be accommodated,
- (v) high gloss surfaces can be achieved,
- (vi) no size restriction.

Disadvantages
- (i) labour intensive,
- (ii) operator sensitive,
- (iii) one finished surface only,
- (iv) slow production rate,
- (v) possibility of high styrene levels requiring costly extraction,
- (vi) excessive skin contact with resin necessitates good personal hygiene.

Despite its disadvantages, hand lay is probably the most versatile and widely used FRP manufacturing method.[2]

References

1. *Polyester Resins Handbook*, BIP Chemicals Limited, Oldbury, Warley, Worcs.
2. T M Gotch, *The Relative Economics of Different GRP Production Methods*, Hands off GRP, PRI Conference, Coventry, July 1979. Plastics and Rubber Institute, London, 1979.

3.8 Injection moulding – thermoplastics

M J FOLKES

One of the most common processing methods for thermoplastics is injection moulding[1-2]. The list of artefacts manufactured using this process is almost endless and includes electric drill casings, gearwheels, business machine housings, telephones and brief-cases. The process has been in existence for well over 100 years, although the development of the technique did not really get under way until the 1920s. In essence, the original concepts of the process were based on the pressure die casting of metals.

In principle, the injection moulding process is very straightforward. The polymer, in either granular or powder form is fed from a hopper into a heated barrel, where it softens and becomes a viscous melt. It is then forced under high pressure into a relatively cold mould cavity. When the polymer in the cavity has had sufficient time to solidify (at least partially), the mould is opened and the fabricated part is ejected. The cycle of operations is then repeated. Correctly controlled, this process is very versatile and is capable of fabricating very complex shaped components with considerable speed and precision.

The earlier injection moulding machines were of the plunger variety and there are still many of these machines in use today. The process of melting the polymer relies entirely on heat conduction from the barrel walls. In view of the low thermal conductivity of polymers, this results in a very non-uniform temperature distribution throughout the melt. Also, there is little or no mixing or homogenization of the melt. This implies that any additives, e.g. pigments or antioxidants remain poorly dispersed with consequential problems in the final artefact. However, for certain applications where it is desirable to minimize deleterious effects on the additives, e.g. when processing fibre reinforced thermoplastics, the simple plunger machine does offer some advantages.

Many of the disadvantages associated with the plunger type injection moulding machine are overcome in the screw injection moulder, which now largely dominates the market. Basically, this machine uses an extruder type screw which acts in a dual role; it is initially used for homogenization of the melt and then reverts to a simple plunger for the injection process itself. The arrangement of the screw in the barrel is shown schematically in Fig. 49. The mould is initially closed and melt is injected into the cavity using the screw as a plunger. The pressure on the screw is maintained until solidification at the gate(s) has occurred. The screw is then allowed to rotate so it conveys fresh polymer towards its front. But the moulded part is still in place, so the screw moves backwards against a predetermined back pressure and the melt is fully homogenized, ready for the next shot. During this operation, the fabricated part has cooled sufficiently so the mould can be opened and the part ejected. The mould then closes and the whole cycle repeats. The overall cycle time is typically a few minutes, although this depends on the dimensions of the part. The briefest part of the cycle is injection, which often takes only a few seconds; much of the time is spent in waiting for the moulded artefact to solidify.

The size of an injection moulding machine is usually defined by its capacity to mould in a single shot, e.g. a medium sized machine would be

Injection moulding – thermoplastics

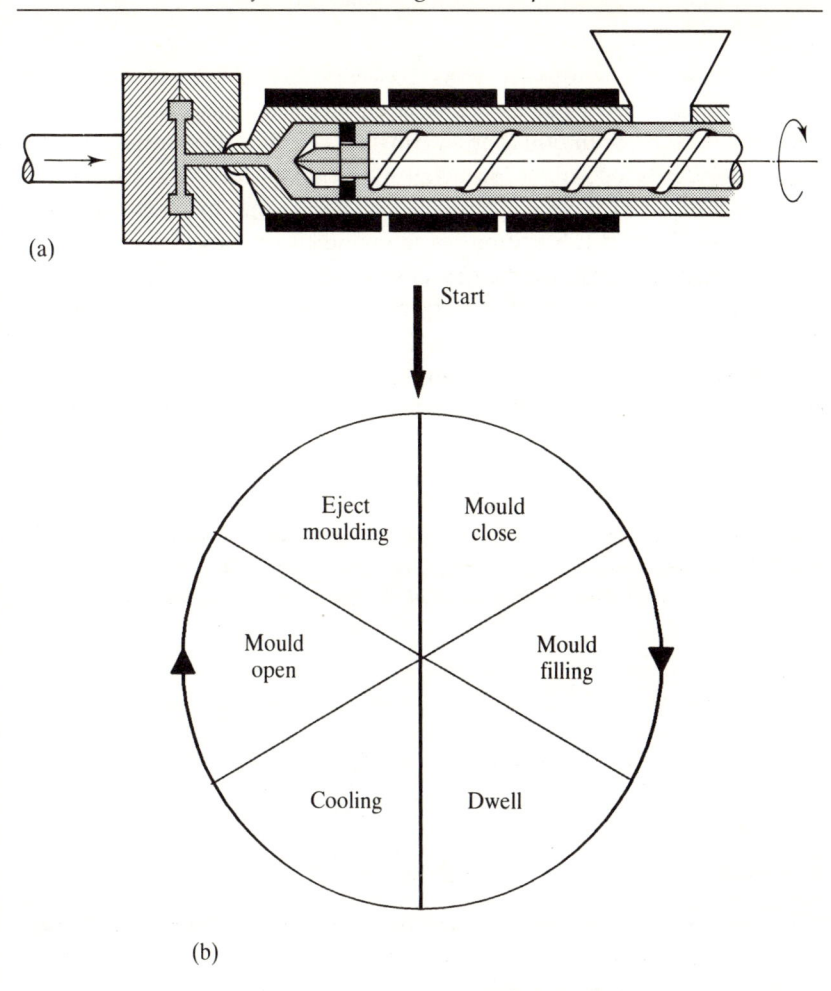

Figure 49 (a) Screw injection moulding machine-inset operations cycle. (b) Cycle of operations in injection moulding

rated 200 g. Machines used for the production of very large items, e.g. rubbish bins, are physically very large pieces of engineering equipment and demand considerable care in operation. The pressure required to inject the polymer melt into the mould can be very high (up to 200 MPa) and this demands the use of high tensile steels for both the injection unit and the mould.

The properties of the moulded part are highly dependent on the processing conditions employed. Barrel and mould temperatures, injection speed and pressure, screw back speed and pressure all influence the properties of the moulded artefact. Especially with some of the newer 'engineering' thermoplastics, minor changes in processing conditions can greatly influence part quality; consequently many of the new generation of injection moulding machines are equipped with 'process controllers'. They allow the continuous monitoring of certain key parameters and enable corrective actions to be taken by means of feedback controllers. Process controllers have done much to develop the injection moulding process into a rapid fabrication route for precision parts.

With discontinuous fibre reinforcement, injection moulding often leads to a skin–core structure of differingly aligned fibres, as discussed in Section 2.2. An example of this is given in Fig. 29. The effect of these aspects on properties is described in Sections 4.14, 4.15 and 4.16. Further new technologies for fibre management are discussed in Section 3.10.

References

1. R J Crawford, *Plastics Engineering*, Pergamon Press, Oxford, 1987.
2. R T Fenner, *Principles of Polymer Processing*, Macmillan, London, 1979.
3. Z Tadmor, C G Gogos, *Principles of Polymer Processing*, John Wiley, New York, 1979.

3.9 Injection moulding – thermosets

A G GIBSON

Materials

Thermoset moulding materials are best regarded as three component systems, containing resin, reinforcement and fillers. The reinforcement provides toughness by a number of mechanisms including fibre pull-out and crack stopping, while the resin acts as a binder for the reinforcement and fillers. The fillers serve a number of important roles in addition to reducing cost. These include modifying mould shrinkage, reducing the cure exotherm, controlling viscosity and improving surface hardness and appearance.

The classes of thermoset moulding material that can be processed by injection moulding are, in order of decreasing cost:

Epoxy
DAP (allylic)
Alkyd
Granular polyester
Bulk moulding compound (BMC-liquid polyester)
Phenolic
Amino

The most expensive materials, the epoxies, are used for encapsulation of electronic components and in other critical electrical applications. The DAP and alkyd resins, which may be cured by a free radical route, are also used in high performance electrical applications. Both these resins employ diallyl phthalate as a reactive diluent, the backbone polymer being an oligomer of DAP and in the former case and an alkyd-type ester in the latter.

The granular polyesters are analogues of the liquid polyester systems. To produce polyester compounds which are solid in the uncured state at room temperature it is generally necessary to replace the styrene reactive diluent with a less volatile monomer. The solid polyesters have some handling advantages over liquid polyesters and can be processed on conventional injection moulding equipment, but they are more expensive and have a higher level of mould shrinkage. They are available in a range of colours and are used in domestic appliances and in some electrical parts.

Bulk moulding compounds, discussed in Section 3.5, are one of the most widely used classes of thermosetting compound. They are based on liquid polyester resins and therefore require the use of specially modified injection moulding equipment. BMCs are also available in a wide colour range. They are the only materials that can be moulded with zero shrinkage.

Phenolic compounds (see Section 2.8) are used in an important range of engineering applications. They are normally cured by the addition of hexamethylene tetramine and other additives. The reinforcement may be glass or a mineral such as mica. Fillers include both mineral fillers and cellulosic ones such as wood flour. Phenolics have excellent heat resistance and low toxicity in fires. A range of mechanical property levels are available from above 100 MPa flexural strength in the case of 'high performance' compounds down to about 40 MPa in the case of 'general purpose' materials. Applications range from automotive, including demanding electrical and engineering under bonnet parts, down to general purpose domestic electrical parts. Phenolics are limited in colour to dark hues, due to the colour of the base resin when cured.

Finally, the amino resins, based on melamine formaldehyde or urea formaldehyde, are the least expensive of the thermoset moulding materials. Although their mechanical properties are modest the availability of these resins in a range of brightly coloured formulations allows them to be used in many domestic and electrical applications.

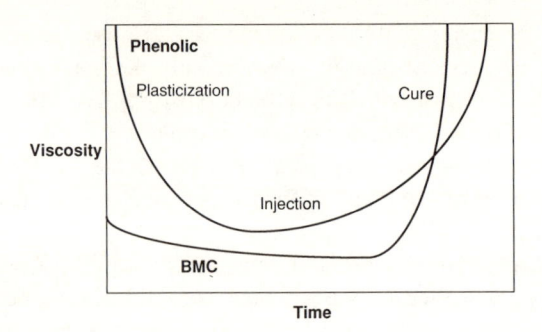

Figure 50 Viscosity variation (schematic) during the thermoset injection moulding cycle

Processing and applications

Thermosets[1] are processed using injection moulding machines fairly similar in basic design to thermoplastic moulding machines (see Section 3.8), the main differences being in the injection unit, which must be capable of removing heat from the compound as well as putting it in, and the mould, which is heated rather than cooled. Most machinery manufacturers offer thermoset versions of their equipment.

The variation in the viscosity of a thermosetting compound during an injection moulding cycle follows a characteristic 'U' shape, as shown schematically in Fig. 50. The viscosity falls initially when the material is heated, then rises due to the effect of the curing reaction. Compounds which are solid at room temperature are heated and plasticized in the barrel of the moulding machine in much the same way as a thermoplastic moulding material, prior to injection. In its low viscosity state the material must be injected into the mould to fill the cavity before the onset of gelation. Finally, after mould filling, the material must remain in the mould until the degree of cure has progressed sufficiently to allow the part to be ejected without damage.

BMC behaves in a slightly different way to the other thermosets: the resin is already liquid at room temperature and the compound only requires to be conveyed along the barrel and warmed a little prior to injection.

The cure chemistry of thermosetting resins always involves a compromise. They must not cure in the injection unit and should remain at low viscosity for long enough to allow mould filling. They are, however, required to cure as rapidly as possible once in the mould. Compounds such as BMC, which cure by a free radical mechanism, can be seen to have more favourable kinetics than materials such as phenolic, which cure by a condensation mechanism. The latter materials not only take longer to cure in the mould, they are less stable in the barrel, and particular care must be taken to ensure that, in the event of the moulding process being interrupted, cure does not take place in the injection unit.

Some of the thermosets, again including phenolics which cure by the condensation route, give off volatile by-products such as ammonia or water. In this case it is necessary to interrupt the cure cycle to allow vapours to leave the mould. Thermoset moulding machines have a 'clamp breathe' facility. During cure the clamping pressure is temporarily removed and the mould partially opened for a short period to allow volatiles to flash off.

Moulds for thermoset injection moulding are usually heated to a temperature of 140–150°C electrically. The cavities are often hard chrome plated to aid mould release and to resist the corrosive effects of some of the volatiles.

In contrast to thermoplastic mouldings, thermoset injection mouldings possess relatively little layer structure because injecting the compound into a hot mould tends to produce an easy slip condition at the mould surface. The principal fibre orientation in reinforced thermosets, which arises largely from the divergent flow that takes places as the compound flows from the gate into the mould, is transverse to mould flow, and the anisotropy of mechanical properties often reflects this.

Reference

1. J F Monk, *Thermosetting Plastics, Practical Moulding Technology*. George Godwin, London, 1981.

3.10 Injection moulding – fibre management by shear controlled orientation

P S ALLAN and M J BEVIS

The application of macroscopic shears to solidifying polymer melts provides a route for the management of reinforcing fibres in thermoplastic and thermosetting matrix polymers. Shear controlled orientation technology (Scortec) can be applied to injection moulding (Scorim) and to extrusion (Scorex). The concept has been applied to a wide range of thermoplastic matrix composites with glass reinforced polypropylene as the principal study material.[1] Thermosetting polyester matrix dough moulding compounds[2] may also respond to the application of macroscopic shears during injection moulding.

Macroscopic shears of specified magnitude and direction, applied at the melt–solid interface provide several advantages.

(i) Enhanced fibre alignment by design in fibre reinforced materials.
(ii) Elimination of mechanical discontinuities that result from the initial mould filling process, including internal weld lines.
(iii) Reduction in the detrimental effects of a change in moulded section thickness on the mechanical performance of fibre reinforced polymers.
(iv) Elimination or reduction in defects resulting from the moulding of thick sectioned components.

To apply macroscopic shears to a solidifying melt within a cavity requires the provision of a multiplicity of live feeds to the cavity, where the pressure applied to each of the feeds can be independently controlled. Another requirement is to be able to displace sufficient molten material within the mould to create a macroscopic shearing of the melt.

A simple embodiment of the concept in the form of a two live-feed device is illustrated in Fig. 51, located between the injection moulding machine screw/barrel, and the mould cavity. Initially, the pistons 1 and 2 are positioned to allow the melt to enter the mould from either one or both of the feeds. When the mould is full, pistons 1 and 2 may be moved back and forth to operate on the solidifying melt as follows.

(i) Pistons 1 and 2 can be pumped back and forth at the same frequency but with a phase difference of 180°.

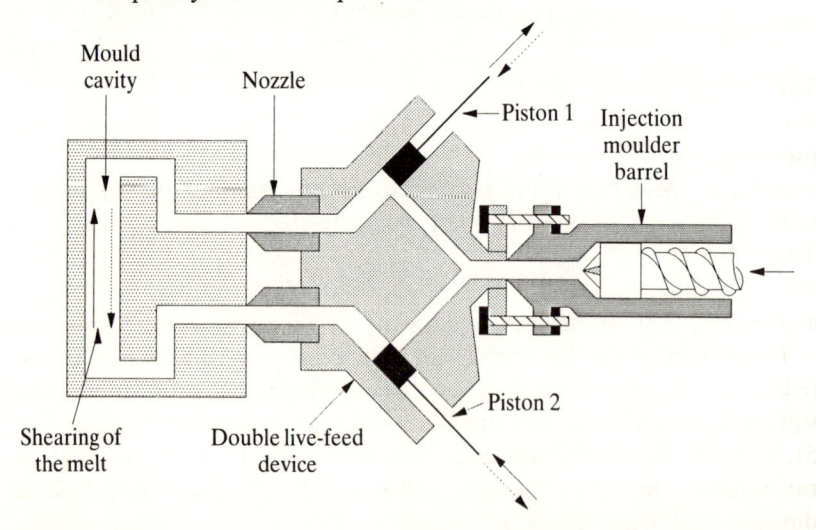

Figure 51 Schematic diagram of a two live-feed device. Reproduced from reference 5 with permission

(b) Edge 2 mm Centre

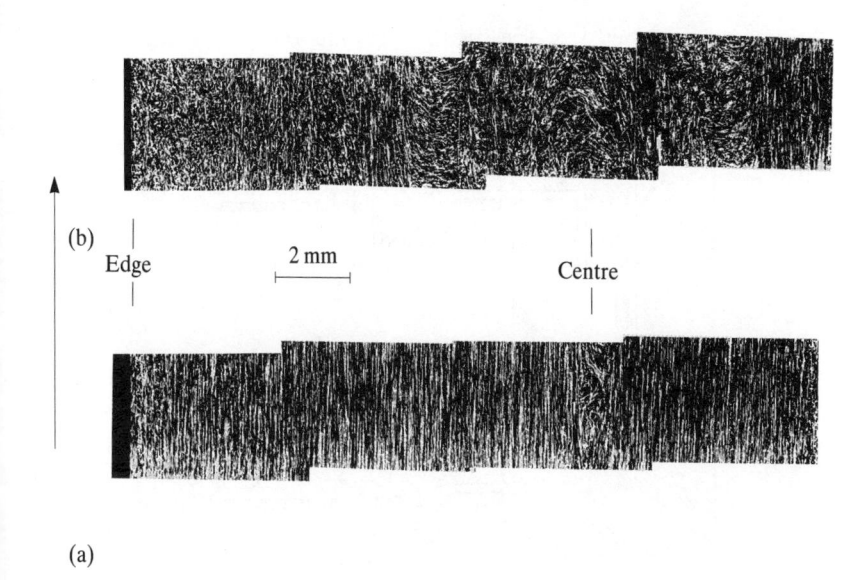

(a)

Figure 52 Contact micrographs showing the preferred orientation of short glass fibres in moulding produced using a) SCORIM and b) conventional injection moulding, combined with oscillating packing pressure to reduce void formation in the very thick moulding. Reproduced from reference 1 with permission

(ii) Pistons 1 and 2 can be pumped back and forth at the same frequency and in phase.
(iii) Pistons 1 and 2 can be held down under a static pressure.

The operation of the first of these three modes induces the preferred alignment of fibres at the melt–solid interface. As the melt–solid interface propagates from the outer to the inner regions of the moulding, the successive application of the out of phase operation of the pistons results in preferred orientation of fibres through the thickness of the moulding. The frequency and displacement of the pistons and the sequencing of the three modes of piston movement are selected to optimize fibre orientation and moulding cycle time.

The enhancement of uniaxial alignment of short glass fibres in a polypropylene matrix that is possible is illustrated in Fig. 52. A 20 mm square cross-section moulding similar in plan to that illustrated in Fig. 51, was produced using Scorim and conventional moulding. Contact X-ray microradiographs taken from sections parallel to the process direction and through the centre of the mouldings are shown in Fig. 52. These illustrate the preferred orientation of glass fibres that result from Scorim and conventional moulding, respectively. The pronounced

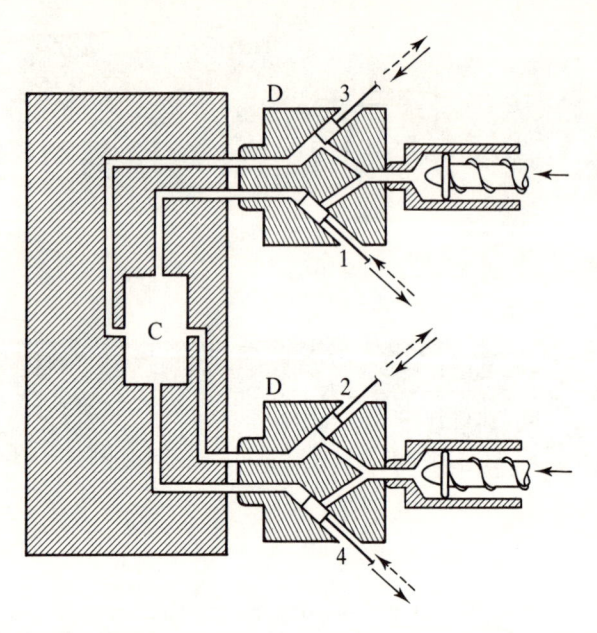

Figure 53 One embodiment of a four live-feed moulding apparatus, based on a two colour moulding machine arrangement. Two live-feed devices (D) are positioned on each of the conventional injection units, to provide four live-feeds to the mould cavity (C). Reproduced from reference 5 with permission

uniaxial alignment of fibres resulting from the application of Scorim corresponds to a substantial increase in tensile modulus when measured parallel to the process direction.

Internal weld lines are major sources of weakness in injection mouldings. They may be erased by the application of a macroscopic shear after mould filling, resulting in substantial enhancement of strength with minimal increase in cycle time. In some cases, the increase in strength of the weld facilitates the ejection of Scorim mouldings from the mould cavity at shorter times than is possible with conventional moulding.

The management of the preferred orientation of fibres is primarily determined by the positioning of the live feeds; these should be positioned where they are needed and to enhance physical properties by design. For example, the arrangement in Fig. 53 provides for the formation of 0–90° laminates.[3] The alternating and out of phase operation of pistons 1, 4 and 2, 3 during progressive solidification of the moulding, produces the preferred orientation in successive layers. Recent application of Scorim using the arrangement shown in Fig. 53 provides for the management of in-plane thermal expansion and impact properties of moulded plaques.

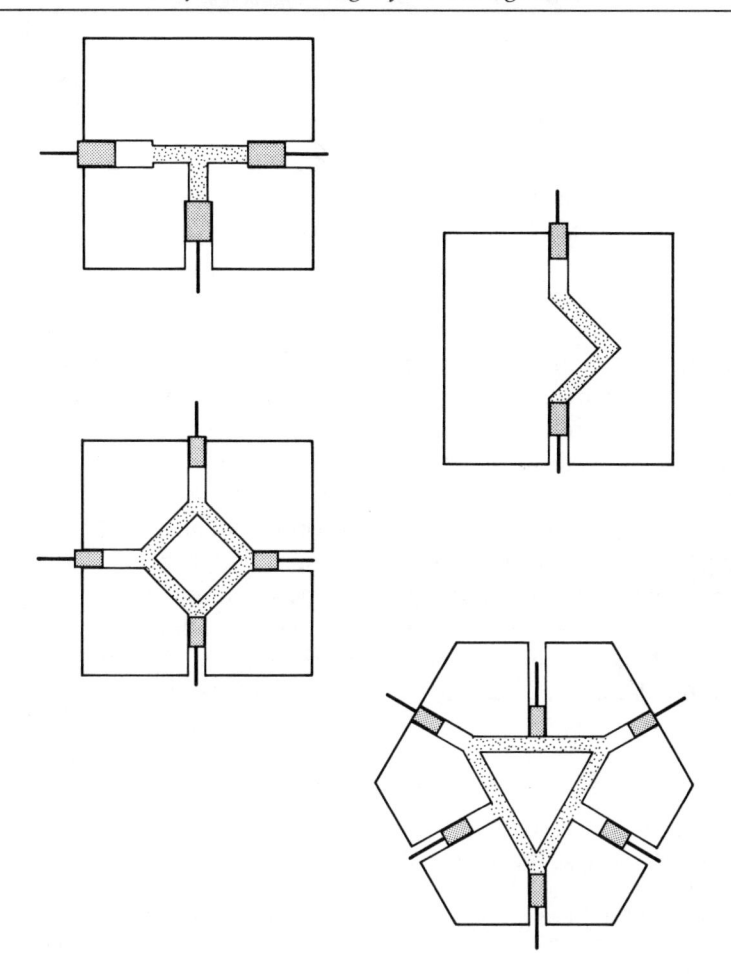

Figure 54 Schematic diagrams of live-feed arrangements used to produce preferred orientation of fibres in a selection of extrudate profiles

Positioning and sequencing of live feeds to provide for the preferred circumferential alignment of fibres in moulded rings have been described.[3] This results in a substantial enhancement of hoop strength and stiffness, and also provides the basis for close tolerance moulding through the control of microstructure throughout the volume of moulded rings.

The second variation of the Scortec technology, Scorex, is a more recent innovation. The basic functions of the Scorex extrusion die have been detailed previously[4] and are listed here.

(i) to pass molten material from an extruder through a one way valve into the die chamber,
(ii) to solidify the material before exit from the die,

(iii) to apply a macroscopic shearing action on the solidifying melt to influence the alignment of reinforcing fibres in extrusions.

The schematic diagrams in Fig. 54 illustrates arrangements for managing the orientation of short fibres in a selection of profiles, with emphasis on the production of transverse orientation. The sequencing of the operation of pistons is selected to produce the desired level of preferred orientation. When applied to a circular die,[4] Scorex can produce a uniform structure in both the hoop and axial directions, resulting in substantial enhancement of the hoop properties of the pipe arising from the controlled circumferential orientation of fibres. The process also provides for excellent control of dimensions and surface finish.

The processes referred to above also apply to the in situ formation and management of liquid crystal polymer fibres, as produced from blends of liquid crystal polymers and specified thermoplastic matrices; and to the management of ceramic fibres in injection moulded metal matrix and ceramic matrix composites,[5] utilizing sacrificial polymer binders to provide the mechanism of alignment before solidification, prior to removal of the binder and subsequent sintering of the component to realize full density.

References

1. P S Allan, M J Bevis, *Plastics Rubber Proc. Appl.*, **7**(1), 3 (1987).
2. J R Gibson, P S Allan, M J Bevis, *Composites Manufacturing*, **1**(3), 183 (1990).
3. P S Allan, M J Bevis, *Composites Manufacturing* **1**(2), 79 (1990).
4. P S Allan, M J Bevis, *Plastics Rubber Composites Proc. Appl.*, **16**(2), 133 (1991).
5. I E Pinwill, F Ahmad, P S Allan, M J Bevis, *Powder Met.*, **35**(2), 107 (1992).

3.11 Long glass fibre reinforced thermoplastic sheet

P M JACOBS

Long glass fibre reinforced thermoplastic sheet[1-4] (RTS) can be divided into two broad categories differentiated by the form of the reinforcement used. Glass mat reinforced thermoplastics (GMT) were the first type of RTS to be introduced in the early 1970s and are reinforced by glass fibre mat, as the name suggests. Structural thermoplastic composites (STC), on the other hand, are reinforced by chopped glass fibres and were introduced some time later. The main advantage of these RTS materials is that they provide a way of producing large strong thermoplastic composite parts at low cycle times in a high volume process. As they evolved from the sheet metal stamping process, it is not surprising that RTS materials have found greatest success in the automotive industry, although they are gaining acceptance in the communication, construction, furniture manufacturing, sporting goods and domestic appliance industries.

The most widely used resin for RTS is polypropylene as it is relatively inexpensive and offers an excellent balance of physical properties. However, resins other than polypropylene are beginning to be used in applications in which increased performance is required. These include the recently introduced polycarbonate/PBT alloy and PET based sheets. Although these resins are more expensive than polypropylene, they offer superior heat resistance and paintability, and are claimed to give improved surface appearance. Nylon has also been considered as an alternative to polypropylene in applications where its superior heat resistance, chemical resistance and paintability would be an advantage.

Manufacturing processes

Semifinished blanks of the two different types of RTS are produced by two quite different manufacturing processes. The extrusion or 'dry' process is used to produce GMT, whereas the wet slurry process is used to produce STC.

Glass mat reinforced thermoplastics

In the 'dry' process, which was the first to be commercialized, random continuous glass fibre mats are laminated between melt extruded films of thermoplastic resin to produce a sheet. A similar effect has also been achieved by the consolidation of glass fibre rovings and resin particles to form an impregnated 'mat' in situ. The advantages of using continuous fibres are that they result in parts with high strength, modulus and impact resistance, and they offer the possibility of preferentially orienting the fibres for specific applications (e.g. bumper beams and snowboards). However, the use of continuous fibres also results in a poor fibre distribution in complex parts and so it is usually necessary to process the glass mat reinforcement in such a way as to increase fibre mobility. This increased fibre mobility is achieved by mechanically degrading the glass mat to reduce the fibre length and is accompanied by a decrease in mechanical properties.

Structural thermoplastic composites

In the wet slurry process, which was developed from traditional paper-making technologies, discrete glass fibres and resin particles are dispersed in water to produce a slurry. This intimate mixture of fibre and resin is deposited on a web to produce a felt, dried and then consolidated to produce a sheet. The dispersion of glass fibres and resin in a foam is a variation on this technique which has also been considered. The main advantages of the wet slurry process are the uniform fibre distribution and high mouldability which result from the use of discrete fibres and allow greater design freedom.

Processing of RTS

RTS is usually processed by either stamping (forming) or flow moulding preheated blanks in the matched metal mould of a hydraulic press. Stamping is normally reserved for RTS, in which the lack of mobility of the glass mat reinforcement limits the material to the manufacture of relatively flat parts. The shape of the blanks used is determined by the shape of the finished part and any holes or cut-outs are either made in the blank before preheating or in an additional post-moulding operation. On the other hand, flow moulding is used to manufacture more complex parts and demands a greater degree of mobility from the glass reinforcement. The preheated material is deformed at high speed, allowing ribs, cut-outs, inserts and variable wall thicknesses to be achieved in a single operation, without compromising fibre distribution. STC, produced by the wet slurry process, is well suited to flow moulding applications as the use of discrete long fibres enables a uniform fibre distribution to be achieved in even the most complex parts.

Table 38. Typical range of properties quoted for commercial 30% fibre glass reinforced polypropylene RTS

Property	Typical values
Density (g cm^{-3})	1.12–1.15
Thermal expansion coefficient (K^{-1})	30–35 $\times$ 10^{-6}
Tensile strength (MPa)	60–90
Elongation at break (%)	1.5–2.5
Tensile modulus (GPa)	4.3–4.8
Flexural strength (MPa)	80–150
Flexural modulus (GPa)	4–4.5
Impact strength, unnotched Charpy (kJ m^{-2})	35–80

Properties

Unlike many injection moulded thermoplastic composites, in which one property is often increased at the expense of another, the long fibre reinforcement of RTS permits a combination of high impact strength and stiffness to be achieved. It exhibits a high energy management capability and is not prone to fracture or splitting. RTS also offers good flexural fatigue resistance, a low creep tendency and excellent dimensional stability over a wide range of temperatures. As a replacement for steel, RTS offers reduced part weight, corrosion resistance and superior thermal, acoustic and electrical insulation. The typical range of properties quoted for a number of commercial 30% fibre glass polypropylene-based RTSs are given in Table 38. The wide differences in some properties are a result of process variations between manufacturers.

Processing advantages include cycle times, which are shorter than other processes such as injection moulding, pressing of thermosetting materials or structural foam moulding. An additional advantage of RTS over themosetting materials is storage life; RTS does not have a limited storage life and does not need to be stored in air conditioned rooms. RTS is also intrinsically recyclable, a further advantage in industries where recyclability of materials is already a major concern (e.g. automotive applications).

Future developments

Since the introduction of RTS, polypropylene/E-glass based systems have been the most widely used, although engineering thermoplastics are beginning to find acceptance as matrix resins in applications where their increased heat resistance is an advantage. Reinforcing fibres other than E-glass may also be considered for applications in which increased mechanical properties, flame retardancy, electromagnetic interference, shielding or antistatic properties are required. However, the economic advantages of using polypropylene/E-glass based systems means they will continue to be the predominant systems for some time to come. In addition to the continuing development of more versatile materials, increased automation and improved part design will enable RTS to be used in an increasing number of applications.

References

1. D Brownbill, *Modern Plastics Intl.* January, 1986, p. 39.
2. M Mahlke, *Eng. Plastics*, **2**(3), 163 (1989).

3. C Price, *Eng. Plastics*, **1**(5), 360 (1988).
4. P Saris, *Composite Polymers*, **2**(2), 134 (1989).

3.12 Pultrusion

A G GIBSON

The pultrusion technique, shown in Fig. 55, is used to manufacture sectional products in continuous lengths and is one of the most economically attractive methods for processing thermoset composite materials.[1] The principle of this continuous process is simple: fibrous reinforcement is impregnated with thermosetting resin and pulled through a heated steel die, which shapes the product section and initiates the crosslinking reaction in the resin, to form a solid product.

Processing

Most often, the reinforcement is impregnated by pulling it through a bath of liquid resin. The different reinforcing elements are then assembled into the shape of the product by a series of preforming dies, often made from ultra high molecular weight (UHMWPE) polyethylene. These dies also squeeze out excess resin, which runs back into the impregnation bath. In a variant of the process the resin is injected into the entry section of the die under pressure.

The processing speed in pultrusion is generally in the range 1–5 m min^{-1}, determined by the time to heat up the incoming material by conduction from the die, plus the time taken for cure to reach an acceptable level before the pultrusion leaves the die. It is not possible to speed up the process by using long dies, as the pull force increases with die length. Most pultrusion dies are between 0.4 m and 1 m long. Some

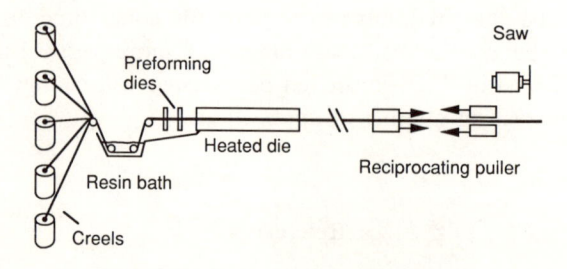

Figure 55 The pultrusion process

improvement in speed may be obtained by preheating the material prior to entering the die, using RF heating or other means.

The product is pulled through the die by a powerful haul-off unit which may be either of the reciprocating type, where the pultrudate is gripped and pulled in a 'hand over hand' fashion or of the caterpillar type. After the haul-off, the product is generally cut by a saw to the required length, although products of small section may be coiled on a drum. A post-curing oven is sometimes placed between the die and the haul-off.

Complex sections can be produced by pultrusion: tubular and hollow shapes require mandrels located inside the die and anchored upstream of it.

Variants of pultrusion are pulforming, used to make curved products or parts of varying section, and pulwinding, where tubes are externally wrapped with reinforcement to improve the circumferential strength.

Materials

Any reinforcement available in continuous form may be processed by pultrusion. Glass fibre (see Sections 1.10 and 1.11) is most commonly used, often as alternate layers of unidirectional rovings and continuous strand mat. Aramid fibres (Section 1.3) are increasingly used for high stiffness applications. Woven reinforcement may also be used.

Unidirectional rovings have the lowest cost and produce pultrudates with high strength and stiffness in the machine direction (see Sections 4.2 and 4.3 for theoretical considerations). Unidirectional E-glass pultrusions, for instance, can have flexural strength levels of up to 1000 MPa. Products of this type are used as tension members, for example in the load-bearing reinforcement at the centre of optical fibre cables. However, there is often a need for better transverse strength than is obtainable with unidirectional material. One way of achieving this is to incorporate fibre mat along with the unidirectional rovings. Continuous fibre mat ('swirl' mat), is used because it can withstand the forces applied to it when it is dragged into the die. However, pultrusions must always contain a certain proportion of unidirectional fibres, to enable them to withstand the force needed to pull them through the process.

Product properties can be tailored by varying the proportions of different reinforcement types. Unidirectional reinforcement is often used in the centre of the section, with swirl mat in the outer layers.

Data[2] show that when such a construction is used with equal amounts of mat and unidirectional fibres in a flat rectangular section, the flexural strength values in the machine and the transverse direction are almost equal, at about 260 MPa.

Most heat-cured resins are amenable to pultrusion, although tailoring of rheology, cure kinetics and shrinkage may be required to ensure processability. Effective internal release agents are essential. Unsaturated polyester resins (see Section 2.10) and vinyl esters (see Section 2.12) are widely used, as are the modified acrylic resins. The last are well suited because of their rapid cure kinetics. 'Modar' products with improved fire performance can be made by incorporating high levels of alumina trihydrate filler (see Section 1.1).

Epoxy resins (see Section 2.4) may be pultruded, but more slowly than polyesters. Phenolic resins (see Section 2.7) are difficult to pultrude due to problems with die corrosion and water evolution, but they have considerable potential because of their low toxicity in fire. There is also interest in the pultrusion of thermoplastic matrix composites (see Section 2.3), but the technology is different to that used for thermosets.[3]

Applications

Like other composites, pultruded products are often used because of their light weight, corrosion resistance and lack of electrical conductivity. Electrical applications include cable trays, live rail protectors and ladders for live line working. In the construction industry, pultrusions are used in a wide range of applications, including frames for lightweight buildings. In the chemical industry they can be used as handrails and components of grid flooring.

In structural applications it should be remembered that pultruded sections have a lower ratio of shear stiffness to bending stiffness than metallic ones. Shear deflections may therefore be larger and torsional stability lower. The improvement in torsional stability obtainable by the use of closed sections makes box section pultrusions more attractive than the traditional I sections, although the latter can be used with care.

References

1. J A Quinn, *Metals and Materials*, **5**(5), 270 (1989).
2. H Engelen, The influence of process parameters on the mechanical properties of pultruded GRP profiles, in *Proceedings of the 24th Journees Europeenes des Composites*, JEC, Centre de Promotion des Composites, Paris 1989.
3. B J Devlin, M D Williams, J A Quinn, A G Gibson, *Composites Manufacturing*, 2(3/4), 203 (1991).

3.13 Recycling of polymer fibre composites

D W CLEGG

Introduction

Polymer fibre composites are based on thermosetting or thermoplastic polymers and contain fibres ranging in length from short to continuous. As well as fibres, polymer composites may contain fillers, pigments, fire retardants and other process aids and property modifiers. Together with the different types of matrix, i.e. thermoset or thermoplastic, and the need to retain fibre length, these fillers impose major constraints on recyclability. However, as an alternative to recovering the original material in a recycling operation, it is possible to recover the energy locked in the material and to use the composite as a source of fuel. Several problems exist in this respect. For instance, one of the most difficult composites to reprocess is tyre rubber. Retreading, pyrolysis and cryogrinding to produce crumb are possibilities. The main problems have been economics of production and the identification of suitable markets for the products. Incineration to recover inherent energy is currently the most promising route. There are similarities between tyres and thermosetting polymer composites. Incineration has been used extensively with plastics for some time. Considerable amounts of domestic refuse containing plastics are incinerated each year. However, the rapid introduction of legislation and public opinion is causing a far reaching analysis of the recycling situation.

Major considerations

It has been estimated that US car makers will double their consumption of all plastics between 1980 and 2000. This will contribute to a significant weight saving per vehicle, translatable into a worthwhile fuel saving. Plastic composites make a valuable contribution but they feature more strongly in low volume, high value production than in mass produced vehicles; this situation is likely to continue. The main components constructed from composites are body panels, structural members, suspension parts and under bonnet components.[1] This situation is even more apparent in the aerospace industries. By the year 2000, it is possible that 65% of the structural weight of commercial aircraft will be composed of composites and most of that is likely to be plastic based. For military aircraft, the figure may be nearer to 85%. The composites used in aerospace applications are generally more sophisticated and have higher property profiles than those used in automotive applications. At the same time, design lives are much longer, for example 25 years compared with 10 years. The total consumption of aerospace composites is lower than automotive composites, therefore we are confronted with high value, low volume composites in aerospace which will not become available for recycling for 25–30 years. On the other hand, the automotive industry will generate large quantities of scrap plastics and composites from a much earlier date, although the inherent value per kilogram will be less. Careful economic analysis is necessary to identify appropriate strategies. In the automotive industries, margins are very narrow and environmental pressures very great. There is a sophisticated metal recycling industry which must be developed to deal with plastics and composites as the quantities involved become higher. It is unlikely to be acceptable simply to use these materials as landfill or to incinerate them in an unsophisticated manner. Separation procedures must be developed so that metals can be separated from plastics and one type of plastic and composite separated from another. It is becoming increasingly apparent that considerable thought must be given to component design in order to render the separation process workable at the recovery and recycling stage.

Component design

For certain components, recycling presents little problem. If they are composed of a single material and can be easily separated then recycling is straightforward. All that is necessary is a simple means of identification such as bar coding or a similar method. It has been proposed that a 'closed loop' approach is adopted where components are reprocessed into similar components at some future time. However, this entails a very long range approach to materials selection, which is not likely to be very practical. Where components are constructed from several polymeric materials, it is sensible to ensure they are compatible. For example, a bumper system composed of a reinforced plastic, energy absorbent core and a flexible cover should be based on compatible polymers, e.g. polypropylene. Then recycling is simplified as the plastic component as a whole can be shredded and compounded to recover a reinforced compound with useful properties.

Incineration

A significant proportion of composite production is in the form of thermosetting materials such as SMC and BMC. The merits of incineration as a means of recovering the inherent energy values of these materials, the energy values available in these materials and the problems involved in the incineration process are being investigated.[2]

Table 39. Comparative calorific values

Material	Approximate calorific value (MJ kg^{-1})
Thermosetting plastics, e.g. polyester, vinyl ester and epoxy	30
Thermosetting fibre composite containing 80% fibre/filler	6
Tyres	31
Polythene	46
Domestic refuse	10
Polymers in domestic refuse (average value)	38

After incineration, a considerable quantity of inert reinforcement and filler is left; one objective of these studies is to find worthwhile uses for it. The calorific values of thermosetting plastics used in polymer fibre composites such as polyesters, vinyl esters and epoxies are approximately 30 MJ kg^{-1}. These values are generally reduced in direct proportion to their filler and reinforcement contents. Many thermosetting composites contain up to 80% fillers and reinforcements, so their calorific values are significantly reduced. But endothermic reactions can absorb considerable amounts of energy; when 1 kg of calcium carbonate decomposes at between 800 and 900°C to yield lime, it absorbs 1.8 MJ. Some fillers and/or polymers may react to form sulphur dioxide, nitrogen oxides, etc. Table 39 shows some typical calorific values for comparison purposes.

Flue gas emissions are a possible problem but initial studies on polymer composites have not revealed major problems. The indications are that particulate and gaseous emissions should be containable within legislative limits.[2] Polyester, vinyl ester and epoxy based composites give low char yields and the ash yield depends on the resin content. Phenolics and urea based composites give much higher values. Work is in progress to find uses for the recovered fillers in, for example, the cement industry, where the energy recovered would also be of considerable use. Lime produced by the decomposition of calcium carbonate described above is potentially useful in this respect.

Reconstitution processes

Short fibre thermoplastic composites can be granulated and melt processed by injection moulding or extrusion. However, some mechanical degradation of fibre length is inevitable together, in some cases, with degradation of the matrix. In-house recycling at the initial production stage is straightforward. In other cases, the situation is less satisfactory although the presence of fibres is often a help in compatibilizing systems comprising mixtures of polymers and composites. Thermosetting

Table 40. Comparison of reclaimed APC-2 and a quasi-isotopic continuous fibre APC-2 composite

Property	APC-2 reclaim with PEEK (40% carbon fibre) injection moulded	Reconstituted single prepreg plies (25.4 mm) hot pressed	APC-2 quasi-isotropic lay-up, hot pressed
Flexural modulus (GPa)	30.4	37.2	40.9
Flexural strength (MPa)	410	440	616
Tensile strength (MPa)	242	290	704

composites can be re-used by shredding them into fine particles then using them as a filler in other polymeric materials. Many novel technical innovations are being made in this area including applications using RRIM polymers.

In the case of long fibre polymer composites, it is an advantage to preserve fibre length as far as possible in a recycling process in order to maximize properties and value. This is difficult to achieve but is worth considering in view of the probable high cost of the fibres and polymer matrix. Studies have been made on recycling APC-2, a thermoplastic composite based on poly ether ether ketone (PEEK) with 68% continuous carbon fibres.[3] In pattern cutting, the prepreg material utilization is unlikely to be more than 75%. Thus re-use of the offcuts is desirable. Studies have been made on the use of controlled size pieces of offcut prepreg followed by reconstitution by hot pressing. Good properties are achievable as can be seen from Table 40. It is conceivable that this approach could be extended to pattern cutting offcuts of thermosetting composites but handling and shelf lives present considerable problems. Injection moulding the reclaimed APC-2 after granulation and dilution with PEEK is also possible and typical results are included for comparison.

Conclusions

Effective recycling strategies for polymer fibre composites require considerable planning right from the component design and materials selection stage. Recycling is easiest with thermoplastic composite components if they are readily identifiable and separable. Thermosetting composites are recyclable as fillers for other polymers but incineration to release energy and reusable fillers may be a more practical solution. Pyrolysis to release low molecular weight products is a possibility for thermosetting polymer composites but is not technologically advanced at present.

References

1. E M Rowbotham in *Proceedings of the 1st International Conference on Plastics Recycling*, London, Sep. 1989, paper 12, Plastics and Rubber Institute, London, pp. 1–10.
2. S J Pickering, M Benson in *Proceedings of the 2nd International Conference on Plastics Recycling*, Mar. 1991, Plastics and Rubber Institute, London.
3. D W Clegg, G McGrath, M Morris, *Composite Manufacturing*, **1**, 85 (1990).

3.14 Reinforced reaction injection moulding (RRIM)

F R JONES

Introduction

Reaction injection moulding (RIM)[1-5] describes the process whereby two liquid reactants are mixed rapidly at a mixing head and injected into a mould at relatively low pressures. RRIM refers to the composite processing route in which the reinforcement, generally in the form of particulates or very short fibres is included in one of the monomer streams. For the most common polymer, the polyurethanes, the polyol would contain the filler. Structural RIM (SRIM) (see Section 3.19) is RIM processing in which the 'reinforcement' is placed in the mould in the form of a preform. This should be differentiated from resin injection moulding or resin transfer moulding, annotated RTM (see Section 3.15) for clarity, where one stream of catalysed resin is used. In-mould polymerization occurs simultaneously with the formation of the moulding. Relatively low viscosity reactants are quantitatively and rapidly dispersed through a mixing device into the mould so that relatively low clamping forces are required even for large surface area components. In comparison to a conventional injection moulding machine (see Section 3.8) at the same clamping force the projected area is approximately 10 times greater. Although the moulds are less expensive and simpler than for injection moulding, RIM moulds are not cheap. They need careful design and good surface finish because expansion during moulding, as a result of exothermicity of the reaction, can cause the surface detail of the mould to be reproduced.

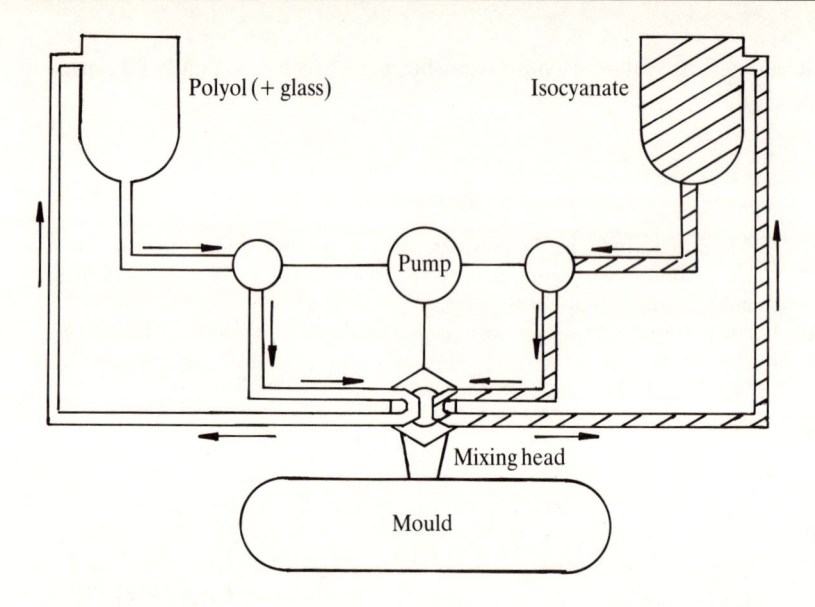

Figure 56 Schematic description of reaction injection moulding (RIM)

Simple pumps and static mixers can be used to disperse the monomers when the rate of reaction is so much slower than the fabrication cycle but effective mixing of fast reacting reagents is best achieved with a self-cleaning impingement mix-head which permits the reactants to continuously circulate in their separate systems (as shown in Fig. 56) until an injection shot is required. The opening and closing of the valve needs to be controlled to accurately deliver the exact quantity of reactants to the mould, otherwise overfilling will cause mould flash and underfilling an imperfect moulding. Typically 2–20 kg of material can be delivered to the mould in 5 s or less and millisecond accuracy is required. It is now generally agreed[2] that good mixing occurs when the Reynolds number R_e in an impingement mixing nozzle is greater than ~200, where

$$R_e = \rho v d / \eta$$

and ρ = the reactant density, v = average stream velocity, d = minimum opening and η is the apparent viscosity of the reactant.

Imperfect mixing leads to poor matrix and hence composite properties, especially for systems where stoichiometric reactant concentrations are needed for complete reaction.

Further mixing occurs during mould filling and more specifically the dispersion of the reinforcement must occur effectively. As with any plastics moulding process, orientation of fibres with high aspect ratio occurs within the flow fields. This is generally less of a problem than with melt processing but increases in viscosity during polymerization can cause problems. More specifically the mould should be filled before gelation occurs.

Generally, heated moulds are not employed to initiate the polymerization but whether or not heated moulds are required, the exothermicity of the process needs to be adequately catered for. This means that moulds (and mouldings) with large surface area are required but this restricts applications to 'thin' sections such as automotive panels. Differing skin–core structures and properties can arise from thermal history and fibre orientation effects. The former can arise from differing crystallite morphologies for crystallizable polymers or from phase separation in toughened or blended systems.

Since the fibre volume fraction is limited to relatively low levels, SRIM provides the best way of providing the properties best associated with composite materials. RRIM technology on the other hand enables the impact strength of the artefact to be fine tuned to the requirements. For example, vertical automotive panels resistant to a defined low speed impact can be manufactured.

Polymer matrix types

RIM was developed originally for polyurethane and nylon 6 (NyRIM). The former is a versatile group of polymers which can be varied across a wide range of properties. It also embraces the related polyureas.

The principles of polyurethane chemistry[1] is the step-growth (or step-addition) polymerization of a diisocyanate (ii) (e.g. TDI–toluene diisocyanate (iii)) with a polyfunctional polyol (i). The polyol is a hydroxy terminated polyester or polyether, polyesterol or polyetherol respectively. For traditional polyurethane foams, the in situ reaction of the isocyanate with water to give CO_2 simultaneously with the catalysed polymerization leads to the formation of a stable foam. The resilience and/or hardness of the resultant polymer (whether as a foam or as a solid) can be adjusted by judicious choice of diisocyanate and polyol.

An example of (ii) is TDI 80/20 which is generally an 80/20 mixture of 2,4 and 2,6 isomers (iii)

$$\underset{80}{\underset{(iii)}{\text{2,4-TDI}}} \qquad \underset{20}{\text{2,6-TDI}}$$

(iii)

An example of (i) is

$$HO{-}(CH_2{-}\underset{CH_3}{CH}{-}O)_n{-}CH_2{-}\underset{OH}{CH}{-}CH_2{-}O{-}(\underset{CH_3}{CH}{-}CH_2)_n{-}OH$$

(iv)

Multifunctional hydroxylic compounds (iv) provide a mechanism for crosslinking and network formation. At the higher concentrations, more rigid materials are created whereas at the lower concentrations they provide resilience to the material. The polyetherol is generally preferred where highly resilient elastomeric properties are required. It also provides for better hydrolytic resistance. In fact, polyesterol based flexible foams are one of the few polymers which can suffer from bacterial degradation in humid environments.

For RRIM based composites the main benefit is the ease with which the properties can be varied across a wide range. Thus for automotive panels, a greater degree of resistance to low speed collision impact can be built-in to the front/rear bumpers than required of the side panels. Because of the relatively low fibre volume fraction which can be used, RRIM car body panels have insufficient stiffness (modulus) for use as horizontal panels (such as boot lids and bonnets). As a consequence the Pontiac Fierro (in USA) and the Scimitar SSII (in UK) employed RRIM vertical panels but SMC horizontal panels. Even with demoulding times of <20 s claimed for RRIM and the low cost of tooling, it has not proved competitive with metal stamping technology for volume car production. However, with the rigid 'space' frame (or cage) type of automobile construction with bolt-on body panels, RRIM technology, which provides the ready customization of a basic car economically, is under continuous evaluation.

Decorative finishes can be applied to RRIM mouldings by painting (automotive panels) or where colour matching between components is less important by self colouring. However, in-mould colour coating of the mould surface prior to RIM provides a better finish.

Other matrices for RRIM

The reactive processing of plastics materials provides for rapid, lower energy cost production. As a consequence, the chemistry required for catalysing rapid polymerizations of epoxy resins (see Section 2.4) modified acrylate (e.g. Modar), unsaturated polyesters and cyclopentadiene by metathesis are all actively being investigated. Furthermore the tailoring of matrix performance using multistreams of monomers is actively being researched.

Anionic ring opening of caprolactam to form nylon 6 (NyRIM) and related polymers is the only other serious matrix for RRIM. NyRIM are block copolymers of hydroxyl terminated polyether or polybutadiene with caprolactam. A magnesium bromide-caprolactam complex is used as a catalyst, in combination with mould temperatures of 130–140°C. Bis-N-adipyl caprolactam is incorporated to omit the induction period often observed in this type of ring-opening polymerization. The requirements for heated moulds and storage tanks are the main disadvantage. However, the polyurethanes clearly form the major matrix for composite materials produced by RRIM technology.

Reinforcement

Short E-glass fibres (<5mm) or glass flake (both E and ECR are available) are the most common reinforcement. Because less energy is put into the material, compared to a preformed thermoplastic melt, during moulding, the fibres as chopped rovings have a strong tendency to remain as discrete bundles within the RIM matrix. As a consequence, the reinforcing efficiency is low because the reinforcing element is a short fat bundle and not a high aspect ratio fibre. This can be circumvented by milling (on a two-roll mill) the fibre bundles, prior to use. While this tends to provide discrete fibres in the matrix, their average length is much lower, as a result of the disintegration which occurs. Nevertheless, the reinforcing efficiency can still be as good because the aspect ratio of the reinforcing element may have increased. Glass flake is often preferred because of the complications explicit in the use of chopped fibres and its ease of handling.

Because of the reaction of isocyanates with absorbed water and/or surface hydroxyl groups, the glass or other particulate reinforcement cannot be incorporated into the isocyanate stream and is dispersed in the polyol stream. Typically only 5–15% by volume of reinforcement can be used in RRIM whereas SRIM can employ up to 50% by volume. Because of the relatively low reinforcing efficiency, other higher modulus fibres are not generally used in RRIM. SRIM technology is likely to provide a more economic route to higher performance. For both RRIM and SRIM the interfacial chemistry of the fibre–matrix interaction could differ from conventional, and this is an area requiring future research.

Conclusion

In a relatively short period RIM and RRIM have become important moulding techniques in the polymer industry. The next decade will see the expansion of SRIM technology. There is also evidence that some RIM polyurethanes can be recycled by 'depolymerization' to useful monomers, and this may provide further impetus to RRIM and SRIM technology in future.

Bibliography

1. C Hepburn, *Polyurethane Elastomers*, Applied Science, London, 1982, Ch. 6.
2. A F Johnson, P D Coates, *Materials World*, **1**, 156 (1993).
3. F M Sweeney, *Introduction to Reaction Injection Moulding*, 2nd edn, Technomic, Westport, USA, 1990.
4. C W Macosko, *RIM–Fundamentals of Reaction Injection Moulding*, Hanser, Munich, 1989.
5. *Concise Encyclopaedia of Polymer Processing and Applications*, Pergamon Press, Oxford, 1992.

3.15 Resin transfer moulding (RTM)

C D RUDD

Resin transfer moulding (RTM) describes a family of processes for the fabrication of composite components. The distinguishing feature of RTM is the transfer of a thermosetting resin from some external supply into a matched mould that contains fibre reinforcement in the form of a mat or preform (Fig. 57). This is facilitated by a pressure difference between the resin supply and the mould cavity.

Resin transfer moulding (RTM)

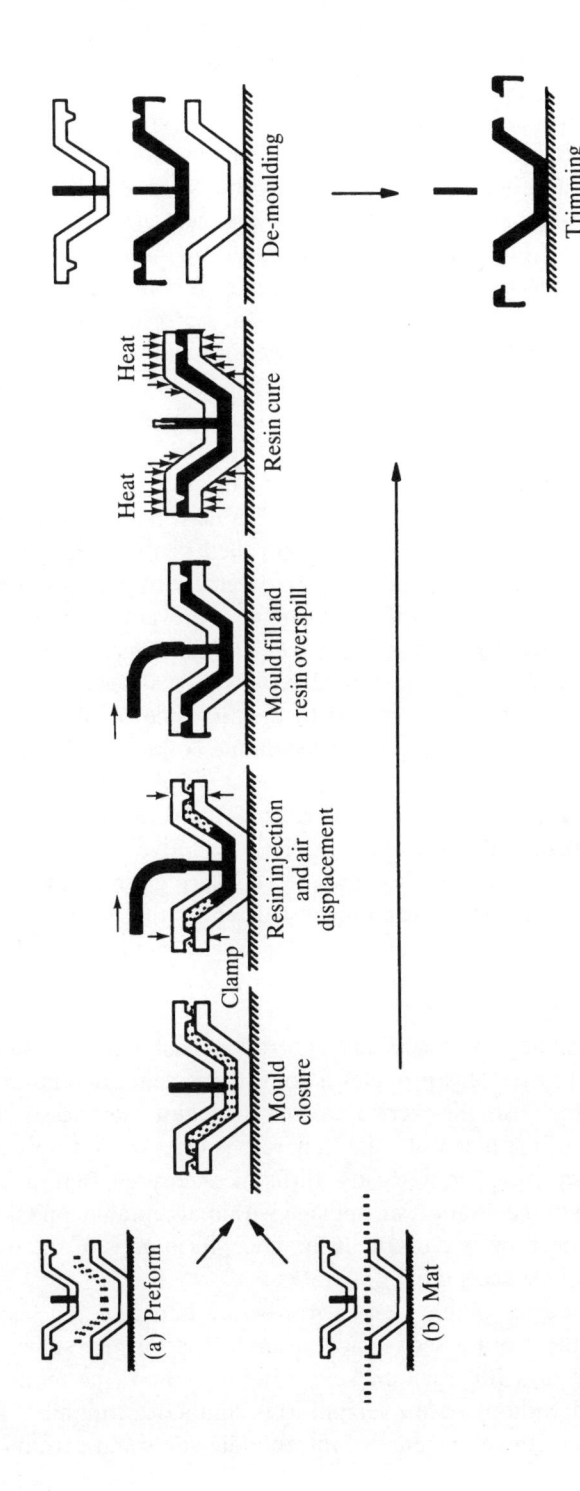

Figure 57 Schematic diagram showing stages of RTM

Process technology

One of the problems to be addressed during the impregnation phase is the removal of the air from the mould cavity.[1] This includes air within the fibre bundles in addition to that which surrounds the bundles. Recent work has confirmed that the liquid flow front proceeds at two levels.[2] The macroscopic front advances at a rate determined by the forcing pressure gradient. The advancement of the microscopic front (which occurs within the fibre bundles) is determined by the capillary pressure and depends upon surface tension. It follows that if the wet-out and wet through flow fronts are not coincident then significant voidage is likely to occur since there is no mechanism to expel the entrapped air once the flow front has passed.

The traditional method of air removal is to use the macroscopic flow front to expel the air. Alternatively, a barrier may be installed at the perimeter of the mould; this provides controlled venting. Perhaps the simplest arrangement is the trapping of reinforcement to produce an area of low permeability or pinch-off; this can also prevent reinforcement movement under the drag or washing force. The difference between the viscosities of air and many commercial resin systems is such that the pinch-off provides negligible resistance to the air while the flow rate of resin is minimized. A variation on this technique is the use of a porous gasket.

An alternative approach is to use a barrier seal in which discrete vents have been manufactured. The vented seal is attractive as it enables a moulded edge to be produced. Because the vents are discrete, they must be sited correctly to avoid air entrapment. Currently, this is an empirical process but the development of computer aided engineering (CAE) techniques to predict resin flow fronts will provide a more reliable route to the siting of vents.

A further possibility is a carefully controlled flash-gap around the perimeter of the cavity. This provides a continuous vent and eliminates the fibre-containing flash, the barrier seal and the maintenance associated with them. As with the pinch-off, the flow rate is inversely proportional to the fluid viscosity. The viscosity difference ensures that a gap, sufficiently narrow to contain resin spillage within acceptable limits, will permit the required flow rate of air. In practice, flash-gaps of the order 0.02–0.10 mm provide acceptable flow rates.

Alternative strategies enable air removal before impregnation begins. The simplest of these processes is vacuum impregnation. By subjecting the mould cavity to a full vacuum before impregnation, the moulding may be produced without resin overspill; this eliminates trimming. It is also likely to reduce the occurrence of intrabundle voids and permits the use of internal runner systems. Runners provide substantial reductions in impregnation times due to the reduced flow path lengths. However, moulds operated at low pressures can result in void formation due to monomer boil, and positive pressures are desirable to eliminate this effect.

Vacuum assisted resin injection or VARI is a further variant that has found several industrial applications. A partial vaccuum is applied to provide mould clamping, reinforcement compaction and an increase in the forcing pressure gradient. The reduced internal pressure provides a useful means of reducing mould deflections, particularly important when producing large area mouldings.

Resin systems

The resin may be a single component, premixed system including catalyst, accelerator, inhibitor, pigment, fillers and additives. This approach is effective for hot-setting polyester and acrylic formations. Multistream systems are also used; they combine just before mould entry in a static mix head and cold-curing formulations are often processed in this way. Reactive processing using an impingement mix head (see Section 3.19) can also be employed with many resins to provide significant reductions in cycle time.

Reinforcements

RTM provides a convenient fabrication route for a variety of applications ranging from high stress components with directional reinforcements to non-structural items using modest levels of random reinforcement. Accurate fibre management is essential for processing

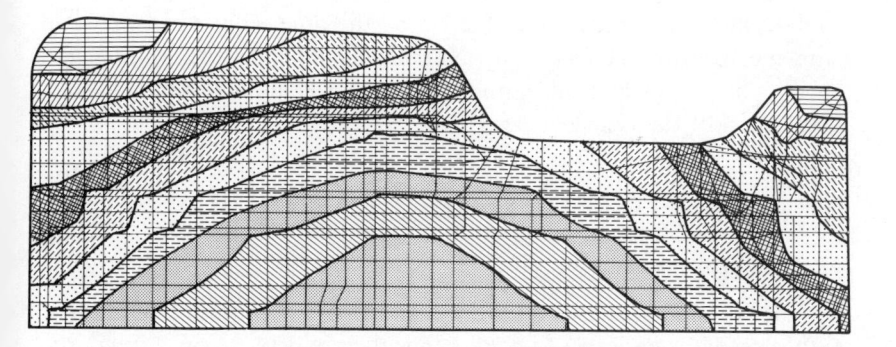

Figure 58 Impregnation modelling for an RTM component. The contours represent lines of equal fill time (isochrones)

reasons as well as structural integrity. Fibre preforms (see Section 1.12) are desirable in all but very low volume applications and current manufacturing methods are limited to assembly from mats or spray-up of chopped rovings. Currently, research efforts are being applied in this area for the manufacture of high quality net shaped preforms.[3]

Impregnation

Prediction of the impregnation phase during RTM is an important part of mould design. Generally, this is often described using the Darcy relationship[4], which relies on Newtonian behaviour of the resin. The flow is normally considered to be two-dimensional and numerical solutions have been developed for this problem using finite difference and finite element methods (Fig. 58). Special consideration needs to be given to preforms containing multiaxial reinforcements where three-dimensional flow can have a significant effect on mould filling. Comprehensive solutions require changes in rheology due to temperature[5] and polymerization to be taken into account.

The resistance to flow of the preform is characterized by its permeability.[6] Preforms are generally anisotropic and accurate mould filling predictions require characterization of materials. Test methods vary considerably and measurements have been made using moulds that promote either radial or rectilinear flow.

Cure

The resin formulation and process variables are selected so that no significant polymerization occurs until the mould cavity has been completely filled. This is achieved by adjustment of supply pressure or flow rate, mould temperature and catalyst/accelerator/inhibitor levels. The overall cycle time is often limited by the time required to heat and cure the resin at the injection point. Predictions for the heating and curing phases can be made using a heat transfer model that includes a description of the resin kinetics.

Mould design

Flow considerations

In order to minimize mould fill times and to avoid problems such as air entrapment it is desirable to predict the advance of the resin through the preform. This may be achieved by numerical solution of the Darcy equation combined with continuity of mass or volume. The location and configuration of the resin inlet gate has several important effects on the impregnation phase. Inappropriate gate design can lead to prolonged fill times, inadequate impregnation and materials wastage caused by excessive venting of liquid resin. Experience has shown that the mould heating circuit needs to be designed subsequent to the location of the inlet gate to optimize mould filling.

Thermal characteristics

In common with many forms of thermoplastic and thermoset injection moulding, the thermal and rheological design of a mould will affect component processing significantly. Elimination of mould quench caused by the cold incoming resin can reduce moulding cycle times appreciably. Preheating the resin before injection is an obvious solution, although precatalysed resin systems require careful handling in this respect. Zone heating can be used to counteract heat losses or to provide additional heat input to specific regions and give further reductions in the cycle time. This should also be linked to the design of the injection gate, as zone heating in this region can provide large (greater than 50%) reductions in overall cycle time.

Mechanical design

One of the major attractions of RTM is the low pressure involved; this enables the use of relatively flexible tooling. Adequate joint support is essential to maintain seal integrity, particularly on large area moulds. It is also important that the structure is designed to withstand appropriate cavity pressures so that seal integrity and part thickness can be maintained. Prediction of the pressure history is a further result of impregnation modelling, which forms an integral part of the design process.

References

1. K F Hutcheon, 'The Application of Resin Transfer Moulding to the Motor Industry.' M. Phil. Thesis 1989. The University of Nottingham.
2. R J Morgan, D E Larive, H Yung, Z Huang Yuan, K P Battjes in *Proceedings of the ASM/ESD Advanced Composites Conference*, Oct. 1990, pp. 229–233.
3. M J Owen, V Middleton, C D Rudd, *Composites Man.* 1(2), 74 (1990).
4. H Darcy, *Les Fontaines Publiques de La Ville de Dijon*, Dalmont, Paris, 1856.
5. C D Rudd, M J Owen, F N Scott in *Proceedings of the 4th International Conference FRC 1990* University of Liverpool, I. Mech. E., London, 1990, Paper C400/012.

6. R Gauvin, M Chibani, in *Proceedings of the 45th Annual Conference*, Composites Institute, Society of the Plastics Industry, New York, Technomic Basel and Westport CT, 1990, Session 9-F, pp. 1–6.

3.16 Resin transfer moulding (RTM) – practicalities

A A ADAMS

There are many opportunities within a diverse range of industries that could realize benefits from RTM processes. The essence of any process and material selection is, or should be, practicality. Some reluctance in industries to adopt RTM processes and materials stems from the inability to understand what can be done in practice.

If the decision maker's background covers a 'need to survive', or they come from a 'how can we achieve results with minimum investment' background, or if the need to have short lead times with low investment is paramount, then the RTM methods, materials and technology may fit the bill.

The process of transferring a polymer resin from a container, through a mixer, into a closed mould that contains a fibre reinforcement is well understood. The chemical transition, whereby the polymer moves from a liquid to a solid state is not quite so fully appreciated.

The important liquid to solid state, resin transition, or gelation to cure time period determines in many ways the practicality of using the RTM technologies in the first place.

The resin polymer has to remain in a liquid state long enough to allow complete impregnation of the fibre reinforcement. The fibre materials are usually bound together with a chemical binder soluble in resin. The binder must be completely dissolved before the polymer sets.

The time taken for tool filling, binder wet-out, resin gelation and cure are related to the size, thickness and complexity of the moulded component. Therefore, small simple mouldings can have very short cycle or tool turn round times. Larger, more complex parts, by nature, require much longer tool cycles.

Practical issues

A large component takes longer, it should therefore be designed to eliminate as many post-moulding operations as possible. If separate parts can be united at the product design stage, a much more practical process will result because costly jigging and panel assembly can be reduced.

Resin transfer moulding (RTM) – practicalities

Of course, as we have said, the larger the moulding, the longer the cycle. However, this slower cycle can be considerably offset when the reduced assembly savings are made, further downstream.

In short, RTM techniques should not be seen only as a means of providing an alternative material, but more as an opportunity to produce comprehensive components with minimal assembly times.

New methods of accurate fibre preforming and placement have added greater control to RTM processing. Applications where high detail definition is essential can be more readily met when the fibre reinforcement is precisely shaped before positioning into the matched tooling. Positive fibre location is most important if consistent products are to be maintained.

With the introduction of the Elan, Lotus replaced its earlier plastic faced tools with electroformed nickel faced tools. The new type of tooling needs much lower maintenance and repair; its life is greater, over 100,000 mouldings, and its heat transfer is greater when tools are heated.

Figure 59 Multiple impression moulding which includes rear upper body, rear trunk lid and two lower body panels

The heated metal faced tools also help the moulder to key into new resin technology. Low temperature, low profile resin systems provide stable, non-shrink panels with a much improved surface finish, easier to paint to Class A quality standards.

Nickel faced tooling is produced directly from epoxy bathmasters. The bathmasters are made from a working set of low cost, plastic faced tools. This allows for design prove-out and confirmation before the nickel shells are made and the additional nickel tooling costs are incurred.

The plastic tools can be used for preproduction mouldings after the bathmasters have been made and while the nickel tools are constructed.

The economic production volumes for RTM processes are dependent upon the size and complexity of the component to be produced. These factors determine cycle times.

Multiple impression tooling (Fig. 59) can increase output and therefore reduce the overall number of individual tools. Lotus aims to reduce the chemical cycle times and five minute cycles could soon become a reality for quite large components using the new technologies and materials. These new ideas are becoming possible using low pressure vacuum-assisted resin injection (VARI) technologies, which do not require high tonnage presses or matched steel tooling.

3.17 Sheet moulding compounds

A G GIBSON

Sheet moulding compounds (SMCs) are pre-impregnated moulding materials that contain glass fibre reinforcement, in the form of chopped strands, along with filler and a thermosetting resin, usually unsaturated polyester.[1] They are one example of a wider class of materials known as polyester moulding compounds. The most commonly used fillers are calcium carbonate and dolomite, which is a mixed calcium and magnesium carbonate.[2] Alumina trihydrate is also used when flammability is critical.

SMCs are processed by hot press moulding. For polyester based SMCs, the curing reaction takes place by the free radical mechanism, described in Section 2.10. Cure is started by the thermally induced decomposition of a peroxide type initiator; a range of initiators covers a range of decomposition temperatures.

For special applications, SMCs based on resins other than polyester are becoming available. These include vinyl ester SMCs, for improved damage and environmental resistance, and phenolic SMCs, for use where exceptionally low flammability is desirable.

With automated processes taking over from the slower, more labour intensive processes in the composites industry, the use of SMCs has increased at a rate of 4.5% per year in the late 1990s. Current SMC usage accounts for about one-tenth of composites processing activity in Europe, although this varies between countries. In Germany, for instance, 41% of all glass fibre reinforced composites are processed as SMCs.

SMC manufacture

The manufacturing process for SMCs is shown in Fig. 60. The resin and filler are first compounded to form a slurry, which is spread in a uniform layer on to two thermoplastic backing films. E-glass strands are chopped and sprinkled, in the required proportions, on to the slurry. Both layers are then squeezed together and passed over a succession of rolls to impregnate the glass strands. After impregnation, the SMC and its layers of backing film are rolled up for subsequent storage and transport.

Because of the nature of the manufacturing process, the type of orientation found in SMCs is usually fairly close to random-in-plane, although there is a version of the SMC process that incorporates unidirectional reinforcement (Section 1.10). The E-glass fibres are chopped to a length of 25 or 50 mm and 20–40% by weight of glass is used in the compounds. Glass fibres for SMC employ a 'hard' size coating, insoluble in polyester resin, so the individual glass strands remain largely intact in the final moulding (Section 1.11).

A typical SMC formulation is given in Table 41. SMC processing involves the use of special additives for thickening and for shrinkage control, as discussed below.

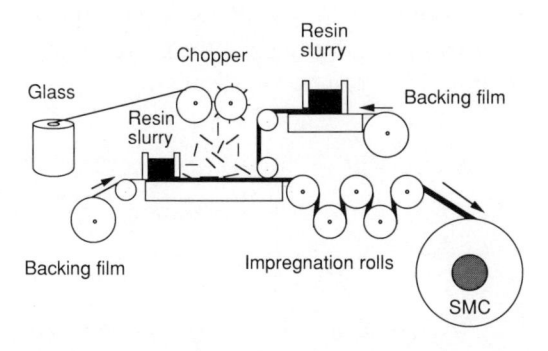

Figure 60 Schematic diagram of the SMC manufacturing process

Table 41. Composition of a typical SMC (quantities are given by weight in parts per 100 parts of liquid resin)

Resin phase	
Unsaturated polyester resin	66.6
Shrinkage control additive	33.3
Cure system	
Tertiary butyl perbenzoate	1.25
Inhibitor	0.01
Thickener	
Magnesium oxide	1.00
Release agent	
Zinc stearate	2.00
Filler	
Calcium carbonate (stearate coated)	180
Reinforcement	
E-glass chopped strands	93

Thickening of SMC

The viscosity of the slurry of resin and filler is required to increase during the manufacturing process. Low viscosity is needed at the outset to achieve good wetting of the glass strands. After wet-out, the viscosity must increase to stabilize the resin in the prepreg and to produce a tack-free handleable moulding material. A certain level of viscosity is also needed to maintain uniform flow of fibres and resin during the moulding process. SMC technology is therefore dependent upon having a reproducible means of 'thickening' the resin.

The most frequently employed thickening technique requires a polyester resin with a controlled number of acid end groups. A small amount of magnesium oxide is added to the resin slurry immediately before spreading. 'MgO' or 'mineral' thickening operates through a reaction with the carboxyl end groups of the polyester chains. The exact mechanism is complex, but can be approximated as follows.

$$\sim\sim\sim\sim\sim COOH + MgO + HOOC \sim\sim\sim\sim\sim$$

$$\downarrow$$

$$\sim\sim\sim\sim\sim COO^- \ Mg^{2+} \ {}^-OOC \sim\sim\sim\sim\sim$$

$$+ H_2O$$

This reaction effectively produces a chain extension of the polyester and results in a considerable increase in the viscosity of the resin slurry. Because the thickening reaction does not immediately proceed to completion, it is necessary to store or 'mature' the SMC for a few days to allow a reproducible viscosity level to be achieved before moulding.

For most applications, MgO thickening provides a very cost effective method of achieving the required viscosity increase. However, there are some disadvantages. In addition to being slow in its approach to completion, the viscosity-increasing reaction can be affected by variables such as moisture content, filler characteristics and ambient temperature, leading to variability in the processing behaviour of the SMC. Some alternative methods of thickening SMC have therefore been developed. These include 'solid state thickening', where a resin additive causes partial crystallization of the polyester component, as well as processes based on urethane crosslinking.

Shrinkage control additives

There is a range of 'shrinkage control' additives that allow the mould shrinkage of SMCs and other polyester compounds to be tailored to any required level or even reduced to zero if needed. These additives are an essential part of SMC technology, especially when accurate parts with a Class A surface finish are required for the automotive industry.

Shrinkage occurs with all thermosets due to the volume change of the resin during cure and the thermal contraction that takes place as the part cools from the cure temperature. This undesirable effect manifests itself both in distortion of the part and in an uneven surface. The surface of the moulding develops an 'orange peel' appearance, due to the differential contraction between the resin phase and the other solids present. The cure contraction and thermal contraction of the resin are hindered to some extent by the presence of fibres and filler particles, so the resin surrounding these particles is left in a state of hydrostatic tension. Near to the surface of the mould, this tension can be relieved by local resin contraction, which leaves visual evidence of the fibres just below the surface of the part.

Shrinkage control or 'low profile' additives are generally solutions of a thermoplastic polymer, such as polyvinyl acetate, in styrene. The mechanism of operation is still the subject of much debate,[3] but it is known that, as the cure reaction advances, the thermoplastic additive is precipitated from solution in a finely dispersed form. At the cure temperature, the thermoplastic is above its glass transition temperature and possesses little strength, so the thermoplastic domains probably act as nuclei for the formation of a network of resin microcracks, which relieve stress and compensate for the resin shrinkage.

Processing

As mentioned above, SMCs are commonly processed by hot press or compression moulding; this involves heated tool steel moulds (often chrome plated for ease of mould release) and large hydraulic presses. The mould temperature is often in the range 140–150°C and the pressures are in the range 0.5–5.0 MPa. The operation of SMC compression moulding involves cutting out a 'charge' of material from the pre-impregnated sheet, removing the thermoplastic backing film and placing it in the mould. The total cycle time for SMC compression moulding is usually in the range of 60–180 s. As compression moulds are expensive, production runs of at least 5000 parts are usually needed for the process to be economically attractive.

One problem with SMC is that, to obtain the best possible surface finish, the parts usually require to be painted after moulding. To overcome this, 'in mould' coating processes have been developed. These processes enable parts to be produced with a hard glossy surface of the required colour direct from the mould. There are two versions of the in-mould coating process. One operates by spraying the pigmented coating resin on to the mould surface before moulding. The other involves injection of the coating resin into the mould between the curing moulding and the mould surface.

Although the bulk of SMCs are processed by compression moulding, a certain percentage of the material is injection moulded. Although SMCs are more expensive than the types of polyester compound normally used for injection moulding (Section 3.5), their longer fibre length offers the possibility for improved mechanical properties. Injection moulding is usually confined to companies with in-house SMC manufacturing facilities.

'Low pressure' SMCs, which have a lower viscosity than conventional ones, have recently been developed. It is claimed these can be used with less expensive steel tooling, which makes their use viable for shorter production runs.

Properties and applications

Some typical properties of SMCs are given in Table 42. Their main advantages are relatively low cost, good surface finish and high heat deflection temperature. SMCs are used in many industries, but their widest use is in the automotive industry. In recent years, SMCs have

Table 42. Typical properties of SMCs

Property	Value
Specific gravity	1.90–1.92
Cost (£ kg^{-1})	1.5–2.5
Young's modulus (GPa)	9–12
Tensile strength (MPa)	65–80
Flexural strength (MPa)	160–190
Unnotched Charpy impact strength (kJ m^{-2})	27–38
Notched Izod impact strength (kJ m^{-2})	1–1.2
Elongation at break (%)	0.7–1.2
Heat deflection temperature (°C)	>240
Expansion coefficient (K^{-1})	12–14 × 10^{-6}

become an established material for the manufacture of panels for the cabs of truck and other commercial vehicles. The ability of the material to withstand the temperatures encountered in the ovens of automotive paint lines also makes SMC attractive for use as car body panels. In Europe, the car and commercial vehicle industries account for about 47% of SMC consumption. The next largest use is in electrical components (23%).

References

1. S Newman, D G Fesko, *Polymer Composites*, **5**, 88 (1984).
2. J V Milewski, *Handbook of Fillers and Reinforcements for Plastics*, Van Nostrand Reinhold, New York, 1978.
3. T Mitani, H Shiraishi, K Honda, G E Owens, in *Proceedings of the SPI/Composites Institute 44th Annual Conference*, Feb. 1989, Dallas, Texas, Society of the Plastics Industry, New York, 1989, Paper 12F.

3.18 Spray deposition

T M GOTCH

Spray deposition of FRP in its simplest form can be considered as a first stage automation of hand lay. In concept, the improvement relates to the material deposition phase; consolidation and cure are similar to hand lay.

The deposition phase consists of spraying down chopped rovings and activated resin from a special purpose gun on to a tool prepared to the gelcoat stage. It is also possible to automate this gelcoat stage by the use of a special version of the spray deposition equipment.

The principal criterion for spray deposition to be efficient and effective is that a satisfactory spray pattern must be developed for both gelcoat and laminate steps without localized excessive material build-up. This is generally a factor of geometry and size of the component. The main disadvantage with this type of process is that, due to the atomization of the raw materials, efficient extraction is required to prevent the build-up of styrene (for polyester resins) and catalyst fumes with their associated health and safety connotations.

The tool is prepared in the conventional way and gelcoat applied by either brush or spray. It is essential that sprayed gelcoat is applied no thicker than brushed gelcoat, otherwise a brittle surface could result. The use of a comb thickness gauge to monitor and control thickness is desirable.

When the gelcoat has reached the 'green' stage, a light resin spray without reinforcement is applied to the tool to provide a key for the resin/reinforcement combination. This is produced by the activated resin and reinforcement impinging and wetting the reinforcement just outside the gun head. For this operation, the reinforcement is used in a different form to hand lay: continuous rovings are employed instead of chopped strand mat. These rovings are chopped in the gun head, to consistent lengths ranging from 25–50 mm such that the final mix is similar to a chopped strand laminate in hand lay. The efficiency of the reinforcement wetting and the spray pattern obtained depend on the gun design and nozzle selection. They are generally chosen by experiment and experience to achieve optimum deposition characteristics for a given tool geometry.

Resin and reinforcement are deposited simultaneously in bands by moving the spray gun from side to side, successive passes overlapping each other to avoid discontinuities and hence weak points at the joint line between passes. For the best results, the gun head movement direction should be at right angles to the plane of the spray pattern.

As in hand lay, the sprayed component should be consolidated after each layer has been deposited until the requisite thickness of laminate is achieved. Unlike hand lay, where the reinforcement type predetermines the weight of reinforcement used, in spray deposition this is much more operator dependent and involves considerable technique and expertise to obtain accuracy commensurate with hand lay. This can be assisted by the use of coloured tracer rovings in the continuous fibre and by the use of comb depth gauges.

The geometry of the component can radically affect the efficiency of deposition. It is important to spray surfaces starting at the bottom of the tool and to consolidate from the top down; this keys the material in position and prevents unconsolidated material coming away from the tool surface. There are two basic types of deposition equipment: twin pot and catalyst injection.

Twin pot

The twin pot technique (Fig. 61) is based on two pressurized containers holding pre-accelerated and precatalysed resin respectively.[1] The resin pots and the continuous reinforcement roving are connected to the gun head. The gun head contains two units: the chopping head and the resin port.

Atomized sprays of the two resin components are usually mixed externally before introduction of the chopped reinforcement. The stream conveys impregnated composite material on to the tool surface. This system obviates the need for a solvent flush; the only limiting factor is the pot life of the precatalysed resin, as the two resin components do not mix in the gun head.

Catalyst injection

As its name implies, catalyst injection (Fig. 62) is the injection of catalyst into a pre-accelerated resin component.[1] Both catalyst and resin component are fed to the spray gun from pressurized containers. As with the twin pot technique, the fibrous element is introduced into the activated system by chopping continuous rovings at the gun head; the resin stream carries the mix on to the tool surface. Two methods of introducing the catalyst are in general use: internal and external mixing.

The internal mix system introduces the catalyst in the gun head chamber and is based on physical mixing, whereas the external mixing method relies on the interaction of atomized streams of catalyst and pre-

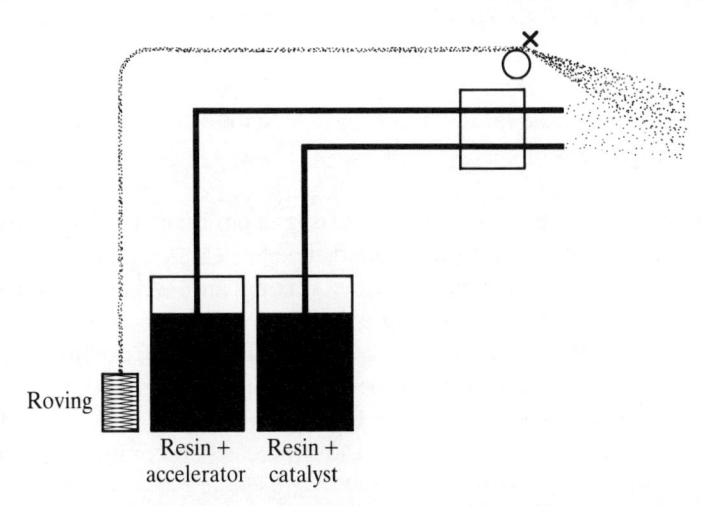

Figure 61 Twin pot system (figure reproduced courtesy of PPG Industries (UK) Ltd)

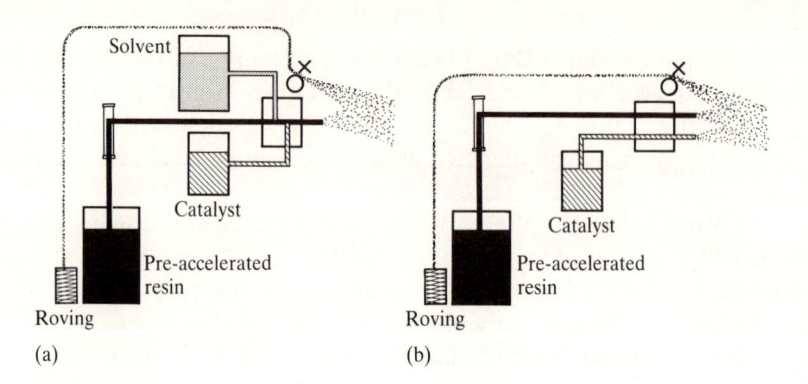

Figure 62 Single pot or catalyst injection, (a) catalyst injection with internal mixing system; (b) catalyst injection with external mixing system (figure reproduced courtesy of PPG Industries (UK) Ltd)

accelerated resin. Choice depends on individual assessment of the level of risk associated with each system. The internal mix system requires solvent flushing, whereas the external mix system has the problems of inconsistent mixing and safe atomization of catalyst under pressure; external systems could spray neat catalyst should the resin supply fail.

Gelcoat spray is based on the catalyst injection principle and generally the gelcoat viscosity has to be adjusted to suit the requirements of the spray process. However, modern equipment is being developed to cope with the viscosities of the hand lay resin systems, reducing the need to stock two forms of the same resin.

Spray deposition offers the potential to produce open moulded components faster than using the basic hand lay technique; other advantages and disadvantages can have a significant effect on process selection.

Advantages	Speed of application is greater than hand lay, giving faster tool turn round time.
	There is no pouring, mixing and weighing of resins during processing.
	Chopping grade rovings can be less expensive than equivalent quantities of chopped strand mat.
	Cheap tooling can be used.
	Large surface area tools are easily accommodated.
Disadvantages	Capital outlay is higher than for hand lay.[2]
	Extraction is essential.

Quality control is entirely in the hands of the operator, particularly in the total weight of material deposited.
Mouldings have only one finished surface.
Shape complexity is more restricted than hand lay.
Incorporation of inserts, cores, etc. is not impossible but considerably slows down the process and hence reduces the potential savings.

Despite its disadvantages, spray deposition finds considerable use, particularly in the manufacture of large, simple shaped mouldings.[2]

References

1. *Spray deposition of GRP*. PPG Industries (UK) Ltd., Wigan, UK.
2. T M Gotch, *The Relative Economics of Different GRP Production Methods*, Hands off GRP, Proceedings of conference, Coventry, July 1979, Plastics and Rubber Institute, London, 1979.

3.19 Structural RIM (SRIM)

P A SHEARD AND M PARMAR

Structural reaction injection moulding (structural RIM or SRIM) is a relatively recent process development for mass producing structural composite components. It provides the opportunity to mould parts suitable for automotive applications in cycle times approaching one per minute.

The process is illustrated schematically in Fig. 63. It is similar to RTM technology (see Sections 3.15 and 3.16) in that a thermosetting resin is injected into a closed mould containing dry reinforcement in the form of mat or woven material (see Sections 1.10 and 1.12). In some cases the reinforcements are preshaped as preforms. The main difference is that structural RIM uses RIM equipment (see Section 3.14) to provide high pressure impingement mixing and rapid injection of resin into the mould. The process therefore combines the advantages of RIM, to provide rapid cycle time, and RTM technology, to give structural composite components.

The two resin components are held in separate storage tanks to maintain the required material temperature. During the pouring stage, resin from each tank is transferred to the mix head under high pressure. The two material components are subjected to high pressure impingement mixing before entering the mould. Although the pressure is high in

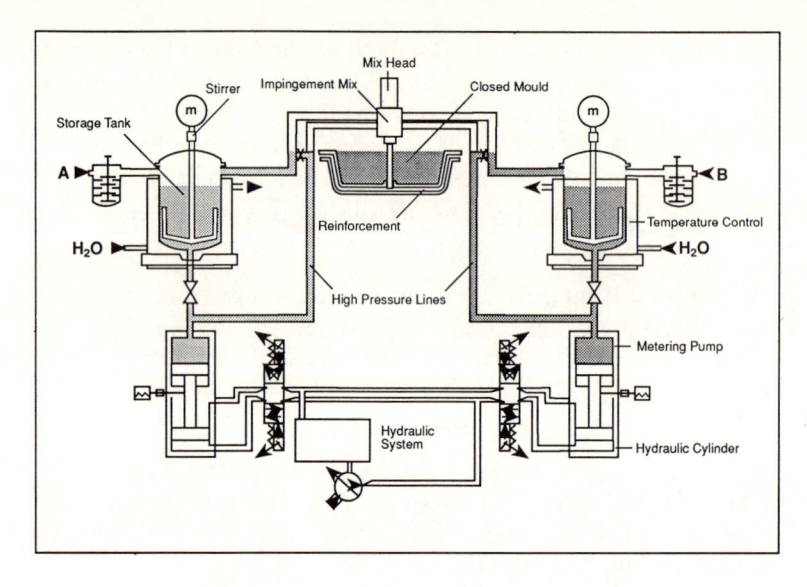

Figure 63 Schematic representation of SRIM

the mix head, the resin pressure can be quite low in the mould, depending on the type and level of reinforcement. Once the mould is filled, the component cures either at room temperature or elevated temperature and is then ejected. Typical properties of SRIM composites[1] are given in Table 43, which also includes properties of neat resin for comparison.

Selection of the appropriate fibre is critical to the process. The fibres must allow fast wet-out and be compatible with the resin system (see Section 1.11). They also need to be stable enough to resist fibre washing

Table 43. Typical properties of SRIM composites and neat resin based on the ARSET HT3500 system[1]

	\multicolumn{3}{c}{Glass percentage}	
	0	22	47
Specific gravity (g cm^{-3})	1.27	1.36	1.56
Tensile strength (MPa)	60	76	179
Young's modulus (MPa)	2620	5760	10760
Elongation (%)	4.0	2.5	2.1
Flexural strength (MPa)	90	180	290
Flexural modulus (MPa)	2415	5035	9800
Izod impact (kJ m^{-2})	160	505	985
DTUL* at 1.82 MPa (°C)	121	204	230
Shrinkage (%)	1.51	0.15	0
Hardness, Shore D	85	–	–
Heat sag at 174°C (mm)	0	0	0

*Deflection temperature under load.

during the resin injection cycle and should have good preformability/drapeability characteristics. Glass is currently the most widely used fibre reinforced material.

The resin system needs to enable complete wet-out of the fabric while having the viscosity and rapid cure schedule to reach all areas of the mould before cure commences.

To date, levels of reinforcement used in the process tend to range from 5 to 55% by weight.[2] Resins generally have two components: isocyanate and polyol. They behave differently to polyurethanes in that the reaction to form the polymer occurs after the resin has penetrated the reinforcement and filled the mould.

Considerable development work is currently focusing on introducing modified resins based on unsaturated polyester, methylmethacrylate and epoxy resins.[3,4] These are believed to overcome some of the limitations associated with conventional SRIM resins. Developments in fibre technology and production of near net shape preforms are continuing to make this process more attractive to a wider range of industries.

References

1. U E Youngs, *Urethanes Technology*, Jun./July, 20–23 (1990).
2. C F Johnson, in *Engineered Materials Handbook*, Vol. 1, ASM International, Metal Parks, Ohio, 1987, pp. 564–568.
3. M Parmar, P Sheard, G S Boyce in *Proceedings of the 3rd International Conference on Automated Composites (ICAC 91)*, Plastics and Rubber Institute, London, 1991, pp. 3211–3218.
4. M Parmar, P Sheard, G S Boyce, *Reinforced Plastics*, **35**(6), 34 (1991).

CHAPTER 4

Mechanical Properties of Composites – micromechanics

4.1 Continuous fibre reinforcement

J E BAILEY

Utilization of continuous lengths of filament in the fabrication of fibre reinforced materials allows for the most efficient form of fibre packing. That is a high volume fraction of fibres can be incorporated. The building blocks are the unidirectional continuous fibre laminae of Fig. 64. For ideal regular hexagonal close packing of fibres of circular cross-section, i.e. when the fibres just touch, the volume fraction of fibre V_f would reach a value of 0.91 and for ideal regular square close packing the value of V_f would be 0.76. In practice, very large bundles or tows of fibres are aligned mechanically and the ideal distribution is only realized in small localized regions. Fibre bunching occurs and matrix-rich (resin-rich) areas occur (Fig. 65). Values of V_f greater than 0.65 are difficult to achieve.

Fibre bunching and heterogeneity in the fibre distribution are often caused by tows or fibre bundles maintaining their identity in the fibre lay-

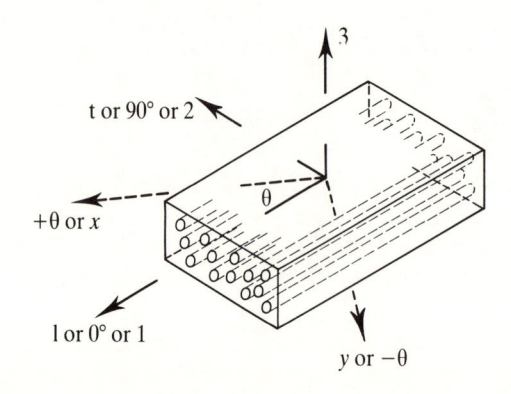

Figure 64 Model unidirectional lamina showing the designated in-plane directions and through thickness, 3, direction

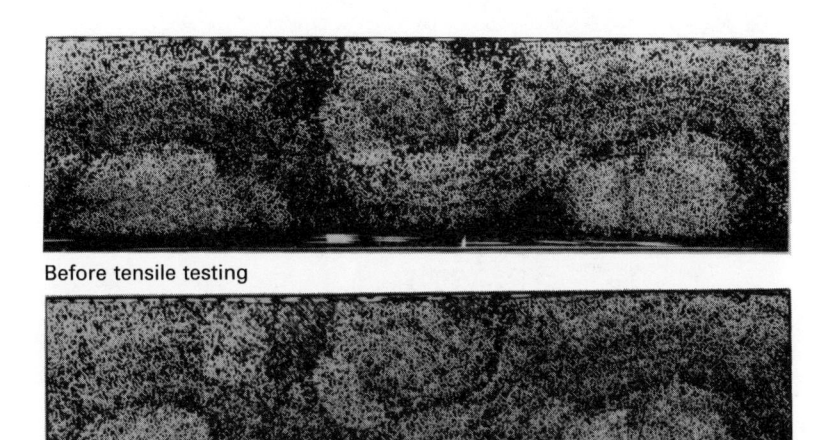

Figure 65 Optical micrographs of the edge of a 90° ply showing the presence of the memory of the fibre tows in a filament wound composite. The location of two transverse cracks is also shown. (Reproduced from Jones (1990)[1], with permission)

up. This is often caused by 'twists' in the tows preventing the splaying out of the fibres, such an effect can cause poor wetting out of the fibres and therefore a poor quality laminate in which pores exist and fibres are touching. It is also interesting to note that even when V_f is comparatively low the distance between fibres is often less than the fibre diameter, i.e. less than 10 μm. This fact may be of significance if the structure of the polymer matrix is affected by the presence of the fibre interface, which in turn may affect the properties developed in the polymer matrix.

Clearly the properties of the unidirectional composites of Fig. 64 will be highly anisotropic since all the fibres are aligned in one direction and we can anticipate that, for example, strength, stiffness, thermal expansion and conductivity properties will be different in directions parallel (often termed 0° or longitudinal, l, direction) and perpendicular (often termed 90° or transverse, t, direction) to the fibres. The anisotropic behaviour will be dealt with in the following articles in this handbook.

The degree of isotropy required can be built into a composite laminate made up of thin unidirectional laminae. If the properties of the thin laminae are known, the properties of a laminate can be designed and fabricated by symmetrically stacking the thin laminae in an acceptable sequence, and rotating them through appropriate angles with respect to one another. Laminate theory (see Sections 4.10 and 5.14) enables the required configuration of laminae to be determined. An example of a

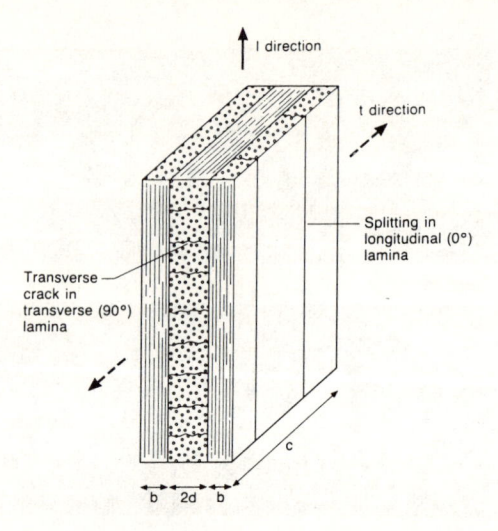

Figure 66 A schematic of a $0°/90°/0°$ crossply laminate illustrating the lay-up of an angle ply laminate. The nature of the microcracking in the form of transverse cracks in the 90° ply and longitudinal splitting of the 0° plies which occurs under an increasing longitudinal stress is also indicated (see 4.10 Laminates-angle plied and 4.12 Laminates-statistics of microcracking)

simple symmetric crossply laminate is shown in Fig. 66. This laminate configuration is termed $0°_x/90°_y/0°_x$, where x and y are the number of individual plies (of prepreg) that make up a lamina. It is clearly possible to laminate the plies at angles other than 90°, θ, to obtain symmetrical angle ply laminates, e.g. $+\theta/-\theta_2/+\theta$, $+\theta/-\theta/+\theta_2/-\theta/+\theta$, etc. The most common arrangement has quasi-isotropy: $0°_2/\pm 45°/90°_4/\pm 45°/0°_2$.

Laminate structured composites of the form discussed above are used to achieve high performance composites for space and aerospace applications and other structural applications where specific strength and stiffness are key factors. Such an approach is generally needed to obtain the economic efficiency necessary to justify the use of expensive fibres and tailored (optimized) polymer matrix materials. There are more detailed presentations of fibre reinforcement.[1-3]

References

1. F R Jones in *Composite Materials in Aircraft Structures*, ed. D H Middleton, Longman, Harlow, 1990, Ch. 6.
2. D Hull, *An Introduction to Composite Materials*, Cambridge University Press, Cambridge, 1981.

3. S W Tsai, H T Hahn, *Introduction to Composite Materials*, Technomic, Westport, Connecticut, 1980.

4.2 Continuous fibre reinforcement – Young's modulus

J E BAILEY

Longitudinal stiffness (Young's modulus) of unidirectional continuous fibre laminae

Consider a load applied to a lamina parallel to the direction of the fibres in Fig. 64. If the fibres are well bonded to the matrix, i.e. no slippage, then the strain ϵ generated by the load on the composite will be the same in the fibre as in the matrix, where the subscripts c, f and m refer to composite, fibre and matrix respectively.

$$^1\epsilon_c = {}^1\epsilon_f = {}^1\epsilon_m \qquad [4.1]$$

Superscripts 1, 2 and 3 refer to direction; they should be applied to the fibre and matrix. It is assumed here for simplicity that both fibre and matrix are isotropic, therefore we can omit the 1, 2 and 3 suffix. This is not correct for all fibres. If the fibre and matrix behave elastically then the stress in the fibre, matrix and composite σ_f, σ_m and σ_c will be given by

$$\sigma_f = E_f \epsilon_f \qquad [4.2]$$

and

$$\sigma_m = E_m \epsilon_m \qquad [4.3]$$

and the composite modulus parallel to the fibres is defined by

$$^1\sigma_c = {}^1E_c {}^1\epsilon_c \qquad [4.4]$$

For the cross-section of the composite perpendicular to the fibre direction, the area fraction of fibre is equal to the volume fraction V_f and the stress on the composite $^1\sigma_c$ will be the sum of the fractional contribution of the fibre and matrix, i.e.

$$^1\sigma_c = \sigma_f V_f + \sigma_m V_m \qquad [4.5]$$

We can substitute for the stresses in this equation using the stress strain relationships (4.2)–(4.4) given above and since the strains are identical (4.1) we arrive at

$$^1E_c = E_f V_f + E_m V_m \qquad [4.6]$$

the so-called 'law of mixtures'.

The analysis assumes (4.2) and (4.3) are valid; this is not strictly true because differing Poisson contractions will result in additional stresses, generally these are not large, <1%. Experimental verification of (4.6) has been obtained for a wide range of composites and it is the key equation for stiffness prediction. When $E_f \gg E_m$ and V_f is high, the contribution of the polymer matrix to the longitudinal stiffness can be insignificant. The volume fraction of matrix $V_m = 1 - V_f$ unless significant porosity is present, and so for good quality composites, the longitudinal stiffness increases linearly with volume fraction of fibres. ^1E_c is often written as E_l and (4.6) becomes

$$E_l = E_f V_f + E_m (1 - V_f). \qquad [4.7]$$

Transverse stiffness (Young's modulus) of unidirectional continuous fibre laminae

If a stress is applied in the direction perpendicular to the fibre direction, i.e. in the 2 or 3 direction in Fig. 64, it is possible to take a similar micromechanics approach to determine the nature of the transverse Young's modulus of the laminae ^2E_c. For this analysis we assume properties are isotropic in the 2,3 plane, i.e. $^2E_c = {^3E_c}$. A more questionable assumption is that transverse stress on the laminae is the same in the fibre as in the matrix, i.e. $^2\sigma_f = {^2\sigma_c}$, and the strains will be

$$^2\epsilon_f = \frac{^2\sigma_c}{E_f} \qquad [4.8]$$

$$^2\epsilon_m = \frac{^2\sigma_c}{E_m} \qquad [4.9]$$

$$^2\epsilon_c = \frac{^2\sigma_c}{E_m} \qquad [4.10]$$

Since the strain on the composite in the 2 direction $^2\epsilon_c$ is given by $^2\epsilon_c = {^2\epsilon_f} V_f + {^2\epsilon_m} V_m$ then if we substitute for the strain using (4.10) we arrive at an expression for ^2E_c namely

$$\frac{1}{^2E_c} = \frac{V_f}{E_f} + \frac{V_m}{E_m} \qquad [4.11]$$

The superscript 2 is left off the fibre and matrix moduli on the assumption that they are isotropic materials. This would be questionable, for example, for carbon or aramid fibres. For simplicity, ^2E_c is often denoted E_t and (4.11) is presented in the form

$$E_t = (E_f E_m)/[E_f(1 - V_f) + E_m V_f] \qquad [4.12]$$

The form of (4.11) is in reasonable agreement with the experimental observations on fibre reinforced polymers for the variation of ^2E_c with V_f, (see Fig. 67), but the absolute values are not predicted. Attempts to take into account Poisson contraction effects, for example by replacing ^2E_m by $^2E_m/(1 - \nu^2)$, apparently improve the fit. But this simple mechanics approach ignores the stress concentration around fibres embedded in the matrix, an unrealistic assumption. The assumption that $^2\sigma_f = {}^2\sigma_m$ is equivalent to replacing the fibres by thin sheets of material of the same modulus sandwiched between sheets of matrix, with the sheets parallel to the fibre axes. More complex expressions have been derived using elasticity theory and finite element analysis; these are more realistic but it is necessary to determine some of the parameters in the resulting equation by fitting them to actual experimental results. The most rigorous approach is that of Hashin and Rosen[3] but the Halpin–Tsai equations[4,5] are generally more applicable and allow for variations in

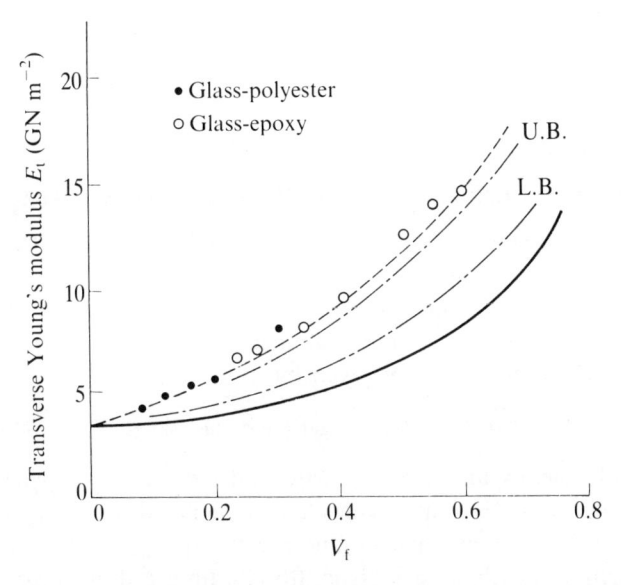

Figure 67 Comparison of experimental and theoretical values of transverse Young's moduli over a range of fibre volume fraction. The continuous line (—) is for the law of mixtures equation (4.12). The dotted line (– – –) is the Halpin–Tsai estimate (4.13). The upper (UB) and lower (LB) Hashin–Rosen bounds are also given. Reproduced from Jones (1990)[1] with permission after original data given by Bailey and Parvizi (1981).[2]

packing geometry and regularity with the factor ξ obtained empirically by fitting the data to

$$^2E_c = E_t = E_m(1 + \xi\eta V_f)/(1 - \eta V_f) \quad [4.13]$$

where $\eta = [(E_f/E_m) - 1]/[E_f/E_m + \xi]$. See also section 4.13, where arguments relevant to the prediction of transverse modulus are also presented.

References

1. F R Jones in *Composite Materials in Aircraft Structures*, ed. D H Middleton, Longman, Harlow, 1990, ch. 6.
2. J E Bailey, A Parvizi, *J. Mater. Sci.* **16**, 649 (1981).
3. Z Hashin, B W Rosen, *J. Appl. Mech. Trans. ASME*, **31**, 223 (1964).
4. J G Halpin, S W Tsai, US Air Force Materials Laboratory Technical Report 67423, 1969.
5. S W Tsai, *Composites Design*, US Air Force Materials Laboratory/Think Composites, Dayton, Ohio, USA.

4.3 Continuous fibre reinforcement – strength

J E BAILEY

Tensile strength parallel to the fibres (longitudinal strength).

The response to stress will depend upon the relative failure strain of the fibre and matrix, ϵ_{uf} and ϵ_{um} respectively. If exceptionally $\epsilon_{uf} = \epsilon_{um}$ then as the strain on the lamina increases and since $\epsilon_f = \epsilon_m$ (see Section 4.2), then fibres and matrix will fail simultaneously and from (4.5) the failure stress of the composite will be given by

$$^1\sigma_{uc} = \sigma_{uf}V_f + \sigma_{um}V_m \quad [4.14]$$

For ductile metals and many plastic matrices $\epsilon_{uf} < \epsilon_{um}$, whereas for virtually all ceramics, some plastics and some metals $\epsilon_{uf} > \epsilon_{um}$.

In (4.14) the fibres are given a unique strength σ_{uf}, whereas in practice the strength of the fibres varies from fibre to fibre and the mean strength of the fibres is equated with σ_{uf}. For brittle fibres, e.g. carbon and glass, the coefficient of variation can be quite high, $\simeq 20\%$; the strength of the fibres is normally determined by the size of the largest effective defect introduced into the fibre during manufacture. Careful processing may reduce the coefficient of variation. However, once fibres are incorporated into the composite material, when the weaker fibres break, in general, the neighbouring fibres are able to take up the additional load and as a consequence the variability of the fibre composite is much less than that of the fibres. Typically $\simeq 5\%$ for the above fibres.

Ductile matrix composites

Glass and carbon fibre reinforced thermoplastic materials are prime examples of composites for which $\epsilon_{uf} < \epsilon_{um}$. When a unidirectional lamina is stressed at low strain, the stiffness is given by (4.6). As the strain increases, the stress–strain curve of the lamina becomes non-linear, due to the decreasing stress contribution of the polymer, although this effect is often difficult to detect because the contribution of the flexible polymer is so small. When ϵ_{uf} is reached, the fibres fail and the stress on the lamina will be

$$^1\sigma_c = \sigma_{uf}V_f + \sigma'_m V_m \qquad [4.15]$$

where σ'_m is the stress on the matrix at the fibre fracture strain. If V_f is high, above some critical value **V**$_{crit}$ the lamina will fail, and the failure strain is given by

$$^1\sigma_{uc} = \sigma_{uf}V_f + \sigma'_m V_m \qquad [4.16]$$

For $V_f < V_{crit}$, the load on the fibres can be borne by the matrix, the matrix strength is given by $\sigma_{um}V_m[=\sigma_{um}(1-V_f)]$, therefore

$$^1\sigma_{uc} = \sigma_{um}(1-V_f) \qquad [4.17]$$

and for $V_f < V_{crit}$, fibre failure (multiple cracking of fibres) would occur before ultimate failure (see Section 4.7). V_{crit} can be determined by equating (4.16) and (4.17). Normally V_{crit} is very small $\simeq 1\%$ and of no technical significance for reinforced polymers. In fact, for most high performance unidirectional polymer composites, the contribution to the tensile strength of the second term in (4.16) is small, normally less than 5% and (4.16) can be written as

$$^1\sigma_{uc} \approx \sigma_{uf}V_f \qquad [4.18]$$

The subscript l is often used to denote longitudinal properties and (4.18) becomes

$$\sigma_{ul} = \sigma_{uf}V_f \qquad [4.19]$$

The micromechanics of strength are shown graphically in Fig. 68a.

Brittle matrix composites

It is possible for the condition $\epsilon_{um} < \epsilon_{uf}$ to occur in some fibre reinforced plastics, for example in some glass and aramid fibre reinforced thermosets. In this case on application of a stress parallel to the fibres in the lamina, the failure strain of the matrix, ϵ_{um}, is reached before that of the fibres, and the stress on the lamina when the matrix cracks will be

$$'\sigma_c = \sigma'_f V_f + \sigma_{um} V_m \quad [4.20]$$

where σ'_f is the stress on the fibre at the matrix failure strain. If V_f is high, i.e. above a critical value V_{crit}, the fibres will be able to support the load originally borne by the matrix; this load is $\sigma_{um} V_m [= \sigma_{um}(1 - V_f)]$. The stress and strain on the lamina will increase until ϵ_{uf} is reached, then the fibres, and therefore the composite, will fail. Thus for $V_f > V_{crit}$, the failure stress of the lamina will be

$$'\sigma_{uc} = \sigma_{uf} V_f \quad [4.21a]$$

or

$$\sigma_{ul} = \sigma_{uf} V_f \quad [4.21b]$$

For $V_f < V_{crit}$, that is when there are insufficient fibres present to bear the load transferred from the matrix, (4.20) determines the failure condition, i.e.

$$'\sigma_{uc} = \sigma'_f V_f + \sigma_{um} V_m \quad [4.22a]$$

or

$$\sigma_{ul} = \sigma'_f V_f + \sigma_{um} V_m \quad [4.22b]$$

V_{crit} can be determined by equating (4.21a) and (4.22a), giving

$$V_{crit} = \frac{\sigma_{um}}{\sigma_{uf} - \sigma'_f + \sigma_{um}} \quad [4.23]$$

If we consider a brittle resin for which say $\epsilon_{um} = 0.02$ and a glass fibre with $\epsilon_{uf} = 0.025$, then $V_{crit} = 0.1$; since for high performance laminae $V_f > 0.5$, matrix cracking will occur for strains on the composite greater than ϵ_{um}. Multiple cracking of the matrix is observed to occur, as illustrated schematically in Fig. 68b. The spacing of the cracks increases as the strain on the lamina increases, if the fibre and matrix are well bonded. The spacing of the matrix cracks can be predicted with reasonable accuracy, by simple consideration of mechanisms by which the load transfers from the matrix to the fibre. The additional load is assumed to transfer to the fibre by a constant shear stress τ at the fibre–matrix interface and the ultimate crack spacing, s, is given by

Continuous fibre reinforcement – strength

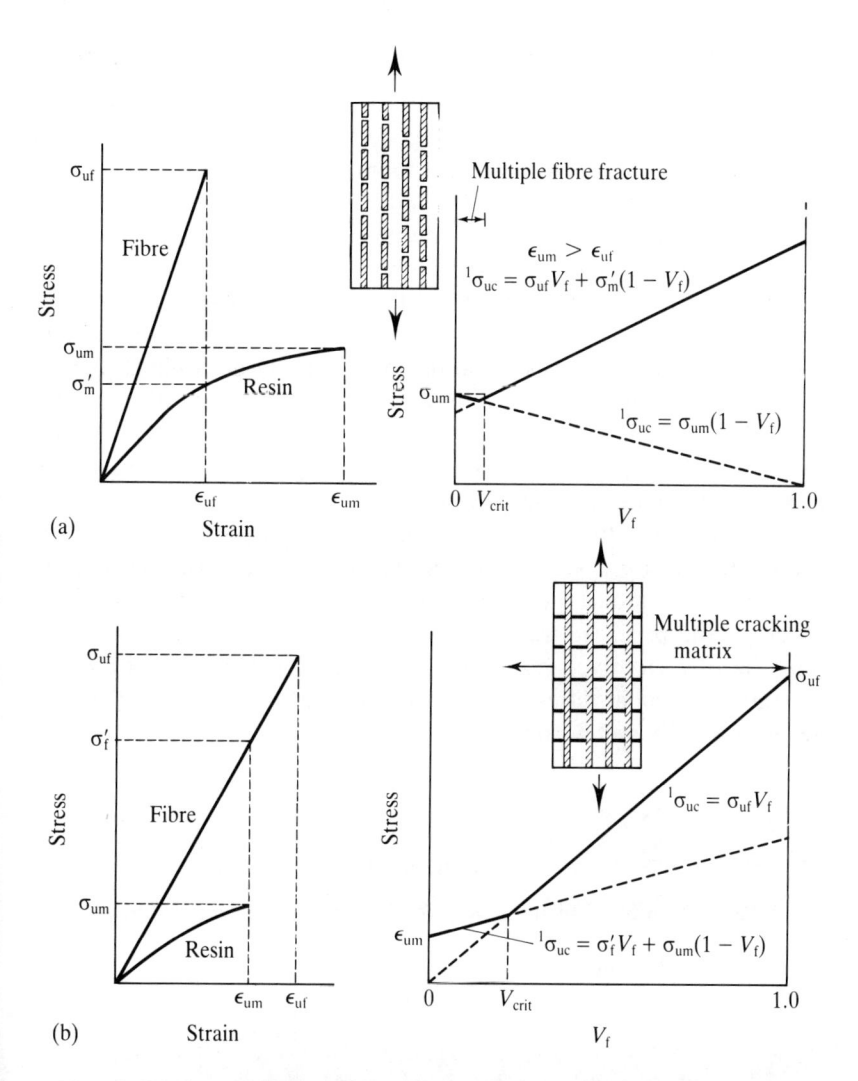

Figure 68 Micromechanics of (a) ductile matrix composite with $\epsilon_{uf} < \epsilon_{um}$, (b) brittle matrix composite with $\epsilon_{uf} > \epsilon_{um}$ showing the fibre volume fraction (V_f) dependence of longitudinal strength and the critical values of V_f (v_{crit}) for reinforcement and the range of V_f where (a) multiple fibre fracture and (b) multiple matrix fracture occurs

$$s = \frac{\sigma_{um} r_f}{2\tau}\left(\frac{V_m}{V_f}\right) \qquad [4.24]$$

where r_f is the radius of the fibre. This equation has been verified for several brittle matrix composites.[1,2]

Tensile strength perpendicular to the fibres (transverse strength)

The transverse strength of the lamina is inevitably low because it has been optimized for maximum strength by concentrating the fibre performance in one direction. The model used to predict the transverse Young's modulus is described in Section 4.2 and Fig. 64. It assumes continuity of stress in the 2,3 direction. If $^2\sigma_c = {}^2\sigma_m = {}^2\sigma_f$ in the transverse direction, then clearly the strength will be determined by the lowest of σ_{um}, σ_{uf} or σ_{ui}, where σ_{ui} is the strength of the fibre–matrix interface. As pointed out in Section 4.2, this model ignores stress concentrations around the fibres. Such stress concentrations are typically determined to be in the range 2.0–3.0. If a strong interface bond exists to strong fibres, σ_{um} will determine the transverse strength and $^2\sigma_{uc}$ will be in the range 0.5 σ_{um} – 0.3 σ_{um}. Direct evidence from experiments on glass fibre reinforced thermosets, which determined values of σ_{uc} for the lamina and σ_{um} for the resin, does support this model but it raises issues about the comparability of σ_{um} values between the bulk resin and the matrix resin. However, this model does lead to a closer agreement between the observed and predicted transverse failure strains of these composites. In the case of a poorly bonded interface when σ_{ui} determines transverse strength, a similar stress concentration argument still applies (see Sections 4.6–4.9 and Section 4.13).

References

1. A Kelly, N H MacMillan, *Strong Solids*, 3rd edn, Clarendon Press, Oxford, 1986.
2. J Aveston, A Kelly, *J. Mater. Sci.*, **8**, 352 (1973).

4.4 Continuous fibre reinforcement – off-axis properties of unidirectional laminae

J E BAILEY

In Sections 4.2 and 4.3, the stiffness and strength of unidirectional laminae are considered. Comparison of the values in the two directions indicates the extent of the anisotropic behaviour. In this section, the off-axis performance is explored.

Orientation dependence of Young's modulus

The prediction of off-axis elastic properties requires the application of anisotropic elasticity theory. In this brief presentation, only the results are quoted. For further detail reference should be made to Hull,[1] Jones[2] and Tsai[3,4] (see also Section 5.14). Four independent elastic or engineering constants define the elastic behaviour of the laminate, i.e. ^1E, ^2E, G_{12} and ν_{12} and the equation for the in-plane off-axis modulus E_x (Fig. 68) is

$$\frac{1}{E_x} = \frac{1}{^1E_c}\cos^4\theta + \left(\frac{1}{G_{12}} - \frac{2\nu_{12}}{^1E_c}\right)\sin^2\theta\cos^2\theta + \frac{1}{^2E_c}\sin^4\theta \qquad [4.25]$$

Similar equations exist for E_y and G_{xy}, the shear modulus. Thus provided ^1E_c, ^2E_c, G_{12} and ν_{12} are known, elastic properties at any angle can be simply calculated. Values for E_x are given in Fig. 69. The most significant aspect of the results to have been confirmed experimentally is the very rapid fall off in stiffness at small values of θ. This result is important from a design viewpoint.

Orientation dependence of tensile strength

Three possible modes of failure can be envisaged in laminae tested at an angle θ to the fibre direction (Fig. 70). For example, failure will occur

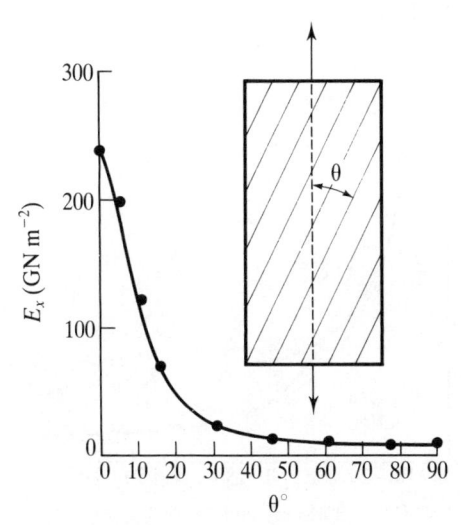

Figure 69 The angular dependence of the modulus of a unidirectional CFRP lamina ($V_f = 0.5$). The continuous line is the prediction of equation 4.25. The points are experimental values. Redrawn from data given by Hull.[1]

when the stress resolved parallel to the fibre direction reaches the value of $^1\sigma_{uc}$, i.e. a longitudinal tensile fracture occurs; the condition is

$$^\theta\sigma_c = {}^1\sigma_{uc} \sec^2\theta \quad [4.26]$$

A second possibility is that the stress resolved in the transverse direction reaches the value $^2\sigma_{uc}$, i.e. a transverse fracture occurs; the condition is

$$^\theta\sigma_c = {}^2\sigma_{uc} \csc^2\theta \quad [4.27]$$

Finally, failure could occur by shear parallel to the fibres when the resolved shear stress reaches the shear strength $^1\tau_{uc}$; the condition is

$$^\theta\sigma_c = {}^1\tau_{uc} \sec\theta \csc\theta \quad [4.28]$$

The quantity $^1\tau_{uc}$ is often referred to as the intralaminar shear strength or in-plane shear strength. The value of $^1\tau_{uc}$ will depend upon the relative shear strengths of the polymer matrix and the interfacial bond.

Clearly the strength at a given angle θ will be determined by the weakest of these failure modes. The three failure modes are plotted in Fig. 70 for a Type 1 carbon fibre epoxy resin system; the anticipated

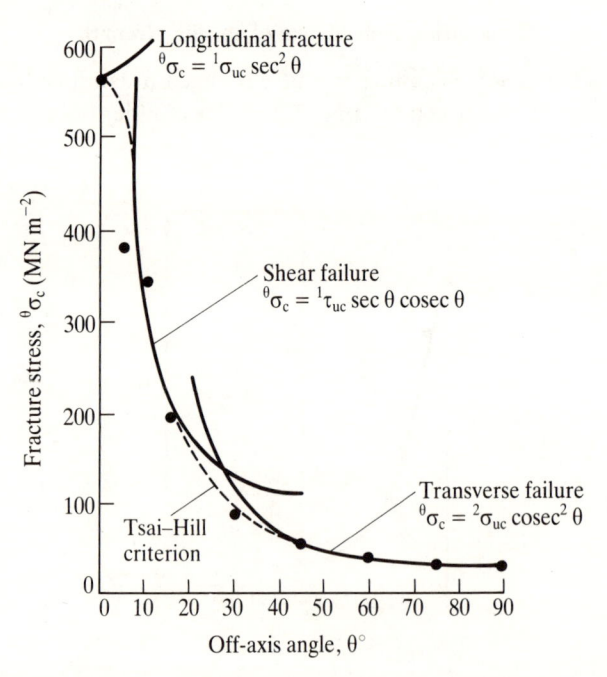

Figure 70 The angular dependence of fracture strength of a CFRP unidirectional lamina ($V_f = 0.5$). Experimental results are those of Sinclair and Chamis.[5] The dotted line is the Tsai–Hill criterion of equation 4.29 presented by Hull.[1]

failure curve shows a transition from longitudinal failure (4.26) to shear failure (4.28) at quite a small angle ($\sim 5°$) and a transition from shear failure to transverse failure (4.27) at quite a large angle (18–42°). This result is characteristic of most fibre–matrix combinations and once again the extreme anisotropy of behaviour is in evidence with the rapid fall off in strength with increasing value of θ. Over the range $\theta = 18$–$42°$, the maximum stress theory has limited predictive ability. The maximum work theory or Tsai-Hill criterion, also included in Fig. 69, is found to be more efficient in predicting the strength of a unidirectional lamina across the whole range of θ

$$^\theta\sigma_c = \left[\frac{\cos^4\theta}{^1\sigma_{uc}^2} + \left(\frac{1}{^1\tau_{uc}^2} - \frac{1}{^1\sigma_{uc}^2}\right)\sin^2\theta\cos^2\theta + \frac{\sin^4\theta}{^2\sigma_{uc}^2}\right]^{-\frac{1}{2}} \quad [4.29]$$

References

1. D Hull, *An Introduction to Composite Materials*, Cambridge University Press, Cambridge, 1981.
2. R M Jones, *Mechanics of Composite Materials*, McGraw-Hill, Washington DC, 1975.
3. S W Tsai, J M Patterson in *Composite Materials in Aircraft Structures*, ed. D H Middleton, Longman, Harlow, 1990, ch. 8.
4. S W Tsai, *Composites Design*, 3rd edn, US Air Force Materials Laboratory/ Think Composites, Dayton, Ohio, USA, 1985.
5. J H Sinclair, C C Chamis in Proceedings of the 34th SPI/RP Antec Conference, Society of the Plastics Industry, New York, 1979, Paper 22A.

4.5 Hybrid effect

M G BADER

The term hybrid effect has been used to describe the phenomenon of an apparent synergistic improvement in the properties of a composite containing two or more types of fibre. The most common use of the term applies to the tensile strength of the composite, which may be higher than would be predicted from a simple application of the rule of mixtures. This is a positive hybrid effect. A negative hybrid effect has also been reported.

A hybrid composite is most commonly one in which two or more types of fibre, e.g. glass and carbon, are embedded within a common matrix. Other forms of hybridization are also possible including hybrid matrix[1] and alternating metal/composite laminates such as Arall and Glare (registered trademarks of Akzo). This article is confined to discussion of the mixed fibre type of hybrid.

The general properties of hybrid composites have been reviewed by Short and Summerscales.[2] The initial concept was to optimize properties by using more than one type of fibre. The driving force is largely economic and the concept of using a relatively small amount of an expensive fibre, such as carbon, in a laminate composed mainly of a cheap commodity fibre, such as E-glass, is attractive. This allows for manipulation of various properties between the limits defined by the properties of an all carbon and an all glass fibre laminate of similar geometry. Early experiments confirmed the stiffness of such laminates conformed to the predictions of the rule of mixtures but the strength could be significantly higher than predicted. Hence, the introduction of the term hybrid effect. According to further work,[3] hybrid laminates in their crudest form often cannot be exploited effectively but may be useful for tailoring properties to service requirements.

There are several possible hybrid configurations. These are illustrated in Fig. 71, which depicts uniaxial hybrid laminates with alternating plies, interspersed tows and intermingled fibres. The alternating ply and interspersed tow configurations are the most commonly used. The cost of preparing intermingled fibres is generally too high for them to be cost effective. Additional possibilities are the use of woven cloths with one fibre type in the warp and another in the weft, and further variations using interspersed tows in woven configurations. Finally the plies containing different fibres, or mixtures of fibres, may be stacked with the fibres aligned in different directions, as in conventional multiaxial laminates.

For simplicity we will consider a composite laminate consisting of alternate uniaxial plies containing E-glass and high modulus carbon fibres at similar fibre fractions in a common matrix resin. Under uniaxial tension, parallel to the fibre direction, the stiffness of the laminate may be

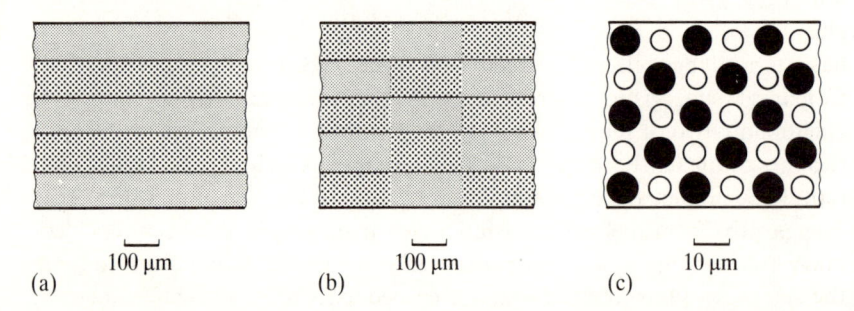

Figure 71 Three basic hybrid laminate configurations. (a) Alternating plies, (b) interspersed fibre bundles, (c) intermingled fibres

accurately predicted from the relative proportions of the two types of fibre

$$E_H = E_1 V_1 + E_2 V_2$$

Where E_H is the modulus of the hybrid laminate and E_1, V_1, E_2, V_2 are the respective moduli and fractions of the component laminae. Experimental measurements are found to fall very close to this prediction and there is no observed hybrid effect for stiffness.

The tensile strength relationship is more complex and cannot be predicted by simple application of a rule of proportionality, i.e. the broken line *AD* in Fig. 72. This is because the strains to failure of the two types of fibre are different. In the case illustrated, the failure strain of the high modulus carbon fibre is about 0.01 (1%), whilst that of the E-glass is 0.028. When a composite containing both types of fibre is extended in tension, the carbon fibres will fail over a range of strains around 0.01 but the glass will survive to the higher strain. However, it is also necessary to consider the relative proportions of the two fibres. If there is only a small proportion of carbon, the glass will be sufficiently strong to continue to bear increasing loads after failure of the carbon; failure of the composite will be at the strain associated with failure of the glass. The carbon will play no significant part in load bearing at strains above 0.01. However, at high proportions of carbon, there will be insufficient glass to carry the load; failure of the whole composite will occur at the load corresponding to failure of the carbon. In Fig. 72 the failure stresses of the all glass and the all carbon plies are the points *A* and *D*. The line *BD* is the stress at which the carbon plies would fail and the line *AE* that for the glass. The intersection *C* marks the change in failure behaviour. At glass contents below *C*, the carbon plies will be expected to fail when the applied strain reaches the failure strain of the carbon. At this point the stress in the composite will be given by the line *AC*, based on the sum of the loads in both carbon and glass plies at the failure strain of the carbon. At these glass contents, there is insufficient glass to carry the load after the carbon has failed, so the whole laminate will fail. At glass contents greater than *C*, the carbon will fail, as before, but there is now more than sufficient glass to bear the load and the stress may be further raised until failure of the glass occurs, as predicted by *CD*. A number of workers[2] have reported failure stresses above the predicted locus, i.e. *ACD* in Fig. 72. This is the hybrid effect and is illustrated by the experimental points taken from the work of Manders and Bader.[3] Note that these points are the stresses in the composite at which the carbon plies fail. The glass plies continue to bear load in the region bounded by *AC*. The basic common observation is that the failure strain of the component with the lower (or

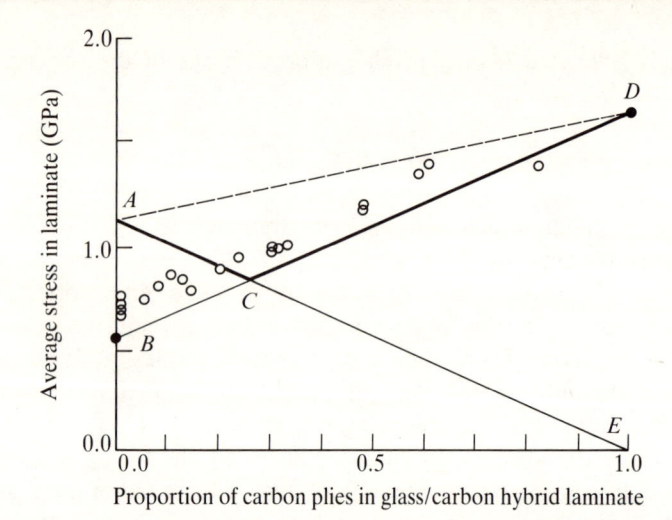

Figure 72 Illustration of the hybrid effect. The line AD is the simple rule-of-mixtures prediction. ACD indicates the predicted strengths based on the failure strains of the two components, and BD is the stress at which the least extendable (carbon) plies are expected to fail. The data points (Manders and Bader[3]), which lie above this predicted locus, indicate a positive hybrid effect

lowest) strain to failure is enhanced when that fibre is in the form of a hybrid.

There is no complete explanation of this phenomenon but there are several contributory factors. The simplest is the contribution of thermal internal stress. Carbon fibres have a small negative coefficient of thermal expansion in the axial direction, whereas glass and the resin both have much larger positive coefficients. When the two fibre types are set in a thermosetting resin, cured at an elevated temperature (usually 100–180°C), then cooled to ambient temperature, the carbon fibre will be forced into compression and axial tensile stresses will be developed in the glass. When an external extension is applied, the precompression in the carbon must first be reversed. This leads to an apparent enhancement in the strain to failure. The effect will be greatest when the proportion of carbon is small but this effect alone is insufficient to account for the whole of the observed hybrid effect.[3]

A more plausible explanation is based on the statistics of failure of fibres and fibre arrays.[3] It is well known that the average strength of short lengths of fibre is greater than that of longer segments. This is based on the relative probability of encountering a severe flaw, which increases as the size of the sample is increased. Likewise, arrays of a small number of fibres are found to be stronger than larger arrays. Therefore, if the high-modulus, low strain-to-failure fibre is subdivided into thin laminae or ligaments and interspersed with a fibre of higher strain to failure, we may expect the average strength of the low strain phase to be enhanced. The effect may be expected to be greater if the low strain fibre is in the form of very thin plies or dispersed ligaments. This is in accord with experimental observations.[3] The mechanics of hybrid composites are discussed in greater detail by Chou.[4]

A further factor concerns the actual mode of failure. In a uniaxial composite containing only one type of fibre, failure in tension is attributed to progressive failure of the component fibres. The fibres have a wide distribution of strengths, so that as the strain is increased, individual single fibres will fail sporadically throughout the composite. At each failure site, there will be a perturbation of the local stress distribution with stress concentration on surviving fibres close to the fracture site. This leads to an enhanced probability of further fibre failures close to previous failure sites, so that regions of several failed fibres develop.[5] Eventually, one of these regions reaches a critical size and a catastrophic succession of fibre breaks travels across the section, leading to complete tensile failure. Now if, in a hybrid, the low strain component is dispersed as a number of thin plies or ligaments, this process will be (temporarily) contained by the higher strain component. This may enhance the strength. It will also radically change the mode of failure. Instead of failure occurring suddenly and catastrophically, there will be progressive ligament or ply failure with associated debonding or delamination. This is often perceived to be a more damage tolerant mode.

The hybrid effect will be greatest when the proportion of the low strain fibre is small and intimately dispersed within the higher strain to failure component. However, the optimum toughness or damage tolerance will generally require a coarser dispersion. This limits the extent to which the effects may be exploited. Clearly, if only a small proportion of a fibre is used, although its failure strain may be substantially enhanced, the effect on the overall properties of the laminate will be small. The more important outcome of the work on hybrids is the recognition that they may be used to control the mode of failure. This has led to the commercial exploitation of mixed fibre hybrids, especially in components where impact and fracture are critical design considerations.

References

1. M A Leaity, P A Smith, M G Bader, P T Curtis, *Composites*, **23**, 387 (1992).
2. D Short, J Summerscales, *Composites*, **10**, 215 (1979); **11**, 33 (1980).
3. P W Manders, M G Bader, *J. Mater. Sci.*, **16**, 2233 (1981); **16**, 2246 (1981).

4. T W Chou, *Microstructural Design of Fiber Composites*, Cambridge University Press, Cambridge, 1992.
5. S B Batdorf, R Gaffarian, *J. Reinf. Plas. Comp.*, **1**, 165 (1982).

4.6 Interfacial bond strength determination – critique

I VERPOEST

The bond strength between the reinforcing fibres and the surrounding matrix is of crucial importance for many mechanical and physical properties of composites. Interface cracks initiate global damage and hence deteriorate the composite. During transverse impact, however, the creation of interface cracks can in some cases dissipate part of the impact energy, and in this way inhibit more serious damage such as delaminations or fibre fractures. In other words, the interface strength is a controlling parameter for other composite properties and the optimum value of the interface strength will be specific for each fibre–matrix combination and for each application.

Optimizing the interface strength requires reliable interface strength data and measurement systems. The difficulty to achieve this is obvious: most fibres used in composites have a very small diameter: it ranges from 5 μm for the new generation carbon fibres to 20 μm for some thicker glass fibres. Measuring the adhesion between these tiny fibres and a polymer or metal in a direct way can only be done when micromechanical techniques, especially developed for this purpose, are used. In these tests, where single fibres are embedded in matrix, the load conditions are such that the stress state at the interface is well controlled and failure will precisely occur at the interface. But without exception, they are laborious and time consuming. Many researchers have tried to develop more handy test techniques on a macro scale. In these test techniques, a macromechanical composite property is measured; this is assumed or proved to be related to the interface strength. Although experimentally more easy, these techniques give indirect information about the interface properties and never generate a pure interface strength, hence their results cannot be used as absolute interface values (i.e. in micromechanical models), they only have a relative qualitative value.

Apart from the direct–indirect distinction, interface strength tests can also be discriminated according to the loading condition applied at the fibre–matrix interface. In practice, interface failure occurs mostly under transverse, and under shear loading, and it seems obvious to simulate these loading conditions both during the direct and the indirect tests. This is perfectly possible for the indirect test methods. The interface strength under transverse loading conditions can be (qualitatively) derived from transverse bending tests on unidirectional specimens, or more accurately from tension tests on crossply laminates. The interface shear strength is related to the shear strength of a unidirectional laminate as measured by the short beam shear test or the Ioscipescu test.

For direct interface test methods, the problem is more complicated. When a single fibre, embedded in a piece of matrix, is transversely loaded, the local stress condition around the circumference of the fibre will be a variable combination of normal and shear stresses; moreover, thermal stresses have to be taken into consideration. A normal interface strength derived from these types of test will only be valid when the interface failure initiates at the point of pure normal stress. Under shear loading conditions, the picture seems clearer as pure shear stresses are present over the whole circumference of the fibre. However, the real stress state along the fibre–matrix interface is quite complex as thermal mismatch between fibre and matrix creates radial normal stresses and as the load introduction in the pull-out and in the micro-indentation test creates an uneven stress distribution along the fibre.

In reality, for the direct interface test methods, the practice of interface strength measurement is more straightforward than the above comments would suggest. Simplified assumptions on the local stress conditions are used in most of the direct test methods but this can lead to erroneous conclusions, as will be shown for the fragmentation test in Section 4.7.

Indirect test methods

The first category of indirect test methods uses unidirectional specimens. The most widely used is the short beam shear test (Fig. 73a). When in a three-point bending test, the span to thickness ratio is reduced to 4 or 5, the shear stress at the mid layer becomes so important that a shear failure occurs in that mid layer, instead of tension failure in the outer layer of the beam. It has been proven by a large number of researchers that the short beam shear strength is well related to the fibre–matrix interface strength. But be warned, the resulting short beam shear strength is highly influenced by the matrix properties, hence an absolute interface shear strength cannot be derived from these experiments. Moreover, the initiation of the shear crack can be influenced by the stress concentrations under the central loading point. Tests can only be considered as valid if no indentation damage is observed.

A more complicated alternative is the Ioscipescu shear test (Fig. 73b), in which the shear stress is concentrated in an area in between the notches. However, the shear stress, assumed constant over this area, is not constant. As the stress profile depends both on the specimen

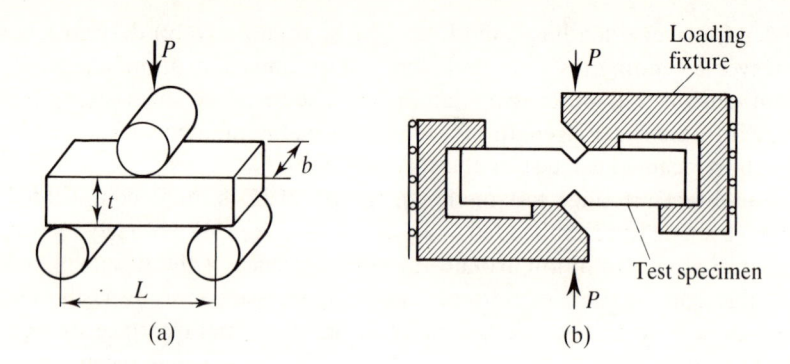

Figure 73 (a) Short beam shear test. (b) Ioscipescu test

geometry and on the elastic properties of the material, extreme care must be taken in interpreting the results.

Angle ply or crossply laminates can also be used to measure indirectly the interface strength. The tensile strength of ±45° specimens is often used for this purpose, but the results are not very reliable for two reasons. First, interface failure can already start at stresses much lower than the ultimate stress, and it is often very hard to detect their initiation. Second, the stress state at the interface is not pure shear; the ratio of normal to shear stress can be reduced by decreasing the angle of the $(\pm\theta)_s$-laminate; the optimum depends on the elastic properties of the material.

The methods mentioned up to now aim at measuring (indirectly) the interface shear strength. Methods for measuring the interface normal (or radial) strength are less developed. Transverse three-point bend tests are possible, but the results (both mean value and scatter) are highly dependent on the accuracy and repeatability of the specimen preparation. Moreover, a very large number of specimens is required.

More recently, a specific tensile test on crossply laminates has been developed; this allows one to derive the transverse strength of a defect free 90°-layer and gives an indication of the optimum interface strength. For this purpose, the stresses at which the transverse cracks initiate have to be detected accurately using piezoelectric transducers. The method is rather laborious, both from an experimental and from a data reduction point of view.[1]

Direct test methods

The disadvantage of the indirect test methods, namely that the result can never be considered as pure interface characteristics, can be overcome by the direct test methods. For interface shear loading, three test methods are currently used.[2]

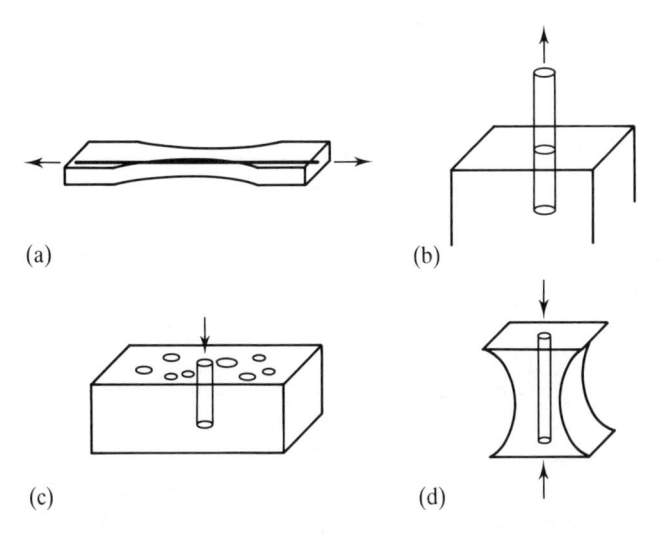

Figure 74 (a) Fragmentation test (b) pull-out test (c) microindentation test, (d) transverse debond test

In the fragmentation test, a single fibre is embedded in a specimen of pure resin, which should have a high strain to failure (Fig. 74a). A tension load is applied parallel to the fibre. When the fibre failure strain is reached, the fibre starts fracturing. The fibre fragment length is related to the interface shear properties. Although simple formulas have been proposed to quantify this relation, the correct data reduction scheme is quite complicated and should take into account other features like debond length, the existence of matrix cracks at the broken fibre ends, residual stresses, the statistical nature of the fibre strength, etc. (see Section 4.7).

The second method seems really straightforward, as it consists of pulling a single fibre out of a plate, bubble or drop of resin (Fig. 74b). The mean pull-out shear stress can be easily calculated if the embedded fibre length is known. Apart from experimental problems, how to grip a single fibre of less than $10\,\mu m$ in diameter, difficulties arise in the calculation of the critical stress condition. Indeed, where the fibre leaves the matrix, shear stresses are locally higher, and radial stresses are superimposed. One way to avoid this problem is to perform pull-out tests at different embedded fibre lengths then extrapolate the mean shear stress to a zero fibre length. This procedure is widely adopted. It does have a limitation for fibre–matrix systems with high interface strength, because in this case the embedded length is limited by the fibre strength.

A third method is the opposite of the previous one. The fibre is pushed into the matrix, and eventually the interface will fail under shear (Fig. 74c). For this test, a well polished section perpendicular to the fibre is needed. The great advantage is that, in contrast to the three previous direct test methods, real composite specimens can be used. Hence, the influence of processing or in-service conditions on the interface shear strength can be quantified. Apart from the stress concentrations, similar to the pull-out test, the main drawback of this method lies in serious experimental difficulties; for some systems, it can be difficult to apply the compressive load exactly in the fibre centre, to measure the load at debonding and to avoid splitting of the fibre. To avoid splitting is difficult for high modulus carbon fibres and almost impossible for polymeric fibres.

For the determination of the transverse interface strength, only one method is used.[3] Proposed in the 1960s, not much progress in its experimental set-up has been reported since then. The specimen (Fig. 74d) is a block of resin, containing a single fibre or a fibre bundle. Parallel to the fibre, a compressive load is applied. Due to the difference in the Poisson's ratio and the Young's modulus, this is translated into tensile stresses normal to the fibre–matrix interface. Thus the transverse interface strength can be directly measured if the failure initiation can be detected. This is the difficult part of the test and explains why it is rarely used.

Conclusions

As a conclusion, one could state that the ideal interface strength measurement method still has to be invented. Indirect methods are experimentally not too difficult but they do not generate absolute interface strength values. Moreover they are influenced by the matrix properties. Direct test methods are always laborious and time-consuming but they can generate absolute results if the correct interface stress state is taken into account. They require accurate and easy to apply data reduction schemes, but so far this need remains unfulfilled. Meantime, it is in the interests of the whole composites community to document interface shear strength experiments as completely as possible, so that experimental results from different authors can be compared and some differences can be traced back to variations in experimental set-up. Indeed, a recent round robin exercise has shown the importance of interlaboratory variations in interface shear strength data measured by direct test methods. The practical aspects of these test methods have recently been reviewed.[4]

References

1. P W M Peters, S I Andersen, *J. Composite Mater.*, **23**, 944 (1989).
2. M Desaeger, I Verpoest, R Keunings in *Controlled Interphases in Composite Materials*, (ICCI 3) ed. H. Ishida, Elsevier, New York, 1990, pp. 653–666.

3. L J Broutman, *Interfaces in Composites*, ASTM STP 452, American Society for Testing and Materials, Philadelphia, 1969, pp. 27–41.
4. M J Pitkethly, I Verpoest, M Desaeger, *et al.*, *J. Composite Sci. Tech.*, **48**, 205–14 (June 1993).
5. P J Herrera-Franco, L T Druzal, *Composites*, **23**, 2 (1992).

4.7 Interfacial bond strength determination – fragmentation

I VERPOEST and M DESAEGER

In the fragmentation test, a single fibre is embedded in a resin specimen. The specimen is loaded in tension parallel to the fibre. When the failure strain of the embedded fibre is smaller than the failure strain of the matrix, the fibre will begin to fracture. During the fragmentation test the strain in the specimen is gradually increased; the strain in the fibre will also increase and reach a new critical fibre strain characteristic of the new fibre fragment length. In this way, the fibre fragments will be fractured again and the process repeats itself. As the strain (or stress) in the fibre is introduced by shear stresses along the fibre surface, the maximum attainable fibre stress must be limited by the fibre length, on one hand, and the maximum achievable shear stress at the interface, on the other.

Experimentally, one can observe that during a fragmentation test, a saturation fragment length or critical fibre length, L_c, is reached at a certain applied strain; further application of loads will not create further fibre fractures. It can hence be assumed that an equilibrium is reached between interface shear stress and fibre strength, at a particular fibre fragment length.

The interface shear stress can be derived from the observed fragment length if one assumes that its absolute value is constant over the whole fragment length (the sign changes at mid fibre length, see Fig. 75). Based on simple equilibrium of forces, one can derive that in each point along the fibre

$$\tau_i = \frac{-d_f}{4} \cdot \frac{d\sigma_f}{dx} \qquad [4.30]$$

where d_f = fibre diameter.

If τ_i is constant, integration over the fibre length, L, yields

$$\tau_i = \frac{\sigma_{f,max}}{2} \cdot \frac{d_f}{L} \qquad [4.31]$$

For further discussion of this analysis, see Section 4.14.

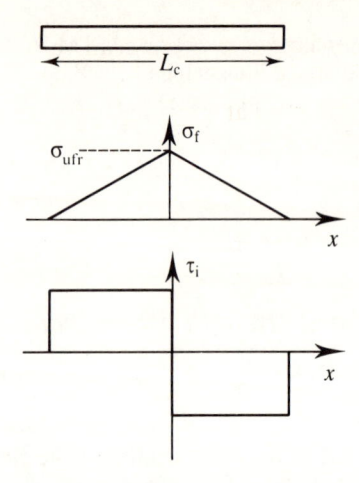

Figure 75 Shear stress and tensile stress distribution in the fibre when the critical length is reached

If we assume that saturation during the fragmentation test occurs when the maximum fibre stress just does not equal the fibre strength, the above formula, where $L = L_c$ and $\sigma_{f,max} = \sigma_{ufr(Lc)}$, the strength of a filament of length L_c, can be used to calculate the interface shear stress. As the interface shear stress is assumed to be constant and characteristic of the fibre–matrix interface, the obtained value is taken to be the interface shear strength. This approach is used by most authors but, as will be shown later, can only be considered as a first, rough estimation.[1]

Although the test seems straightforward, several difficulties both experimental and theoretical have to be taken care of.

First, during specimen preparation, one single fibre has to be taken out of the fibre bundle and the actual fibre diameter has to be measured. The fibre has then to be placed in a mould and the resin has to be poured over it. Wrinkling of the fibre can occur during inaccurate mould filling and during curing. In thermosets, cure shrinkage can be so important that small diameter fibres (like the intermediate modulus carbon fibres) buckle under the longitudinal compressive stresses. The buckling tendency is enhanced when the curing cycle is not optimized and when silicone moulds are used, as they heavily shrink during cooling and exert compressive forces on the specimen. This effect can be avoided by adapting the cure cycle, by reinforcing the silicone mould with a steel wire grid, by using stainless steel moulds or by attaching small lead balls at the fibre ends, so the fibre is strained continually.

Second, the fragment lengths at saturation have to be measured. When the matrix is transparent, the fragmentation process can be followed with a light microscope and the fragment length as well as other features (debonding, matrix cracking or yielding) can be monitored as a function of the applied strain or load. If, however, the matrix is not transparent, it has to be chemically dissolved at a certain strain before the fragments are counted. Acoustic emission can be used to count the fibre breaks but does need careful calibration.

In most fibre–matrix systems, the fragment length will reach saturation when the applied strain is approximately equal to three times the mean fibre failure strain. As a consequence, the failure strain of the matrix should be at least three times that of the fibre.

Third, the fibre strength at the critical length, $\sigma_{ufr(Lc)}$, has to be known. Indeed, the mean fibre strength is a function of the fibre length. As the fragment lengths range from a few tenths of a millimetre to a few millimetres, experimental data are not easily available, and extrapolation from tests on longer fibres have to be carried out. This invariably involves direct measurement of the strength statistics of a large number of fibres of differing lengths or of a fixed length and the application of Weibull statistical methods for the calculation of $\sigma_{ufr(Lc)}$. Methods have recently been suggested to derive the fibre strength indirectly from the fibre fragment length distribution.[2]

Concerning the data reduction route, it was already mentioned that the formula suggested to calculate the interface strength is oversimplified. In fact, when the stress transfer between fibre and matrix is fully elastic, the shear stresses are not constant (Fig. 76a), and (4.31) cannot be used.

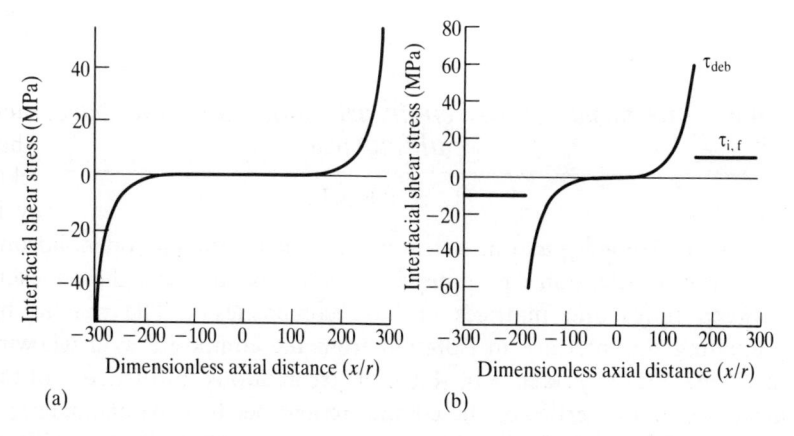

Figure 76 (a) Stress distribution predicted by the elastic model, (b) stress distribution predicted by taking into account the debonding phenomena

Moreover, two other phenomena often happen at the fibre ends. Either the matrix starts cracking under the influence of the stress concentration created by the broken fibre ends or the peak shear stresses create a decohesion at the fibre–matrix interface. In none of these cases, the interface stress state is constant along the fibre length, hence the assumption on which (4.31) is based is violated. For the case where debonding occurs, the interface stress follows the profile depicted in Fig. 76b. A procedure has recently been suggested in which the interface debond strength and the interface friction stress can be derived from the fragment length and the debond length. This new procedure is actually verified.[1]

As a conclusion, one can state that the fragmentation test when carried out correctly cannot directly yield one value for the interface strength. The saturation fragment length in itself is a reliable parameter which for fibres with equal strength gives a good indication for the stress transfer capability of the interface. Absolute values of the two 'interface strengths' (i.e. debond strength and frictional stress) can only be obtained when the actual stress state is taken into account. When using (4.31), one should be aware that the basic assumptions for it are violated in reality and that the obtained 'interface strength' will lie somewhere between the real interface debond strength and the interface frictional stress.

References

1. M Desaeger, T Lacroix, B Tilmans, R Keunings, I Verpoest, *J. Composite Sci. Tech.*, **43**(4), 379 (1992).
2. H D Wagner, A Eitan, *Appl. Phys. Lett.*, **56**(20), 1965 (1990).

4.8 *Interfacial bond strength determination – microdebond pull-out test*

L S PENN

The single filament pull-out test, sometimes called the microdebond test, has received attention for some years as a way to assess the adhesion between fibres and matrices in fibre composites.[1,2] The practice of evaluating the interface in isolation from the composite as a whole is justified; the interface can have a tremendous influence on the macroscopic properties of the composite and needs to be characterized independently in some way. The single filament pull-out test provides a direct measure of interfacial adhesion and can be used with both brittle and ductile matrix resins.

Interfacial bond strength – microdebond test

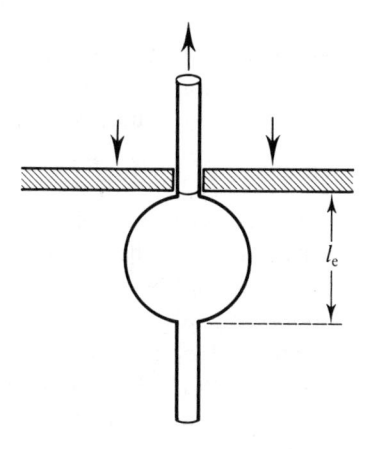

Figure 77 Geometry of the single filament pull-out test. Detailed geometry of the matrix may vary, i.e. from spherical to cylindrical to cubic

Figure 77 shows a widely used test configuration; the matrix is a sphere of resin, deposited as a liquid onto the fibre and allowed to solidify. The top end of the fibre is attached to a load-sensing device, and the matrix is contacted by load points affixed to the crosshead of a load frame or other tensioning apparatus. When the load points are made to move downward, the interface experiences a shear stress that ultimately causes debonding of the fibre from the matrix.

Figure 78 shows a typical load versus displacement curve for an individual specimen. The single most important piece of information from the curve is the maximum load, P_m, which, once determined, can be used in different computations (see below). The sudden drop in load after P_m corresponds to an instantaneous vertical displacement of fibre with

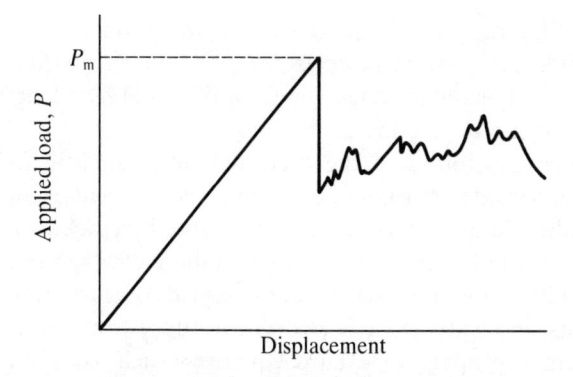

Figure 78 Typical load versus displacement curve for an individual single filament pull-out test

respect to matrix upon completion of debonding.[3] The lower load thereafter is simply due to the friction of the debonded fibre being pulled up through the matrix. The frictional load is proportional to original embedded length and declines to zero as the fibre end exits the matrix completely.

The analysis that should be applied to the single filament pull-out test depends on whether interfacial failure occurs by yielding or by crack propagation, a distinction that cannot be made from the curve in Fig. 78. The simplest and most popular analysis has been based on interfacial yielding, where the shear stress is assumed to be distributed uniformly over the interface from top to bottom. According to this, the interfacial shear stress increases uniformly until, suddenly, every location in the interface gives way simultaneously. For this type of failure, it is justifiable to compute an interfacial shear strength, τ, by dividing maximum load, P_m, by original interfacial area, A.

On the other hand, a more complicated analysis is required if specimen failure occurs by propagation of a crack that starts at the top and travels downward until the interface is completely debonded. In this situation, the stress distribution at the interface is decidedly non-uniform and is usually approximated by a hyperbolic trigonometric function.[4] Fracture mechanics analyses using the energy balance approach have been applied,[1,3] resulting in equations that relate P_m to specimen dimensions, elastic constants of fibre and matrix, initial crack length, and interfacial work of fracture, W_i. One such equation is

$$P_m = \frac{2\pi r_f \sqrt{r_f W_i E_f}}{\sqrt{1 + \operatorname{cosech}^2\left\{\left[\frac{E_m}{E_f(1+\nu_m)\ln(R/r_f)}\right]^{1/2}\frac{(l_e - a)}{r_f}\right\}}} \quad [4.32]$$

where r_f = fibre radius, R = matrix radius, E_f = fibre modulus, E_m = matrix modulus, ν_m = matrix Poisson's ratio, W_i = interfacial work of fracture, l_e = original embedded length of fibre and a = length of small initial crack at top of interface.

Other equations found in the literature differ slightly from (4.32), reflecting the various approximations made by different authors to deal with the multitude of factors involved in failure by crack propagation.

Although it would seem that a computation of τ, as described for specimen failure, by interface yielding would be far more desirable because of its simplicity, there is evidence to suggest that failure actually occurs by crack propagation. Thus equations such as (4.32) are more appropriate, whereas computation of an interfacial strength, τ, is strictly speaking not correct. Equation (4.32) and its variants provide for the computation of W_i, a quantity characteristic of the interface alone and independent of geometry and elastic constants.

Further insight into the failure mode is provided by plotting P_m versus embedded length, l_e. For failure by yielding, one would expect P_m to be directly proportional to embedded length, exhibiting linear behaviour for all l_e. For failure by crack propagation, one would expect P_m to rise non-linearly from zero then reach a plateau indicated by (4.32). The typical pull-out test data in Fig. 79 display a plateau,[3] thereby supporting the identification of crack propagation as the failure mode: of W_i, not τ, as the meaningful quantity. Unfortunately, the computation of W_i is not without problems; P_m and l_e are determined by direct measurement but values for ν_m are often unavailable and values for initial crack length, a, are practically impossible to determine, due to the small size of the specimens.

Fortunately, the above described difficulties do not prevent the use of the single filament pull-out test for comparative evaluation of interfaces. A second look at the rising region of the plot in Fig. 79 shows that, given the scatter in the data, it is indistinguishable from a linear plot. This implies that the slope of the rising region (equivalent to an averaged P_m/A) can be used as a performance index for different interfaces (e.g. those obtained by different fibre surface treatments) from a single fibre–matrix pair. An example is given in Table 44, which compares aramid fibre–epoxy matrix interfaces modified by attachment of molecular chains to the fibre surface.[2]

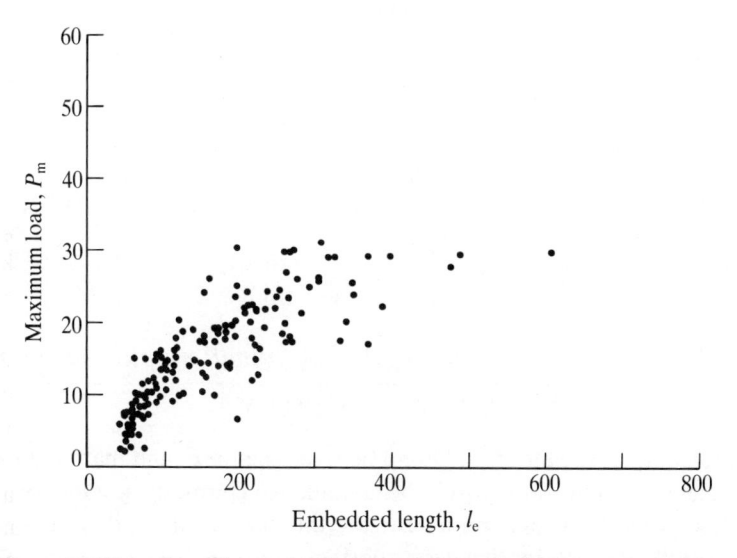

Figure 79 Plot of maximum load, P_m, versus embedded length, l_e where each point represents an individual test. The data appears to rise and then to reach a plateau

Table 44. Comparison of interfaces derived from different fibre surface treatment for aramid fibre–epoxy resin matrix

Chains attached by fibre surface treatment	Performance index, P_m/A (MPa)*
None	21.6 ± 0.51
–(CH$_2$)$_6$–NH$_2$	25.8 ± 1.01
–[(CH$_2$)$_6$–NH–C(=O)–NH]$_2$–(CH$_2$)$_6$–NH$_2$	30.7 ± 0.79
–[(CH$_2$)$_6$–NH–C(=O)–NH]$_6$–(CH$_2$)$_6$–NH$_2$	35.6 ± 1.26

*Average ± standard error mean.

The scatter in single filament pull-out data is typically high, sometimes reaching 30%. A cause for concern among researchers, it has been attributed to such things as inherent variability in the specimens[3] or uncontrolled positioning of the load points on the specimen.[5] However, the testing techniques in common use today are sufficiently consistent to keep load point variation low within a given laboratory. Thus, it is likely that most of the observed scatter arises from inherent variability of the fibre, matrix or interface.

References

1. P S Chua, M R Piggott, *Composites Sci. Tech.*, **22**, 107 (1985).
2. C T Chou, L S Penn, *J. Adhesion*, **35**, 127 (1991).
3. L S Penn, S M Lee, *J. Composites Tech. Res.*, **11**, 23 (1989); **12**, 164 (1990).
4. L B Greszczuk in *Interfaces in Composites*, ASTM STP 452, American Society for Testing and Materials, Philadelphia, 1969, pp. 42–58.
5. P J Herrera-Franco, V Rao, L T Drzal, M Y M Chiang, *Composites Engng*, **2**, 31 (1992).

4.9 Interfacial bond strength determination: pull-out test

M J PITKETHLY

The strength of the interfacial bond between the fibres and matrix is a key parameter in controlling the behaviour of composites. One of the methods used to assess and investigate the interfacial bond is the single fibre pull-out test. In this test,[1] a single fibre is taken and partially embedded in a drop of uncured resin placed on a holder. The resin is then cured with the fibre held upright. The holder, with resin and fibre, is placed in a universal tensile test machine. The holder is secured; the fibre is held in a grip attached to the crosshead then pulled out from the resin. The load and, if possible, the displacement are recorded during the test. If the fibre extends through the block, this is the microdebond test covered in Section 4.8. The basis for all subsequent analyses was that of Greszczuk,[2] whose model is shown in Fig. 80. He utilized the shear lag equations so that the force pulling the fibre out of the resin is balanced by the shear stress at the fibre–resin interface holding the fibre in place. His analysis showed that the maximum shear stress occurs as the embedded length tends to zero, and is given by

$$\tau_{max} = \tau_{av} \alpha l_e \coth(\alpha l_e) \qquad [4.33]$$

where $\tau_{av} = F/2r_f l_e$ and $\alpha = (2G_i/br_f E_f)^{1/2}$.

F is the pullout force, r_f the fibre radius, l_e the embedded length, G_i the shear modulus of the interface, b the effective width of the interface and E_f the tensile modulus of the fibre. The quantity α is a constant for a particular fibre–matrix system and is unknown since neither b or G_i are known, but τ_{max} may be estimated either graphically by plotting τ_{av} against l_e and extrapolating to zero embedded length, or analytically using the analysis of Pitkethly and Doble,[1] whereby if certain conditions are met, α can be eliminated and τ_{max} found using only F, r_f, and l_e.

A number of factors are not considered in this analysis, such as residual radial thermal stresses, the effect of bonding over the end of the fibre and the effect of a meniscus at the point of entry of the fibre into the

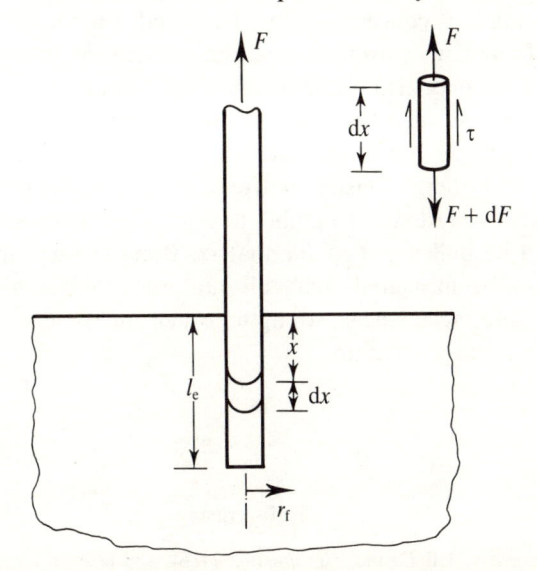

Figure 80 Model for matrix stresses around a partially embedded fibre (after Greszczuk[2])

matrix. Various extensions to Greszczuk's theory have been made to accommodate these effects but practical data reduction schemes utilizing these extensions have not yet been developed. The embedded length has a significant effect on the value for τ_{max}; as the embedded length becomes smaller, the value of τ increases so the widest possible range of embedded lengths should be measured, in order to produce a result with confidence.

Depending on the fibre–matrix system, a wide range of information may be obtained from this test. Four possible regions on a load–displacement trace have been identified that may be associated with different interfacial phenomena; an elastic region, a plastic deformation region, a region in which dynamic recoil of the fibre occurs after fibre–matrix debonding and a region where frictional forces predominate and a stick–slip mechanism may occur. Depending on the fibre–resin system under investigation, not all these regions may be present or identifiable. In systems where strong stiff fibres have a strong interfacial bond, failure is often catastrophic and only a value for τ_{max} may be obtained. At long embedded lengths, fibre failures may occur within the embedded length, resulting in a high apparent pull-out load for a short embedded length; this will distort the calculated value for τ_{av}. As the bond strength becomes greater, the ability to successfully pull a fibre from the matrix drop without the fibre failing becomes increasingly harder and problems may be encountered when testing systems with very high interfacial shear strengths. In other systems, where the interfacial bond strength is weak, debonding may occur before the fibre is pulled out of the matrix. In this case, information concerning the frictional properties may also be obtained. In systems with thermoplastic matrices or low modulus fibres, plastic yielding may also occur before bond failure.

A wide range of fibre–matrix combinations may be evaluated, including carbon, glass, aramid and polyethylene fibres, in both thermoset and thermoplastic matrices. The limiting factor on which materials may be tested is the ability to grip the fibre sufficiently to enable the fibre to be pulled out of the matrix. Because very small embedded lengths may be measured, the pull-out test provides a value for the maximum interfacial shear strength; other micromechanical methods evaluate an average strength.

References

1. M J Pitkethly, J B Doble, *Composites* 21(5), 389 (1990).
2. L B Greszczuk, *Interfaces in Composites*, ASTM STP 452, American Society for Testing and Materials, Philadelphia, 1969, pp. 42–58.

4.10 Laminates – angle plied

S L OGIN

Laminates: crossply, angle ply

In most structures, load is applied in more than one direction. A single composite lamina or ply has excellent mechanical properties, tensile strength and Young's modulus in the fibre direction, but poor properties transverse to the fibres (see Sections 4.2 and 4.3). Hence the need for using a laminate in which the fibres in each layer of the composite are oriented in different directions.

The laminate is usually constructed from thin sheets (about 0.125 mm thick) of partially cured resin (prepreg) containing continuous, unidirectionally aligned fibres. It is fabricated by bonding the prepreg during the curing of the resin under a combination of heat and pressure (see Section 3.12). Laminates can also be manufactured from prepreg containing short aligned or randomly oriented fibres. In addition, it is possible to vary the fibre type, and even the matrix properties, from ply to ply to produce a hybrid laminate. The following discussion will be restricted to laminates with plies having continuous unidirectional fibres of the same type in a common matrix.

The proportion of fibres in any direction within the laminate is chosen on the basis of the magnitude and direction of the applied loads. Additional requirements need to be considered because the behaviour of the laminate depends on the sequence of plies. Laminates are classified according to whether they are balanced (i.e. have as many $+\theta$ as $-\theta$ plies), symmetric (i.e. symmetric about the mid-plane in both geometry and material properties), or both. A laminate that is not symmetric will tend to warp on cooling after fabrication (see Section 4.12). More surprisingly, and in contrast to isotropic materials, laminates can show unexpected coupling between loading and response. For example, a laminate that is not balanced can show an in-plane shear when subjected to a simple in-plane tensile load, not a response expected of most engineering materials. A laminate that is not symmetric will tend to bend out of plane when subjected to a simple in-plane tensile load (see Section 5.14). In most applications, laminates that are both balanced and symmetric are used. But in some circumstances, designers make use of the special flexibility in design which composites can provide.

A convention is used to specify the laminate stacking sequence. A ply with fibres parallel to the long dimension is a 0° ply. A ply with fibres at an angle θ to this direction is a θ ply. Individual adjacent plies are separated by a slash if their angles are different, or by a multiplying subscript if their angles are the same. The plies are listed in sequence from one laminate face to the other. In the case of a symmetric laminate, the code for one-half of the laminate is given, with the subscript S outside the bracket indicating that the other half is symmetric about the mid-plane. For example, the sequence 0°, 0°, +45°, −45°, 90°, 90°, −45°, +45°, 0°, 0° becomes $[0_2/\pm 45/90]_S$. See Agarwal and Broutman,[1] for example, for further details of the convention.

In angle ply laminates, the angle of fibres in adjacent plies alternates. Crossply laminates are a special case of angle ply laminates in which the fibres are aligned in the 0° or 90° directions.

Constraint

The response of a laminate to load depends on its geometry and material properties. A crossply laminate will be discussed here for simplicity. Under tensile loading, which is the simplest case, the features of significance will be the initial modulus of the laminate, the stress or strain at which damage initiation occurs, the residual properties of the damaged laminate if unloaded before failure, and the tensile strength (Fig. 81).

The initial modulus can often be estimated using a rule of mixtures approach based on the effective modulus of each ply in the loading direction and the volume fraction of each type of ply. However, this

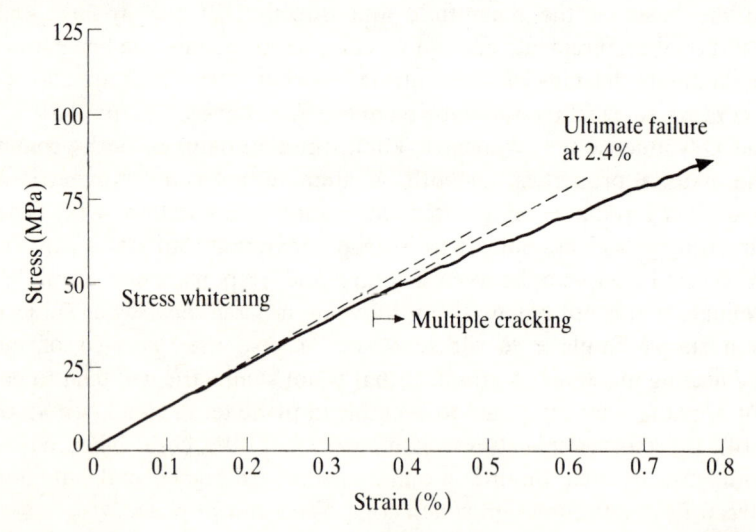

Figure 81 Low strain portion of stress/strain curve for 0°/90°/0° cross-ply laminate. Changes in gradient are associated with a rapid increase in stress whitening, and with the beginning of multiple cracking (after Manders et al.[7] Reproduced by permission of Chapman & Hall Ltd, London.)

approach takes no account of the orientation dependence of Poisson's ratio from ply to ply and discrepancies with measured values can be serious, particularly in angle ply laminates. Laminated plate theory takes the change in Poisson's ratio into account to give a better prediction of the undamaged laminate modulus.[2]

Generally, the first observable damage to occur is matrix cracking, i.e. cracking of the plies at an angle to the loading direction. In this type of damage, the cracks run parallel to the fibre direction and extend across the thickness of the ply. Matrix cracking is preceded by the development of fibre–matrix debonds but these are difficult to detect without careful microscopy (the 'stress whitening' in GFRP laminates) and do not affect the macroscopic stress–strain response significantly. Matrix cracks, on the other hand, usually give rise to a pronounced 'knee' in the stress–strain curve as a consequence of the reduction of the load-carrying capacity of the ply. This knee is obscured in laminates where the contribution of the 90° plies to the laminate modulus is low (particularly in CFRP laminates with thin transverse plies).

Figure 81 shows a typical load–strain curve for a crossply glass–epoxy resin GFRP laminate with a thick transverse ply. The onset of matrix crack formation (the transverse ply cracking threshold strain) at about 0.4% strain coincides with a knee in the stress–strain curve. Further straining produces an accumulation of cracks (see Fig. 66) and a reduction in the laminate modulus. Under tensile loading, crack accumulation saturates, with the crack spacing approximately the same as the transverse ply thickness; this is the 'characteristic damage state' or CDS.[3]

The transverse ply cracking threshold strain is an important parameter since it determines the onset of macroscopic damage in the laminate and is sometimes used as a design strain. Surprisingly, it is not independent of the ply thickness. This is a consequence of constrained cracking and it is of technological importance. The strain for the onset of matrix cracking increases with decreasing ply thickness in crossply[4] and angle ply laminates.[5] However, at 'large' ply thicknesses (typically >0.5 mm) in crossply GFRP laminates, a plateau occurs where the threshold strain for cracking is equal to the strain to failure of a 90° ply tested alone, after correction for thermal strain (see Section 4.12).

The reason for the phenomenon of constrained cracking can be understood using a fracture mechanics argument[6] based on the stress intensity factor associated with a flaw in the transverse ply and the fracture toughness of the ply. In an unconstrained ply (large ply thickness), debonds between fibre and matrix initiate and link to form a flaw during loading (Fig. 82). The flaw grows by the linking of debonds to a size sufficient that the combination of flaw size and stress in the

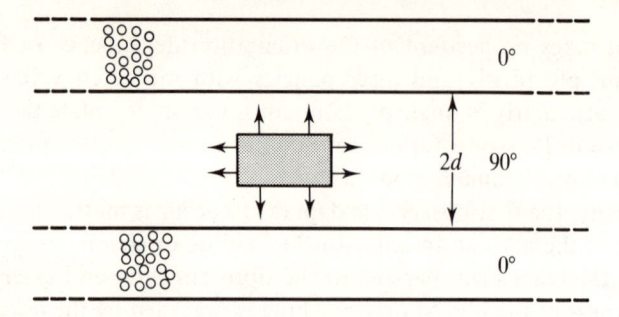

Figure 82 A flaw in a transverse ply (ply thickness, 2d) viewed from the loading direction

transverse ply is greater than the fracture toughness of the ply. At this point, the flaw grows catastrophically into a transverse ply crack spanning the ply thickness and the ply width. In the case of a constrained (thin) ply, the extent of growth across the ply thickness by the linking of debonds is limited by the adjacent 0° plies. More strain has to be applied to achieve the necessary combination of stress and flaw size to enable the crack to grow catastrophically across the width into a transverse ply crack. This mechanism-based approach explains the transition from thin ply to thick ply behaviour and why the strain for the onset of matrix cracking in the thick ply case is the same as the strain to failure of an unconstrained ply. A complementary approach, based on the energy release when a crack forms, can also predict the thin ply dependence but cannot by itself explain the transition from constrained (thin ply) to unconstrained (thick ply) behaviour.

References

1. B D Agarwal, L J Broutman, *Analysis and Performance of Fiber Composites*, John Wiley, New York, 1990.
2. R M Jones, *Mechanics of Composite Materials*, Scripta Book Co., Washington D.C. (McGraw Hill Book Co., New York), 1974.
3. J E Masters, K L Reifsnider, *An Investigation of Cumulative Damage in Quasi-isotropic Graphite/Epoxy Composites*, in ASTM STP 775, American Society for Testing and Materials, Philadelphia, 1982, pp. 40–62.
4. A Parvizi, K W Garrett, J E Bailey, *J. Mater. Sci.*, **13**, 195 (1978).
5. D L Flaggs, M H Kural, *J. Composite Mater.*, **18**, 339 (1984).
6. S L Ogin, P A Smith, *ESA J.*, **11**, 45 (1987).
7. P W Manders, T-W Chou, F R Jones, J W Rock, *J. Mater. Sci.*, **18**, 2876, (1983).

4.11 Laminates – crossply cracking

F R JONES

Introduction

The mechanism by which high performance composites fail is complex. This is a consequence of the anisotropy and heterogeneity of the material both on the gross and microstructural scales. (This ignores the likely molecular heterogeneity which results from interphase formation and the presence of chemical interfaces.) Transverse cracking (also referred to as matrix cracking), fibre fracture and fibre–matrix debonding are of particular importance to the initiation and accumulation of damage. Transverse cracking has received detailed analysis and consideration since the first systematic studies of Bailey and coworkers[1,2] because of its significance to the prediction of performance and design criteria. Thus matrix cracking can lead to, for example, the reduction in stiffness of high performance laminates or to a permeation mechanism for a pressurized fluid contained in filament wound containers or pipes.

The shear-lag multiple cracking theory[1,2]

As indicated in Fig. 66 the first damage to crossply laminates under a tensile stress occurs in the plies at 90° to the applied load. If the 0° plies are very thin and cannot withstand the additional load, complete failure will occur. However, in practical laminates fracture does not occur. Providing the individual plies remain elastically bonded, and that delamination does not occur, with the crack(s) pinned by the 0° plies, the additional stress, $\Delta\sigma$, on the 0° plies decays as one moves away from the crack in the longitudinal direction over the distance y. The value of $\Delta\sigma$ is given by

$$\Delta\sigma = \Delta\sigma_0 \exp(-\phi^{1/2} y) \quad [4.34]$$

where $\Delta\sigma_0$ is the maximum additional stress in the 0° plies in the plane of the transverse ply crack, and $\phi = E_c G_t (b+d)/E_l E_t b d^2$ where $E_c E_t E_l$ are the initial Young's modulus in the y-direction, the Young's moduli of the individual transverse (or 90°) and longitudinal (or 0°) plies. G_t is the shear modulus of the transverse ply in the y-direction, b and d are the 0° and semi-90° ply thicknesses respectively.

The corresponding load F transferred back into the transverse ply at a distance y is given by

$$F = 2bc\Delta\sigma_0(1 - \exp(-\phi^{1/2} y)) \quad [4.35]$$

where c is the width of the laminate. The transverse ply can crack again when the load F equals $2\epsilon_{utl}E_t dc$ where ϵ_{utl} is the transverse ply cracking strain. This will occur at infinity when the applied stress σ_a is just at the value at which the ply first cracks.

For a further crack to form in the 90° ply, σ_a and hence $\Delta\sigma_0$ must be increased to a value such that the load in the transverse ply reaches $2\epsilon_{utl}E_t dc$ at the furthest end of the specimen. Similarly the next cracks will form midway between the existing cracks when $\Delta\sigma_0$ is large enough to produce a load of $2\epsilon_{utl}E_t dc$ in the transverse ply midway between two cracks of spacing t.

$$\Delta\sigma_0 = \epsilon_{utl}E_t d/b[1 + \exp(-\phi^{1/2}t) - 2\exp(\phi^{1/2}t/2)]^{-1} \qquad [4.36]$$

$\Delta\sigma_0$ is the additional stress on the 0° plies at $y = 0$ and is related to σ_a by

$$\Delta\sigma_0 = \sigma_a(b+d)/b - E_1\sigma_a/E_c \qquad [4.37]$$

The term σ_a/E_c is the strain in the composite in the absence of cracks and the first crack occurs at $\sigma_a = \epsilon_{utl}E_c$.

The solution to this equation is a stepped curve, as illustrated in Fig. 83, which predicts the range of crack spacing for a laminate of particular length. However, it has been suggested that since the transfer distance y is small by comparison with the crack spacing, the first cracks form statistically at the weakest point in the 90° ply. Kimber and Keer[3] showed that the average crack spacing t_{av}, is related to the lower bound of the shear lag analysis t

$$t_{av} = 1.337t. \qquad [4.38]$$

The reasons for such complications are clear in that defects at the edges of cut coupons can lead to a reduction of the transverse cracking strain. Furthermore, the residual thermal strain (see Section 4.12) which is formed as a consequence of cooling from either a fabrication or service temperature and which reduces the first cracking strain (ϵ_{utl}) relative to the ply failure strain (ϵ_{ut}) is also a function of fibre volume fraction. Thus for practical laminates a memory of the tow dimensions within the cured matrix (see Fig. 65) and slight variations in ply thickness can lead to areas of the inner ply which have a higher residual thermal stress and a variable cracking strain. Sheard and Jones[4] found that the locations of the first 10 transverse cracks were affected by the geometry and curing conditions. As the post-cure temperature was increased the locations of these early cracks changed from high fibre volume fraction areas to areas of low fibre volume fraction, as a consequence of increased thermal stresses in resin-rich areas. Variations in inner ply thickness were also significant.

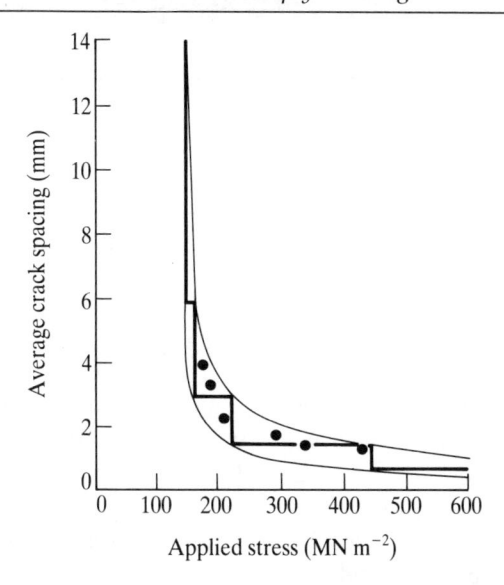

Figure 83 Comparison of experimental results (●)[1] with theoretical curves of crack spacings as a function of applied stress for 190 mm long glass fibre reinforced epoxy crossply specimens with a transverse ply thickness of 1.2 mm. The stepped curves show the crack spacing when the first crack occurs in the middle of the specimens of this particular length, and the upper and lower continuous curves indicate the range of crack spacings for specimens of any length with an arbitrary position of the first crack. (Reproduced from Parvizi and Bailey[1] with permission of Chapman and Hall Ltd, London)

These types of microstructural effects can be expected to be more likely when the number of the filaments in a tow is large and the fabrication process does not enable full fibre dispersion. Heterogeneity of this type means that both dependent and independent cracking can occur. The latter refers to an isolated crack whose stress field is unperturbed by the presence of other transverse cracks. However as the crack density increases dependent cracking occurs because the stress fields associated with adjacent cracks overlap.[4] For highly heterogeneous plies (such as woven cloth or poor quality material) this can occur locally and be more significant.

Other theories of multiple microcracking (see Sections 4.10 and 4.13)

Nairn[5] has recently reviewed the theories of transverse or matrix cracking. However, some brief comments here will enable the importance of this phenomenon to be stressed. Wang[6] has also developed theories of matrix or transverse cracking using two-dimensional finite element analysis, which depends on the presence of gross laminate defects rather than micromechanical fluctuations. The major problem with this argument is that the effective flaw dimensions bear no relationship to real observable flaws, and are equal to the thickness of one ply.

Peters (Section 4.13) has developed the statistical argument further to provide a technique for determining the maximum strength of a transverse ply, which has been assigned to the weakest element of a perfect 90° lamina, namely the interfacial strength.

In Section 4.10 Ogin discusses the general nature of damage accumulation mechanisms and has indicated that the ply thickness has a profound effect on the cracking strain. The energy released when a crack forms enabled Bailey et al.[2] to derive (4.39) which predicts the thickness dependence of ϵ_{utl} and identifies the benefits of thin plies. However, the fracture mechanics approach of Ogin can better predict the strength of thick (>0.5 mm) 90° plies when this becomes the dominant effect (see Sections 4.10 and 4.13).

$$\epsilon_{ut}^{min} = [2bE_l\gamma_t\phi^{1/2}/(b+d)E_tE_c]^{1/2} \qquad [4.39]$$

where ϵ_{ut}^{min} is the minimum transverse failure strain and γ_t is the surface free energy of the transverse ply.

It should be remembered that the residual thermal strain in the 90° ply will reduce the cracking strain by an equivalent amount (see Section 4.12) according to

$$\epsilon_{ut} = \epsilon_{utl} + \epsilon_{tl}^{th} \qquad [4.40]$$

$$\epsilon_{utl}^{min} = [2bE_l\gamma_t\phi^{1/2}/(b+d)E_tE_c]^{1/2} - \epsilon_{tl}^{th} \qquad [4.41]$$

where ϵ_{tl}^{th} is the thermal strain in the longitudinal direction of the transverse ply and ϵ_{utl}^{min} is the minimum cracking strain of the transverse ply.

Longitudinal splitting

As indicated in Fig. 66, 'transverse cracking' or splitting of the longitudinal or 0° plies can occur when the Poisson-induced strain (see Section 4.12) exceeds the transverse failure strain of the 0° plies. Analogously to the 90° plies, the transverse residual thermal strain in longitudinal plies (ϵ_{lt}^{th}) also contributes to the stress level in the 0° ply.

Consider the 0°/90°/0° laminate in Fig. 66 under a tensile load then the Poisson strain at the onset of splitting ϵ_{lts}^p, is given by

$$\epsilon_{lts}^p = [(\epsilon_{lt}^{th})^2/16 + \gamma_t dE_l\phi^{1/2}/(b+d)E_{ct}E_t]^{1/2} - 3\epsilon_{lt}^{th}/4 \qquad [4.42]$$

Since from a consideration of force equilibrium the poisson strain in the longitudinal ply in the transverse direction is given by

$$\epsilon_{lt}^p = [E_l d/(E_l d + E_t b)][\epsilon_{ll}\nu_l - \epsilon_{utl}^i \nu_t] \qquad [4.43]$$

where ϵ_{ll} is the strain in the longitudinal direction of the longitudinal ply, ν_l and ν_t are the Poisson ratios of the longitudinal and transverse plies respectively. ϵ_{utl}^i is the transverse ply cracking strain of the inner ply, E_{ct} = Young's modulus of 0°/90°/0° composite in the transverse direction which is the Young's modulus of a 90°/0°/90° composite. By combining (4.42) and (4.43), the minimum applied composite strain (ϵ_{lls}^{min}) for longitudinal splitting is given by

$$\epsilon_{lls}^{min} = \frac{1}{\nu_l}\left[\frac{E_{ct}(b+d)}{E_l d}\left\{\left[(\epsilon_{lt}^{th})^2/16 + \gamma_t dE_l \phi^{\frac{1}{2}}/(b+d)E_{ct}E_t\right]^{\frac{1}{2}} - \frac{3}{4}\epsilon_{lt}^{th}\right\} + \nu_t \epsilon_{utl}^i\right]$$

[4.44]

The value of ϵ_{lls} for a glass fibre composite is typically $> 1.3\%$, which is less than the fracture strain of the composite. Thus sufficient Poisson strain can be generated to cause splitting. However, for carbon fibre composite laminates, insufficient Poisson strain can be induced to cause splitting before the laminate fails.

Conclusions

Transverse (or matrix) cracking of composite laminates can be preceded by fibre matrix debonding. However, it represents the first damage mode and therefore justifies the efforts to examine the microstructural parameters and derive predictive equations.

References

1. A Parvizi, J E Bailey, *J. Mater. Sci.* **13**, 2131 (1978).
2. J E Bailey, P T Curtis, A Parvizi, *Proc. R. Soc. Lond.* **A 366**, 599 (1979).
3. A C Kimber, J G Keer, *J. Mater, Sci. Lett.*, **1**, 353 (1982).
4. P A Sheard, F R Jones, in *Sixth International Conference and Second European Conference on Composite Materials, ICCMVI/ECCM2 London*, eds F L Matthews *et al.*, Elsevier Applied Science, London, 1987, Vol. 3. pp. 123–135.
5. J A Nairn, S Hu, in *Damage Mechanics of Composite Materials*, ed. R Talreja, Elsevier, Barking, UK, 1992.
6. A S D Wang, *Comp. Tech. Rev.* **6**, 45 (1984).

4.12 Laminates – residual thermal and related strains

F R JONES

Thermal strains are generated in composite materials on the micro and macro scales. The former arise from a mismatch in the thermal expansion coefficients of the fibre and resin, whereas the latter arise as a result of differing thermal expansion coefficients of the longitudinal and transverse plies in an angle ply laminate. These macrostrains are an order of magnitude larger than the microstrains that occur in a unidirectional laminate and are of major importance since they reduce the failure strains of the individual plies in multidirectional ply laminates. The extreme case, the crossply or $0°/90°/0°$ laminates are considered here to illustrate the material consequences.

Consider the crossply laminate given in Fig. 66, the residual thermal strains that develop in the plies on cooling from temperature T_1 to temperature T_2 are

$$\epsilon_{tl}^{th} = \frac{E_l b (\alpha_t - \alpha_l)(T_1 - T_2)}{E_l b + E_t d} \quad [4.45]$$

$$\epsilon_{lt}^{th} = \frac{E_l d (\alpha_t - \alpha_l)(T_1 - T_2)}{E_t b + E_l d} \quad [4.46]$$

where ϵ_{tl}^{th} is the thermal strain in the transverse ply in the longitudinal direction and ϵ_{lt}^{th} is the thermal strain in longitudinal ply in the transverse direction; E_l, E_t are the longitudinal and transverse moduli, sometimes referred to as ^1E_c and ^2E_c, respectively (see Section 4.2); α_l, α_t are the linear coefficients of thermal expansion of the longitudinal and transverse plies; b, d are the outerply and semi-inner ply thicknesses, respectively.

The residual thermal strains are positive in magnitude in the transverse directions of the plies because the higher modulus in the fibre direction constrains the larger contraction 90° to the fibres and this is matrix dominated. Typical values of thermal strain for carbon fibre and glass fibre composites are given in Table 45.

The values for CFRP are much larger than those for GRP because of the differences in fibre expansion coefficient. That of glass is $5 \times 10^{-6} \text{K}^{-1}$ where the radial (transverse) expansion coefficient of carbon fibre is $10-20 \times 10^{-6} \text{K}^{-1}$. Whereas glass is isotropic, carbon fibres are anisotropic and in the axial direction α_f is negative (-0.4 to $-1.1 \times 10^{-6} \text{K}^{-1}$).

From (4.45) it is possible to calculate the value of ϵ_{tl}^{th} when all the terms are known. In many cases the values of α_t, α_l and T_1 are not known. T_1 is not necessarily the temperature from which the laminate is cooled but the

Table 45. Thermal strains (%) in 0°/90°/0° laminates from epoxy resin based CFRP and GRP with varying semi-inner ply thickness d (mm), $b = 0.5$ mm

	CFRP		GRP	
d	ϵ_{tl}^{th}	ϵ_{lt}^{th}	ϵ_{th}^{tl}	ϵ_{lt}^{th}
2.0	–	–	0.053	0.115
1.5	–	–	0.062	0.112
1.0	–	–	0.075	0.107
0.5	0.332	0.322	0.094	0.094
0.25	0.332	0.303	0.107	0.075
0.15	–	–	0.113	0.059
0.125	0.337	0.271	–	–
0.0625	0.340	0.224	–	–

temperature at which the residual strain is first built into the laminate. It will be determined by the viscoelastic response of the resin and will be close in magnitude to the matrix glass transition temperature but not necessarily equal to it. To overcome these uncertainties, the value of $(\alpha_t - \alpha_l)(T_1 - T_2)$ can be obtained directly from the curvature of an unbalanced 0°/90° beam.[1] The mismatch in expansion coefficients of the two plies causes a thin beam to curve in the manner of a bimetal strip. Timoshenko[2] has shown that

$$(\alpha_t - \alpha_l)(T_1 - T_2) = \frac{b+d}{2\rho} + \frac{E_l b^3 + E_t d^3}{6\rho(b+d)} \left[\frac{1}{E_l b} + \frac{1}{E_t d} \right] \quad [4.47]$$

where ρ is the radius of curvature of the beam.

For a laminate, the curvature will develop in both directions to form a saddle shape, which can flip between two geometric forms.

The quantity $(\alpha_t - \alpha_l)(T_1 - T_2)$ is needed for substitution into (4.45) and (4.46); it can be obtained from (4.47) by knowing the radius of curvature of the beam, ρ. This technique is very simple and takes into account any stress relaxations that occur on cooling and eliminates the need for measurements of α_l, α_t and T_1 individually. But measurement of ρ can be tedious; it involves finding a curve of known radius to match the curvature. Jones et al.[3] have applied simple trigonometry to derive (4.48) and (4.49), which enable ρ to be calculated from the beam displacement, δ, and the chord length $2x$,

$$\rho = \frac{\delta^2 + x^2}{2\delta} \quad [4.48]$$

It is often informative to monitor the temperature dependence of ϵ_{tl}^{th} using this technique. In which case, it is difficult to measure $2x$ at temperatures other than ambient. Equation (4.49) enables ρ to be computed iteratively from the length of the beam, L,

$$\frac{1}{\rho} = \frac{2}{L} \cos^{-1}(1 - \delta/\rho) \qquad [4.49]$$

Since L is relatively insensitive to temperature it can be calculated from the room temperature values of δ and x according to

$$L = \frac{\pi}{360} \sin^{-1}\left(\frac{2\delta x}{\delta^2 x^2} \cdot \frac{(\delta^2 + x^2)}{2\delta}\right) \qquad [4.50]$$

The experimental temperature dependence of ϵ_{tl}^{th} for a glass fibre polyester composite, given in Fig. 84, is compared to those predicted from known values of α_t and α_l. There is good agreement between experimental and calculated values. This figure also illustrates the other advantage of the 0°/90° unbalanced beam technique. It is seen that the beam displacement becomes zero (i.e. the beam becomes flat) at the strain free temperature. Thus a direct measurement of T_1 can be obtained.

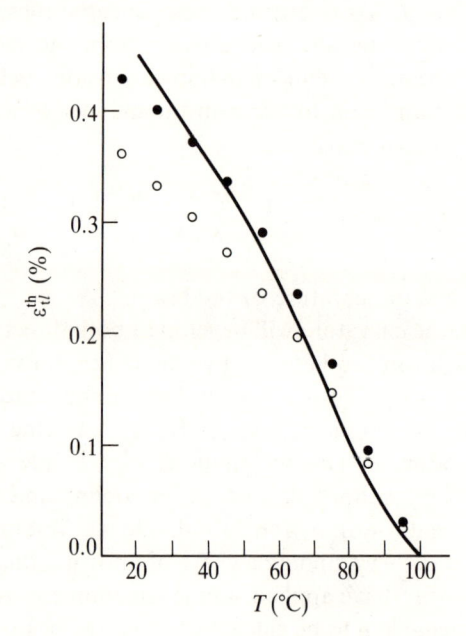

Figure 84 Variation of thermal strain, ϵ_{tl}^{th}, with temperature for an unsaturated polyester crossply laminate post-cured at 130°C. Experimental values obtained from (0°/90°) beams (continuous line), predicted values using known dry laminate properties (○) and from properties of a laminate containing 0.15 % H$_2$O by weight (●). Redrawn from Jones et al.[4] with kind permission of Elsevier Science Ltd, The Boulevard, Langford Lane, Kidlington, OX5 1GB, UK.

The equations of Schapery[5] can be used and have been shown to accurately predict the laminate expansion coefficients from those of the components.[2]

$$\alpha_l = (E_m \alpha_m V_m + E_f \alpha_f V_f)/(E_m V_m + E_f V_f) \quad [4.51]$$

$$\alpha_t = \alpha_m V_m (1 + \nu_m) + \alpha_f V_f (1 + \nu_f) - \alpha_l \nu_c \quad [4.52]$$

where the symbols are conventionally defined, with ν as the Poisson's ratio. The subscripts refer to composite, c, fibre, f and matrix, m.

Effect of moisture

Absorption of moisture into the resin, either as a result of a humid service environment or immersion in water, swells the resin matrix and tends to negate the thermal strains. For a polyester resin based composite, a reduction in thermal strain by 0.2% occurs with a 0.6% increase in water content. These results were obtained by recording the change in curvature of the 0°/90° beam with time of immersion in water. In this case, a thermal residual strain is still present in the wet laminate but in other systems, where the water absorption is higher, ϵ_{tl}^{th} can be totally negated to give an effectively stress free laminate.

Matrix swelling that occurs on absorption of moisture is analogous to the thermal expansion that occurs on heating. As a consequence, an analogous equation (4.53) can be used to assess the swelling coefficients of the individual plies and to predict changes in the residual stress state during environmental conditions.

$$\epsilon_{tl}^s = \frac{E_l b (\beta_t - \beta_l) \Delta M}{E_l b + E_t d} \quad [4.53]$$

where ϵ_{tl}^s is the swelling strain in the longitudinal direction of transverse ply; β_l, β_t are the swelling coefficients in the longitudinal and transverse directions, respectively; ΔM is the change in moisture content. For a dry laminate brought to equilibrium moisture content, $\Delta M = M_\infty$.

The temperature dependence of the resin expansion coefficient after water absorption does not necessarily remain the same. In many cases, the resulting increase in $(\alpha_t - \alpha_l)(T_1 - T_2)$ is not offset by the reduction in T_1 (or T_g), which may also occur. As a consequence, subjecting a wet laminate to a thermal cycle,[2] e.g. post-curing after water absorption by the as-cured laminate or a thermal spike,[6] can lead to an enhanced thermal strain. The changes that can occur are illustrated in Table 46 and Fig. 84, where the as manufactured beam contained $\approx 0.15\%$ water, which caused an enhancement of thermal strain after post-curing.

Table 46. Residual thermal strain ϵ_{tl}^{th} induced by post-curing a wet glass fibre polyester laminate

Post-curing temperature (°C)	Residual thermal strain, ϵ_{tl}^{th} (%)	
	Dry	Wet
50	0.15	0.34
80	0.22	0.40
130	0.26	0.45

Therefore the residual stress state can fluctuate under conditions where moisture absorption and excursions to higher temperatures occur. This aspect is considered further in Section 6.6.

Effect of residual thermal strains on laminate properties

The transverse ply of a crossply laminate is placed under tension by the residual thermal stress, therefore the transverse cracking strain, ϵ_{utl} the strain at which first failure of the inner ply occurs, will be lower than the ply failure strain, ϵ_{ut}. Thus

$$\epsilon_{ut} = \epsilon_{utl} + \epsilon_{tl}^{th} \qquad [4.54]$$

In Section 4.3, ϵ_{ut} is defined as $^2\epsilon_{uc}$. A typical ply failure strain for glass fibre reinforced plastic is 0.6%. For the fully dry post-cured polyester laminate in Table 46, ϵ_{utl} will be reduced to 0.15%. For some carbon fibre epoxy laminates, ϵ_{ut} can reach 1.0%; this reduces the effect of thermally induced residual strains on the propensity for microcracking. The tendency of crossply laminates to microcrack is affected by a number of factors not least the magnitude of ϵ_{tl}^{th} and the degree of constraint which can be induced by ensuring that the fibres are dispersed into a number of thin rather than thick plies (see Sections 4.10 and 4.11).

Thermally induced strains in the longitudinal direction

The tensile strains that develop transversely to the fibres, ϵ_{tl}^{th} and ϵ_{lt}^{th}, must be counteracted by equivalent compressive stresses in the fibre direction. The thermal stresses in the longitudinal direction of the longitudinal ply ϵ_{ll}^{th}, and the transverse direction of the transverse ply, ϵ_{tt}^{th}, are given by

$$\epsilon_{ll}^{th} = -\frac{E_t}{E_l} \epsilon_{tl}^{th} \qquad [4.55]$$

$$\epsilon_{tt}^{th} = -\frac{E_t}{E_l}\epsilon_{lt}^{th} \qquad [4.56]$$

Poisson induced strains

Analogous equations can be derived for the effect of Poisson induced contraction that occurs on loading. Consider the simple crossply laminate in Fig. 66, on loading in the longitudinal direction a positive strain will develop in the outer longitudinal ply in the transverse direction, ϵ_{lt}^P, as a result of the lower Poisson ratio of the transverse ply, ν_t, compared to the longitudinal ply, ν_l, and

$$\epsilon_{lt}^P = \frac{E_l d \epsilon_{ll}(\nu_l - \nu_t)}{E_l d + E_t b} \qquad [4.57]$$

This equation only applies for $\epsilon_{ll} < \epsilon_{utl}^i$. At larger values of composite strain $\epsilon_{ll} > \epsilon_{utl}^i$, i.e. when the inner ply cracks,

$$\epsilon_{lt}^P = [E_l d/(E_l d + E_t b)][\epsilon_{ll} d\nu_l - \epsilon_{utl}\nu_t] \qquad [4.58]$$

The magnitude of these strains are significant and can exceed the transverse cracking strain of the longitudinal ply with the observation that splitting occurs before ultimate failure (Fig. 66). These strains add to the thermal strain, ϵ_{lt}^{th}, that is present before loading

$$\epsilon_{lts} \geq \epsilon_{lt}^P + \epsilon_{lt}^{th} \qquad [4.59]$$

where ϵ_{lts} is strain in the longitudinal ply in the transverse direction for splitting. Splitting before failure has been observed in the longitudinal plies of glass fibre laminates but has not been observed for carbon fibre composites. This is because insufficient Poisson strain can be induced before failure as failure occurs at lower composite strains. A detailed discussion is given in Bailey et al.[1]

Conclusions

The generation of thermally induced strains in the individual plies of composite laminates arise from differences in the linear expansion coefficients in the longitudinal and transverse direction on cooling from an elevated temperature. It has been suggested that monomers that expand during curing can be used to reduce these effects. Thermally induced strains arise from a constrained contraction on cooling, therefore ϵ_{tl}^{th} cannot be negated in this way. There is no evidence to suggest that the thermal strains add to a curing strain. The overwhelming argument is that any curing strain, which arises from a constrained contraction during polymerization at room temperature, can be annealed out during post-curing. Most measured thermal strains are consistent with predictions, provided any anomalous expansion behaviour is fully understood.

References

1. J E Bailey, P T Curtis, A Parvizi, *Proc. R. Soc. Lond.*, **A366**, 599 (1979).
2. S Timoshenko, *J. Opt. Soc. Amer.* **11**, 233 (1925).
3. F R Jones, M Mulheron, J E Bailey, *J. Mater. Sci.*, **18**, 1522 (1983).
4. F R Jones, M Mulheron, J E Bailey, *Composites Sci. Tech.*, **25**, 119 (1986).
5. R A Schapery, *J. Composite Mater.* **2**, 380 (1968).
6. P M Jacobs, F R Jones in *Composites Design, Manufacture and Applications, (ICCM VIII)*, Vol. 2, eds S W Tsai, G S Springer, SAMPE, Covina, USA, 1991, Ch. 16.

4.13 Laminates – statistics of transverse cracking

P W M PETERS

Analysis

Transverse cracking is generally the first damage mechanism to occur in multidirectional fibre reinforced polymers. This is the case for glass, carbon and aramid fibre reinforced plastics (GFRP, CFRP and AFRP). Generally a transverse ply in a multidirectional laminate shows multiple fracture before the laminate finally fails.

There are several approaches to describe transverse fracture as indicated in Fig. 85.

In this section, statistical approaches for the description of *transverse fracture* will be discussed. A transverse ply in a multidirectional laminate

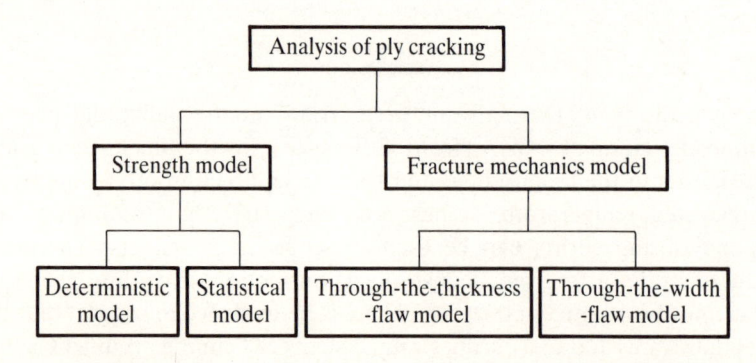

Figure 85 Flow chart for the analysis of transverse cracking

shows multiple fracture due to (i) its low failure strain compared with other ply orientations; and (ii) the fact that stresses are reintroduced again in the transverse ply at some distance from a crack.

The development of transverse cracks is thus influenced by the stress recovery in the transverse ply near cracks. The stress recovery can be described by:

(i) a shear lag analysis,[1]
(ii) finite element or finite difference methods,[1]
(iii) a variational approach.[2]

In a shear lag analysis (the element of which in the case of a $0°/90°/0°$ laminate is given in Fig. 86) the stress recovery is realized by shear stresses in a shear transfer zone (between 90° and neighbouring plies). This zone only experiences a shear stress, whereas the different plies bear only tensile stresses that do not vary over the thickness of the plies. The tensile stress in the transverse ply as a function of the distance x to the crack is given by:

$$\sigma_{2,x} = \sigma_{2,\infty} \left(1 - e^{-\gamma x}\right) \qquad [4.60]$$

where $\sigma_{2,\infty}$ is the undisturbed stress (at $x = \infty$) and

$$\gamma = \left[\frac{G}{b}(1/E_1 a_1 + 2/E_2 a_2)\right]^{\frac{1}{2}} \qquad [4.61]$$

where G and b are the shear modulus and thickness of the shear transfer zone and E_1, E_2, a_1, a_2, the modulus and thickness of the surface plies (ply 1,3) and of the transverse ply (ply 2), respectively. The shear transfer

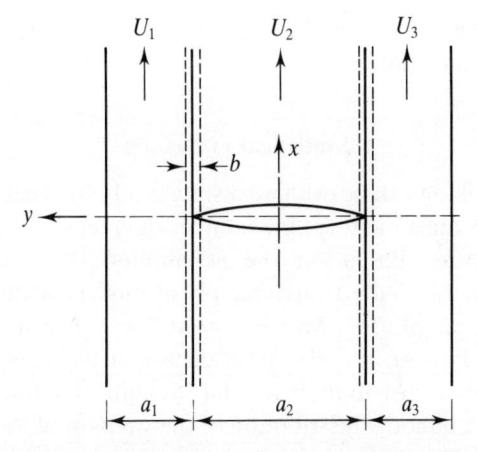

Figure 86 Element of the applied shear lag analysis (displacement $U_1 = U_3$, $a_1 = a_3$)

zone is considered to be a resin-rich zone between transverse and surface plies (thus G is matrix modulus) with a thickness of twice the fibre diameter.[1]

At a small distance to the crack, the stress is low and gives rise to only a small probability of failure for other cracks to occur in this region. This is the reason for a rather regular crack spacing found under static as well as under dynamic loading conditions. Another limitation on the degree of transverse cracking can be imposed by the shear stress in the shear transfer zone. This shear stress measures[3]

$$\tau_{max} = \frac{G}{b}\left(\frac{\epsilon_{2,\infty}}{\gamma}\right)\left(1 + \frac{E_2 a_2}{2E_1 a_1}\right) \quad [4.62]$$

as a maximum at the crack tip. Thus, if the transverse ply stiffness, E_2, and/or thickness, a_2, is large, the maximum shear stress can exceed the interlaminar shear strength, especially if the fibre–matrix bond strength is poor or the matrix shear yield stress is low. This leads to cracking or yielding in the shear lag zone and thus limits the re-introduction of stresses in the interrupted ply. At increasing applied load, the stresses in the interrupted plies do not increase as strongly; this leads to a reduction of cracking and eventually to a saturation of cracking, i.e. no more cracks occur at increased load level.

The scatter in fracture data of a transverse ply can be described with the aid of a two-parameter Weibull distribution. According to this distribution, the probability of transverse fracture F at an applied strain of ϵ is given by

$$F = 1 - \exp\left[-(\epsilon/\epsilon_0)^a\right] \quad [4.63]$$

where ϵ_0 and a are the Weibull scale parameter and shape parameter, respectively. There are several procedures to find this distribution based on transverse crack data.

Statistical evaluation

Transverse cracking can be evaluated statistically by the determination of (i) the crack distance;[4] and (ii) the number of cracks,[1,3,5] as a function of the applied stress. Based on the assumption that the strengths of constitual volumes of the transverse ply are independent of each other and identically distributed, Manders et al.[4] described the crack spacing according to a Poisson process. With the aid of this analysis and a two-parameter Weibull distribution for the strength of a transverse ply, the authors found a bimodal distribution of the strength of the transverse ply in a $0°/90_2°/0°$ laminate ($a_2 = 1.1$ mm). The second mode probably is the result of saturation of cracking.[3]

The second approach is based on the assumption that in a certain range close to an existing crack no more cracking occurs.[1,3,5] For CFRP, this range was arbitrarily chosen to be the range in which the stress is < 90% of the undisturbed stress. The transverse ply is now considered to consist of elements of this length, all of which can break. Substitution of $\sigma_{2,x} = 0.9\,\sigma_{2,\infty}$ in (4.60) delivers half the length of this element and thus the number of elements in the specimen's gauge length.

The number of cracks in the gauge length can be determined by visual inspection of a polished specimen edge, after loading and unloading the specimen to increasing levels of strain. Use of a piezoelectric transducer makes it possible to detect every single transverse crack, so that for every occurring crack, the respective failure strain can be determined.[1] As an example for the last procedure, the two-parameter Weibull distribution for transverse fracture in a specimen from the $0°/90°_4/0°$ laminate tested at $-40°C$ is given in Fig. 87a.

The experimental data, x, with the probability of failure, F, calculated by $F = j/N + 1$ (valid for $N > 20$, j = crack order number, N is number of elements) and the strain, ϵ (consisting of a thermal and a mechanical component), was found to follow very well a two-parameter Weibull distribution (indicated by the straight line). This distribution was

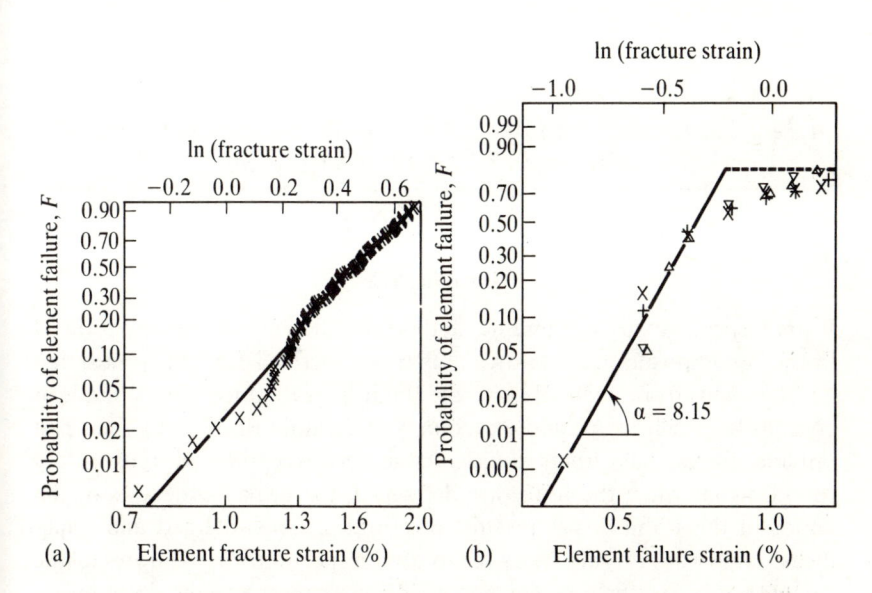

Figure 87 Two parameter Weibull distributions for transverse cracking based on experimental data (a) for CFRP, data points of all occurred cracks (T800/R6376, $0°/90°_4/0°$, $-40°C$); (b) for GFRP, data at intervals of applied strain (glass–epoxy $0°/90°/0°$, $a_2 = 0.303$ mm) ▽ spec 1, △ spec 2, + spec 3, × spec 4

determined with the aid of the experimental data, making use of linear regression and least squares analysis.

An example of the statistical evaluation of transverse cracking, based on visual inspection of a polished specimen edge, is indicated in Fig. 87b. It indicates that GFRP specimens with a thicker transverse ply show a reduced probability of failure at increased applied strain; a result of the shear stress exceeding the interlaminar shear strength (leading to saturation of cracking). With the aid of these statistical distributions for transverse cracking, it is possible to study the influence of different important parameters on transverse cracking, e.g. the fibre–matrix bond strength and the defect distribution in the laminate.

References

1. P W M Peters, *J. Composite Mater.*, **18**, 545 (1984).
2. Z. Hashin, *Mechanics of Materials*, **4**, 121 (1985).
3. P W M Peters, T W Chou, *Composites*, **18**, 40 (1987).
4. P W Manders, T W Chou, F R Jones, J W Rock, *J. Mater. Sci.*, **18**, 2876 (1983).
5. P W M Peters, S I Andersen, *J. Composite Mater.*, **23** (9) 944 (1989).

4.14 Short fibre reinforcement – direct measurement of fibre strain using Raman spectroscopy

R J YOUNG

Introduction

Raman spectroscopy is a unique method of following the deformation of fibres in composites. It is known[1,2] that well-defined Raman spectra can be obtained from virtually all the high performance fibres used in composites, such as aramids, carbon and ceramic fibres, and rigid-rod polymer fibres; glass fibres are about the only exception. Moreover, it is found[1,2] that when the fibres are deformed, the peak frequencies of the bands in the Raman spectra shift position in a well-defined and easily measurable way. It is also possible to obtain spectra from the fibres inside a composite,[1,2] so the transfer of stress from the matrix to the fibres inside the composite can be observed directly.

Raman scattering occurs when a beam of laser light of frequency ν is scattered by a fibre. The majority of the radiation is scattered elastically at the same frequency (Rayleigh scattering) but a small amount of the radiation is scattered inelastically at a different frequency $\nu \pm \Delta\nu$ (Raman scattering). It is this Raman scattered light that carries information about molecular vibrations in the fibre; when the fibres are subjected to simultaneous deformation, the effect of macroscopic stress or strain upon the molecules can be monitored. In the case of high performance fibres, it is possible to focus the laser beam to a spot of the order of 1–2 μm in diameter and effectively obtain a spectrum from a point on the fibre during deformation.[1] Since transparent resins are often used in high performance composites, it is sometimes possible to obtain Raman spectra from a point on the fibre inside the composites, otherwise it is usually possible to grind and polish the composite and obtain the spectrum from a fibre at the polished surface.

Fibre deformation

The use of Raman spectroscopy to monitor molecular deformation in polymer fibres has been reviewed extensively elsewhere,[1,2] so only a brief account will be given here using the Kevlar 49 aramid fibre as an example. In a full Raman spectrum from a single filament of Kevlar 49, obtained using a low power laser beam and a Raman microprobe, there are six well-defined Raman bands in the region 1100–1700 cm^{-1}. It is found that all the bands shift to lower frequency when the fibre is deformed in tension.[1,2] An example of this is shown in Fig. 88, where the peak frequency, $\Delta\nu$, of the most well-defined Raman band at about 1610 cm^{-1} is plotted against strain, ϵ, and it can be seen there is an

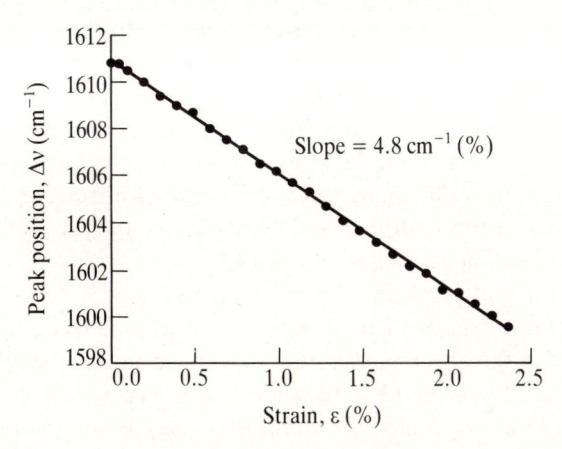

Figure 88 Kevlar 49 fibre in air. Dependence of the peak position of the 1610 cm^{-1} Raman band upon tensile strain for a Kevlar 49 fibre (reprinted from Andrews et al.[3] with kind permission from Elsevier Science Ltd, The Boulevard, Langford Lane, Kidlington, OX5 1GB, UK.)

approximately linear decrease in peak position with strain. It is found[1,2] that the slope of the line in Fig. 88, $d\Delta\nu/d\epsilon$, is proportional to the fibre modulus, E_f; this has been shown to indicate that the Raman technique is a direct measure of the level of axial molecular stress in the fibres. Similar behaviour is found for many other high performance fibres; in general there is an approximately linear dependence of the peak position of the Raman bands upon strain, as in Fig. 88.

Composite micromechanics

The relationship in Fig. 88 is essentially an optical strain gauge calibration for a Kevlar 49 fibre such that, if the fibre is subjected to any state of deformation, the strain at a point on the fibre can be determined from the peak frequency of the 1610 cm^{-1} Raman band. This simple observation has revolutionized our ability to investigate the micromechanics of composite deformation; an example of this is given below.

Single fibre model composite specimens are used widely to investigate composite micromechanics in a variety of systems through, for example, fragmentation tests. Unfortunately, such investigations normally suffer from the problem of not being able to measure fibre stress or strain directly and have to rely upon indirect methods of stress determination. With the advent of the Raman technique, it is possible with a transparent matrix to focus the laser beam at a point on the fibre, obtain a spectrum and, using a relationship such as that in Fig. 88, map out the strain along a fibre. An example of this is shown in Fig. 89 for a single filament of Kevlar 49 in an epoxy resin matrix subjected to different levels of axial matrix strain, ϵ_m.[3] The general form of the variation of strain along the fibre is identical to that predicted by the Cox type shear lag model developed by Kelly.[4] In the undeformed state, $\epsilon_m = 0$, there is no fibre strain but as ϵ_m is increased, the fibre strain rises from the fibre ends then settles at a plateau value along the middle of the fibre such that the fibre strain is equal to the matrix strain. Figure 89 is a vivid example of the power of the technique and gives a direct illustration of the transfer of stress from the matrix to the fibre in the composite.

But deviations from the Cox type behaviour can be seen. For example, it is assumed in the Cox analysis[4] that there is no traction across the ends of the fibres so the value of fibre strain is assumed to fall to zero in the end regions. In a real single fibre composite, there is adhesion at the fibre ends and Fig. 89 shows that higher levels of matrix strain produce a finite value of strain at the fibre ends. It is also possible to monitor other deviations from ideal behaviour, such as matrix yielding, breakdown of the fibre–matrix interface and fibre fracture, from the change in the

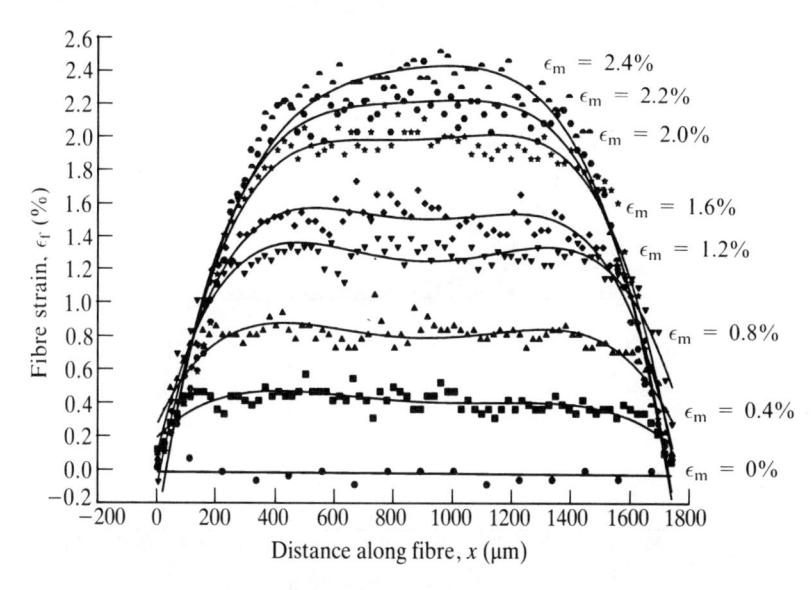

Figure 89 Kevlar 49 single fibre composite. Variation of fibre strain with distance along the fibre for a model Kevlar 49 single fibre composite subjected to different levels of matrix strain (reprinted from Andrews et al.[3] with kind permission from Elsevier Science Ltd, The Boulevard, Langford Lane, Kidlington, OX5 1GB, UK.)

appearance of plots such as Fig. 89.[1,2] A neater way of doing this is by using the data in Fig. 89 to calculate the variation of interfacial shear stress, τ, with distance, x, along the fibre using the relationship[3]

$$\tau = E_f \frac{r}{2} \left(\frac{d\epsilon}{dx} \right) \qquad [4.64]$$

where r is the fibre radius and $d\epsilon/dx$ is determined from the variation of fibre strain with distance along the fibre, i.e. from the slopes of the lines in Fig. 89. The derived variation of τ with x is shown in Fig. 90 for the eight different levels of matrix strain. Again, the form of the behaviour is qualitatively very similar to that predicted by classical shear lag theory.[4] However, it can be seen that above about 2% matrix strain, the value of τ at the end of the fibre reaches a maximum value and does not continue to rise as ϵ_m is increased further. This is an indication that the matrix has undergone shear yielding in the region of high stress concentration at the fibre ends.

The example shown above represents a very brief description of the possibilities that exist in using Raman spectroscopy to follow composite micromechanics. It can also be used to follow fibre compression,[1] to measure strain in off-axis fibres[4] and for carbon fibre epoxy composites.[5]

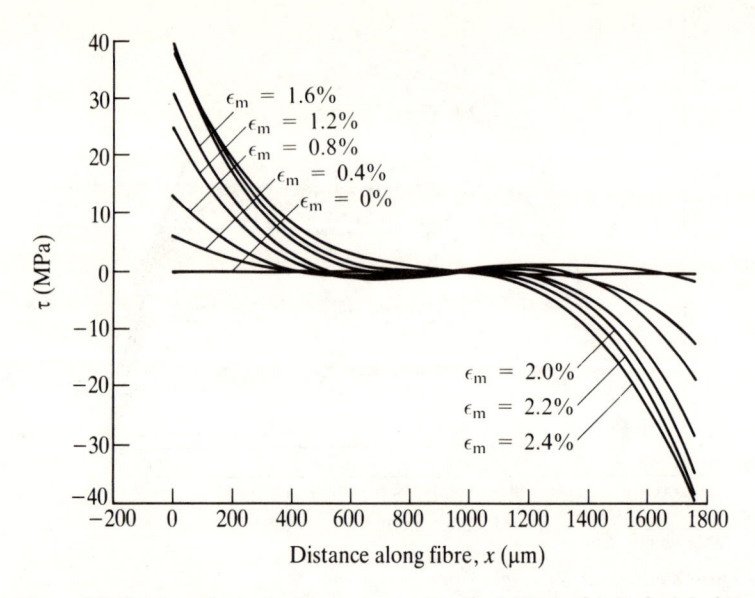

Figure 90 Kevlar-49 single fibre composite. Variation of interfacial shear stress with distance along the fibre at different levels of matrix strain determined using [4.64] for the Kevlar-49 epoxy composite in Fig. 89 (reprinted from Andrews et al.[3] with kind permission from Elsevier Science Ltd, The Boulevard, Langford Lane, Kidlington, OX5 1GB, UK.)

There is no doubt that the use of the technique[3] to study a wide range of phenomena for a large variety of fibre and composite systems will increase rapidly in years to come.

References

1. R J Young in *Advances in the Characterisation of Solid Polymers*, ed. S J Spells, Elsevier Applied Science, London, in press.
2. R J Young, in *Polymer Surfaces and Interfaces 2*, eds W J Feast, H Monroe, R W Richards, Wiley, Chichester, 1993, p. 131.
3. M C Andrews, R J Day, X Hu, R J Young, *Composites Sci. Tech.*, **48**, 255 (1993).
4. A Kelly, N H MacMillan, *Strong Solids*, 3rd edn, Clarendon Press, Oxford, 1986.
5. P Melanitis, P L Tetlow, C Galiotis, C K L Davies in *Interfacial Phenomena in Composite Materials*, eds I Verpoest, F R Jones, Butterworth-Heinemann, Oxford, 1991, p. 288.

4.15 Short fibre reinforcement – stress transfer in discontinuous fibre composites

M G BADER

A discontinuous fibre composite is one that contains relatively short discrete lengths of the fibre dispersed within the matrix. The fibres may be aligned in one direction but are more usually in a random, or semirandom configuration. When an external load is applied to the composite, the fibres are loaded as a result of stress transfer from the matrix to the fibre across the fibre–matrix interface. The degree of reinforcement that may be attained is a function of the fibre fraction (V_f), the fibre orientation distribution (FOD), the fibre length distribution (FLD) and the efficiency of stress transfer at the interface. In general the reinforcement is more effective when V_f is high, the fibres are long, the fibres are aligned in the principal stress direction and the interface is 'strong'.

There are two well-accepted but simplistic, models for stress transfer.[1,2] These will be considered briefly. Cox[1] models the composite as a pair of concentric cylinders, Fig. 91. The central cylinder represents the fibre and the annular outer region the matrix. The ratio of the diameters (r/R) is adjusted to the required V_f. Both fibre and matrix are assumed to be elastic and the cylindrical bond between them is considered to be perfect. It is also assumed that there is no stress transfer across the ends of the fibre. If, as is usually the case, the fibre is much stiffer than the matrix, an axial load applied to the system will tend to induce more strain in the matrix than in the fibre and leads to the development of shear stresses along the cylindrical interface. Cox used a shear lag analysis, which leads to the following expressions for the tensile stress in the fibre, σ_f, and the shear stress at the interface τ,

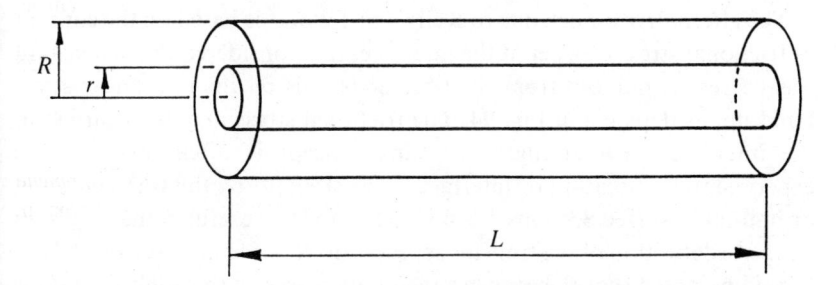

Figure 91 The representative element used in the Cox analysis. The inner cylinder represents the fibre and the outer annulus the matrix

$$\sigma_f = E_f \epsilon_m \left[1 - \frac{\cosh \beta (R_a - x_r)}{\cosh \beta R_a} \right] \quad [4.65]$$

$$\tau = E_f \epsilon_m \left(\frac{G_m}{2 E_f \ln V_f^{-1/2}} \right)^{1/2} \frac{\sinh \beta (R_a - x_r)}{\cosh \beta R_a} \quad [4.66]$$

$$\beta = \left(\frac{2 G_m}{E_f r^2 \ln V_f^{-1/2}} \right)^{1/2} \quad [4.67]$$

where E_f and G_m are the tensile modulus of the fibre and shear modulus of the matrix, ϵ_m is the applied strain, R_a is the aspect ratio of the fibre, $L/2r$, and x_r is the distance from the fibre end measured in units of the fibre diameter.

The resulting stress distributions along the fibre are shown schematically in Fig. 92. The tensile stress in the fibre rises from zero at the end of the fibre to a plateau region along the middle portion. The rate of stress build-up from the fibre ends is greater when the parameters G_m/E_f (shear modulus of the matrix/Young's modulus of the fibre) and r/R (which is related to V_f) are greater. The overall reinforcement efficiency is given by the ratio of the area under the curve of Fig. 92a and the enclosing rectangle, this tends to unity as the fibre aspect ratio tends to infinity. In practical systems, fibres of aspect ratio of ≈ 1000 are virtually as efficient as continuous fibres, as indicated in Fig. 93.

One shortcoming of the Cox approach is that a maximum shear stress is predicted at the end of the fibre; in fact, this is where the shear stress must be zero. However, the stress profile predicted by Fig. 92a has been shown to approximate very closely to that determined by experimental measurements using Raman spectroscopy.[3] The other problem is that interface strength is not considered at all, as the model assumes a perfect bond and only elastic interactions.

The alternative model, due to Kelly and Tyson,[2] is based on the concept of frictional stress transfer at the interface. It is considered that a constant shear stress is induced from the fibre ends; this results in a linear stress build up, as depicted in Fig. 94. The frictional stress may be regarded as the interface shear strength (τ_i); this concept is often used for the experimental estimation of interface shear strength by the fragmentation or pull-out test (see Sections 4.6–4.9). The model is useful in that it allows a precise definition of the transfer aspect ratio R_t, or length (see Fig. 94). It should be noted that this transfer region increases as the applied strain is increased. If the fibre aspect ratio exceeds $2R_t$, there will be a plateau region along the central portion of the fibre. The reinforcement efficiency

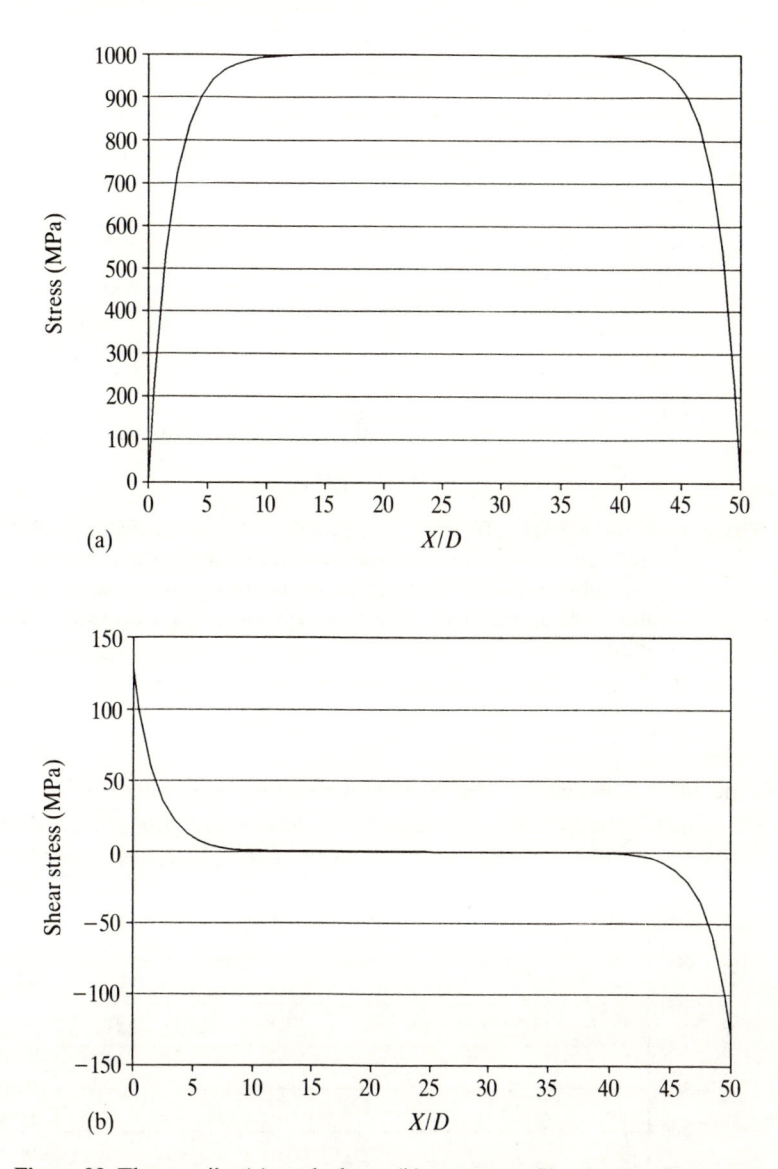

Figure 92 The tensile (a) and shear (b) stress profiles in the fibre and interface according to the Cox model. The values used in the equations are typical of those for an E-glass fibre in an epoxy rein. The transfer length is of the order of 10 fibre diameters. The parameter X/D is the distance from the end of the fibre in units of diameter

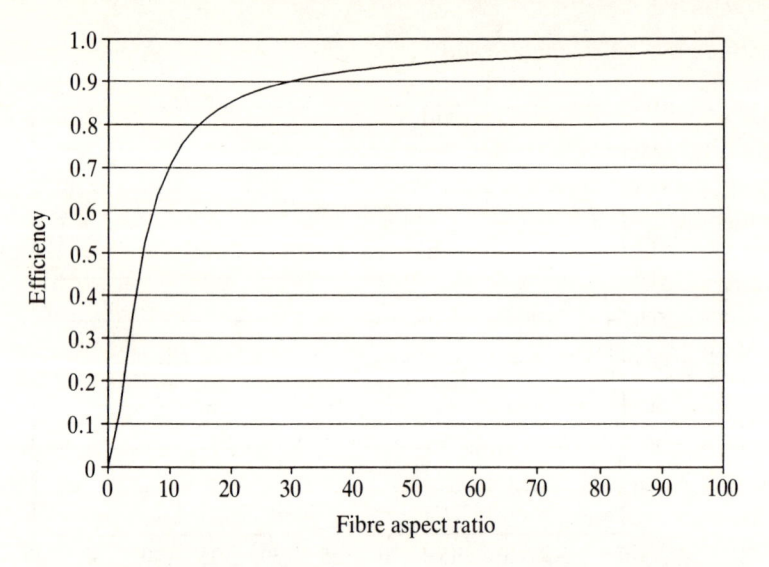

Figure 93 Reinforcement efficiency as a function of fibre aspect ratio for an E-glass/epoxy resin aligned discontinuous system. Continuous fibres would have an efficiency factor of 1.0. In the system illustrated an efficiency of 0.95 is achieved at an aspect ratio of about 60

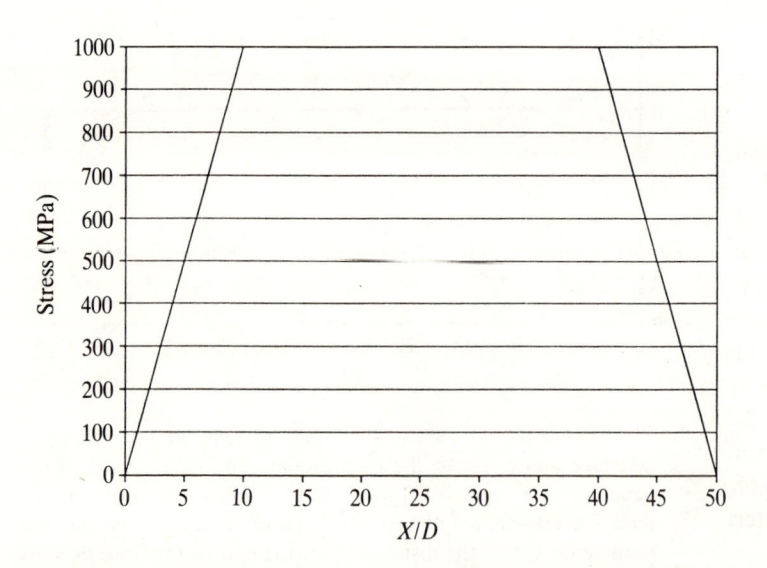

Figure 94 The tensile stress profile in an E-glass fibre in an epoxy resin according to the Kelly–Tyson model. This should be compared with Fig 92a

is obtained from a similar ratio of areas, as for the Cox model (Fig. 92a), and efficiency increases with fibre aspect ratio in a similar manner. Each of the models considered provides an insight into some aspects of reinforcement mechanics but both are inadequate in other respects. The reality is a combination of both concepts but no simple model has emerged to offer significant practical advantages. One of the reasons for this is that real short fibre systems are much more complex than the models. The two main considerations are the length and orientation distributions mentioned at the beginning of this article.

Most discontinuous fibre composites contain fibres with a wide distribution of lengths distributed in a semirandom orientation, most often close to planar random. However, flow processes during fabrication often result in more complex distributions, dependent on component geometry and moulding parameters. To treat the problem of length variations, or more properly aspect ratio variations, it is first necessary to determine the fibre aspect ratio distribution. Then, if the interface shear strength is known, the efficiency of each fibre may be determined using the Kelly–Tyson model and the overall efficiency computed. Likewise, if the fibre orientation distribution is determined, a further efficiency factor may be computed using standard angle ply theory. This has been attempted by several workers[4] who have obtained quite reasonable predictions for the stiffness of short fibre systems (see also Section 4.16). The simplest procedure is to use the aspect ratio and orientation data to produce two constants which may be applied to the Voigt equation

$$E_c = \eta_o \eta_l E_f V_f + E_m V_m \qquad [4.68]$$

where η_l and η_o are the length and orientation constants, respectively. They have a maximum value of unity (long aligned fibres) when the reinforcement efficiency is equal to that of continuous fibres. For short fibre reinforced thermoplastic moulding compounds, values of 0.3–0.5 may be considered typical. A more detailed review of the mechanics of short fibre systems may be found in Chou.[5]

The Bowyer–Bader methodology[6] can be used to predict the stress–strain response of short fibre reinforced plastics. Following the analysis of Kelly and Tyson, described above, the stress on the composite, σ_c, at a given strain, ϵ_c, can be computed by fitting the response to a form of (4.68) with two parameters: the fibre orientation factor, c_θ, and the interfacial shear strength, τ_i,

$$\sigma_c = c_\theta \left[\sum_0^{R_c} V_x R_x \tau_i + \sum_{R_c}^{\infty} \epsilon_c E_f V_y \left(1 - \frac{\epsilon_c E_f}{4 R_y \tau_i}\right) \right] + \epsilon_c E_m (1 - V_f) \qquad [4.69]$$

where $R_c = L_c/d_f$ is the critical aspect ratio of the fibres, $R_y > R_c > R_x$ and L_c is the critical fibre length. Since the fibre distribution is large and R_c is a function of ϵ_c, $\sum_0^{R_c} R_x V_x \tau_i$ is the contribution from the subcritical fibre population and $\sum_{R_c}^{\infty} \epsilon_c E_f V_y (1 - \frac{\epsilon_c E_f}{4 R_y \tau_i})$ the contribution from the supercritical fibre population. The matrix contribution is $\epsilon_c E_m (1 - V_f)$. Values of c_θ and τ_i are obtained by an iterative fitting routine. An initial value of c_θ can be obtained from microscopic studies.

References

1. H L Cox, *Brit. J. Appl. Phys.*, **3**, 72 (1952).
2. A Kelly, W R Tyson, *J. Mech. Phys. Solids*, **13**, 329 (1965).
3. C Galiotis, H Jahankhan, I Melanitis, D Batchelder in *Developments in the Science and Technology of Composite Materials* (*ECCM 3*), eds. A R Bunsell, P Lamicq, A Massiah, Elsevier Applied Science, London and New York, 1989.
4. R K Mittal, V B Gupta, P K Sharma, *J. Mater. Sci.*, **22**, 1949 (1987).
5. T W Chou, *Microstructural Design of Fiber Composites*, Cambridge University Press, Cambridge, 1992.
6. W H Bowyer, M G Bader, *J. Mater. Sci.*, **7**, 1315 (1972).

4.16 Short fibre reinforced thermoplastics

M G BADER

The principal applications for short fibre reinforced thermoplastics are as injection moulding compounds (Sections 3.8, 3.9 and 3.10) and as sheet feedstock for thermoforming operations (Section 3.11). Sheet feedstock materials are often based on random mats of continuous fibre and are not further discussed in this article. A more comprehensive review is given by Bader.[1]

Short fibre reinforced thermoplastic moulding materials are very widely used in industry. They are attractive because the addition of short fibres to the thermoplastic results in some very cost effective property improvements whilst retaining the processability of the thermoplastic. The main applications are for relatively small intricate, load bearing components. These vary from small mechanical details, such as gearwheels, pawls and levers, to casings for electric hand tools and electronic equipment.

There are now two main classes of these moulding compounds, those containing very short fibres and those containing longer fibres (Section 2.2). The fibres are most frequently E-glass of 10–20 μm diameter, the short fibres are typically less than 1 mm long in the moulding compound (i.e. before final moulding), the long fibres are typically 10 mm long at this stage. In both cases, some fibre breakage occurs during moulding so the lengths will be further reduced. The short fibre material is manufactured by mixing either prechopped E-glass roving or continuous glass roving (Sections 1.10 and 1.11) with the thermoplastic in a twin-screw compounding extruder. The most effective technology is to introduce the continuous roving at a venting port directly into the molten thermoplastic. The high shear in the extruder ensures the fibre and plastic are intimately mixed; in this type of material the fibres are dispersed as individual filaments instead of bundles or strands, as is usual in bulk or sheet moulding compounds, for example. The compound is then extruded through a 'spaghetti' die then either chilled and chopped into pellets or chopped directly with a hot face cutter. The resultant pellets are dried and shipped to the injection moulder. They are typically about 4 mm in diameter and length and contain up to 30% by volume of fibre (up to 50% by weight according to the density of the polymer).

The long fibre material is made by passing continuous roving through liquid polymer using a fluidized bed of polymer powder or a crosshead die on an extruder. The material is passed through a die to force the polymer to infiltrate the roving. Finally it is chopped into pellets, typically 10 mm long and 4 mm in diameter. The fibre runs the full length of these pellets, as shown in Fig. 28. When moulded, some fibre breakage occurs but the overall fibre length distribution will be much higher than for the short fibre material. This allows greater reinforcement efficiency to be attained, together with greater relative toughness.

The attraction of these materials lies in a combination of enhanced mechanical properties coupled with good processability. In general, stiffness is strongly enhanced in proportion to the stiffness of the fibre used, the fibre fraction, the length distribution and the orientation distribution achieved in the moulding. Strength is also enhanced but not as much as the stiffness; this is due to a reduction in the fracture strain. This low fracture strain, often only 1–2%, is sometimes an embarrassment but the impact energy (Charpy or Izod) is often enhanced by fibre reinforcement. A better balance of properties is often achieved by a judicious combination of fibre reinforcement and rubber particle toughening. There are two further important effects: the heat distortion temperature is raised and the coefficient of thermal expansion is reduced. This allows the fibre reinforced grade to be utilized at higher service temperatures, reduces in-mould shrinkage and generally improves dimensional stability. Dimensional stability is very important for intricate components, which need to be manufactured to close tolerances. Some typical properties are given in Table 47.[2]

Table 47. Typical properties of selected short-fibre reinforced thermoplastics

Polymer	Fibre	V_f	Relative density	E (GPa)	X_T (MPa)	Charpy impact (kJ m^{-2})	HDT (°C)	CTE (10^{-5} K^{-1})
Polypropylene	None	0	0.91	1.9	39	2.7	60	15
Polypropylene	Glass	0.20	1.14	7.5	110	8.0	155	2.7
Polyamide 6.6	None	0	1.14	3.2	105	>25	100	8.5
Polyamide 6.6	Glass	0.20	1.46	10	230	40	250	2.5
Polyamide 6.6	Carbon	0.20	1.28	20	250	10	255	1.9
Polycarbonate	Glass	0.20	1.45	9.0	135	40	160	2.3
Polyoxymethylene	Glass	0.20	1.58	9.0	140	9.0	165	3.5
Polyphenylenesulphide	Glass	0.20	1.65	11	155	20	263	2.7
Polyphenylenesulphide	Carbon	0.20	1.45	17	185	20	263	1.1
Poly ether ether ketone	Carbon	0.20	1.45	16	215	–	310	–

E = Young's modulus, X_T = tensile strength, HDT = heat distortion temperature, CTE = linear coefficient of thermal expansion. Data has been abstracted from several sources and normalized to a standard fibre volume fraction of 0.20 (20%). These values should be taken as guidelines and not used for design purposes.

Processability is generally very good. Although the addition of the fibres tends to increase the melt viscosity, and hence the mouldability, the pseudoplastic nature of thermoplastics means their apparent viscosity decreases with increased shear rate. The presence of rigid fibres in the melt has the effect of increasing the local shear rate so mouldability remains good up to quite high fibre loadings. The main processing problems arise from flow induced fibre orientation. In regimens of converging flow, the fibres tend to align in the direction of the flow; in regimens of divergent flow, the opposite is the case. Practical mouldings tend to be composed of thin shell-like sections so good heat transfer to the mould can be achieved with consequent reduction in the moulding cycle time. In these sections, the orientation is modified by interactions at the (cold) mould surfaces. The overall effect is to orient fibres in the principal direction of melt flow at the surfaces, but normal to melt flow in the centre of the section; this leads to a sandwich structure, Fig. 95. For an illustration of this phenomenon see Fig. 29 in Section 2.2.

This, of course, will affect the mechanical properties of the moulding, but the more important implication is that it can lead to distortion due to differential shrinkage between sections with different fibre orientations. This can destroy the advantages of the reduced coefficient of thermal expansion mentioned above. A further effect is the formation of knit or weld lines where melt fronts intersect. This occurs wherever there is bifurcated flow, e.g. around a cored feature, or when multiple gates are used (Fig. 96). These can be controlled by multilive feed technology (see Section 3.10).

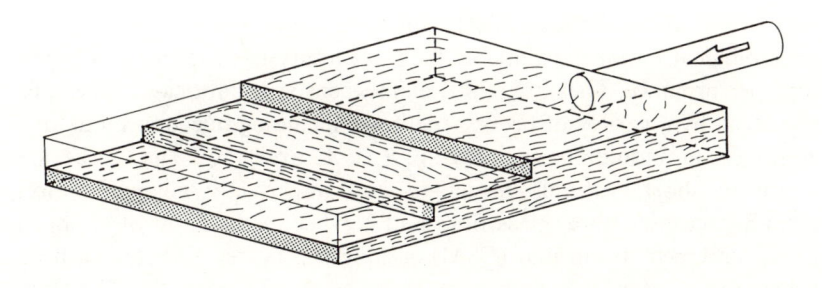

Figure 95 Illustration of the 'sandwich' fibre orientation in an injection moulded component. The fibres are aligned parallel to the mould fill direction at the surfaces and normal to that direction in the centre of the section

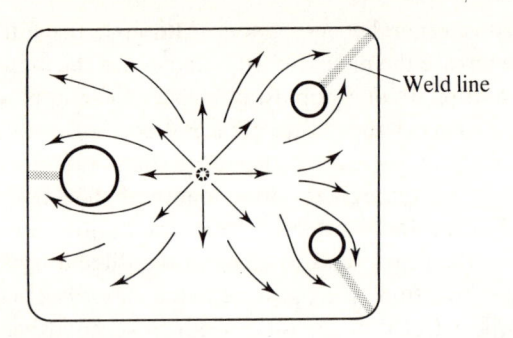

Figure 96 Weld or 'knit' lines are formed wherever melt fronts rejoin after flow around a cored feature. In this case the flow is illustrated for a centre gated plate with three circular holes

References

1. M G Bader in *Handbook of Composites*, Vol. 4, eds A Kelly, S T Mileiko, North Holland, Amsterdam, 1983, Ch. 4.
2. M G Bader in *Structure and Properties of Composites*, ed. T W Chou, Vol. 13 of *Materials Science and Technology*, eds R W Cahn, P Haasen, E J Kramer, VCH Publishers, Weinheim, Germany, 1993 in press.

4.17 Short fibre reinforced thermosets

F CHEN

Introduction

Short fibre (discontinuous fibre) reinforced thermosets are increasing in technological importance due to their versatility in properties and their high performance. They are relatively inexpensive to fabricate and can be produced by a variety of process methods, including injection moulding, compression moulding, wet lay-up, (see Chapter 3). The fibres are relatively short, variable in length and imperfectly aligned. Fibres including carbon fibre, glass fibre and Kevlar, in the form of chopped fibres, chopped strand mat (CSM), among others (see Chapter 1), have been used to reinforce thermoset matrices (see Chapter 2). The fibre orientation in the moulding is affected by the processing method and the nature of the reinforcement, for instance wet lay-up of CSM and compression moulding of sheet mould compound (SMC) and dough moulding compound (DMC) can produce composites with in-plane randomly oriented fibres. However, the fibre orientation in injection mouldings can vary from aligned to three-dimensional random, depending on the injection moulding conditions and the mould geometry. Skin–core structures with differing fibre orientation can also form (see Fig. 95).

Unlike continuous fibre composites, the mechanical behaviour of short fibre reinforced thermosets is often dominated by complex stress distributions due to the fibre discontinuity, in particular the stress concentration at the fibre ends. A number of different approaches have been used to predict the mechanical properties of short fibre thermosets, including modification of the 'rule of mixture' theory originally developed for continuous fibre composites, and the probabilistic theory. Though a full description of the relevant strength properties is not yet possible, the current theories for the prediction of the mechanical performance, with particular reference to the tensile strength, are considered in this section.

Critical fibre length

The stress transfer to the fibres in short fibre reinforced thermosets occurs by shear at the fibre/matrix interface. The stress in the fibre builds up from zero at each end to a value given approximately by[1]

$$\sigma_f = \frac{2x\tau}{r} \quad [4.70]$$

where x is the distance from the fibre end, r the fibre radius and τ interfacial shear strength.

For any given value of composite strain ϵ_c there will be a characteristic fibre length l_ϵ where the fibre stress just reaches the maximum at the centre of the fibre. In this case since

$$\sigma_f = E_f \epsilon_c \quad [4.71]$$

$$l_\epsilon = \frac{E_f \epsilon_c r}{\tau} \quad [4.72]$$

where E_f is the Young's modulus of the fibre.

There is also a special case when the fibre stress is allowed to build up to the fibre fracture stress, σ_{uf} when

$$l_c = \frac{r\sigma_{uf}}{\tau} \quad [4.73]$$

The quantity l_c is termed the critical fibre length. However, it should be noted that there is effectively a critical fibre length for any value of ϵ_c, as given by (4.72).

The critical fibre length of fibre reinforced thermosets has been determined by a number of different methods. It was reported that the critical length for glass fibre reinforced epoxy is 12.7 mm, and from a few millimetres to 15 mm for glass fibre reinforced phenolic composites.[1] It is much longer than that for glass fibre reinforced thermoplastics, which are less than 1 mm. In order to achieve the same reinforcing efficiency as in thermoplastics, the fibre reinforcement in thermosets has to be much longer.

Aligned short fibres

This type of composite can be produced by injection moulding with special mould geometry, and also by wet lay-up with special fibre reinforcement arrangement.

Brittle thermoset matrix

In most of the cases the failure strain of the thermosets is smaller than that of the fibre, i.e. $\epsilon_{um} < \epsilon_{uf}$, for instance, glass fibre reinforced phenolic composites and polyester composites. The brittle matrix will fail when the strain of the composite reaches the failure strain of the matrix, and the load will transfer to the fibres. Depending on the fibre length and volume fraction, the composite either fails immediately or can be further loaded until the fibre fractures or pulls-out.

Supposing that fibres have a uniform length and are longer than l_c, provided the matrix can transfer the load to the fibres, the load on the composite can still be increased until the fracture strength of the fibres is reached. During this time, the matrix will break further into more segments, and this is termed the multiple matrix fracture phenomenon (see Section 4.3). The average stress in the fibre is as follows[1]

$$\bar{\sigma}_f = \sigma_{uf}\left(1 - \frac{l_c}{2l}\right) = \sigma_{uf}\left(1 - \frac{r\sigma_{uf}}{2\tau l}\right) \qquad [4.74]$$

And the tensile strength of the composites, σ_{uc}, will be

$$\sigma_{uc} = \sigma_{uf}V_f\left(1 - \frac{l_c}{2l}\right) = \sigma_{uf}V_f\left(1 - \frac{r\sigma_{uf}}{2\tau l}\right) \qquad [4.75]$$

where V_f is the volume fraction of fibres of length l.

If the fibre is shorter than l_c, but with sufficient fibre volume fraction, the composite will not fail when the matrix fracture strength is reached, instead the resistance to pull-out of the fibres which bridge the matrix crack is sufficient to stand the load on the composite. The strength of the composite is equal to the frictional force needed to pull-out the fibres, and is given by[1,2]

$$\sigma_{uc} = V_f \left(\frac{\tau l}{2r}\right) \quad [4.76]$$

If the fibre length in the composites has wide variations, the fracture strength of the composite will be

$$\sigma_{uc} = \sigma_{uf} \sum_{l_i > l_c} \left[V_{fi}\left(1 - \frac{r\sigma_{uf}}{2\tau l_i}\right)\right] + \sum_{l_j \leq l_c} \left[V_{fj}\left(\frac{\tau l_j}{2r}\right)\right] \quad [4.77]$$

where l_i and l_j are the supercritical and subcritical fibre lengths with fibre volume fractions of V_{fi} and V_{fj} respectively.

In the above analysis, it was assumed that the build-up of the stress in the fibre is linear from the fibre ends (Kelly–Tyson theory), and therefore the interfacial shear stress, τ, is constant. Cox[3] developed a shear-lag theory in which the interfacial shear stress follows

$$\tau = \frac{r\beta}{2}(E_f \epsilon_f)\left(\frac{\sinh[\beta(l/2 - x)]}{\cosh(l\beta/2)}\right) \quad [4.78]$$

and

$$\beta = \left(\frac{2G_m}{E_f r^2 \ln(R/r)}\right)^{1/2} \quad [4.79]$$

where G_m is the shear modulus of the matrix, R one half of the centre-to-centre fibre spacing, and ϵ_f the strain of the fibre.

The average tensile stress in a fibre is

$$\bar{\sigma}_f = E_f \epsilon_f \left(1 - \frac{\tanh(l\beta/2)}{l\beta/2}\right) \quad [4.80]$$

And the fracture strength of a composite with $l > l_c$ is given by

$$\sigma_{uc} = \sigma_{uf} \sum_{l_i > l_c} V_{fi}\left(1 - \frac{\tanh(l_i \beta/2)}{l_i \beta/2}\right) \quad [4.81]$$

Aveston et al.[2] used a different approach and proposed that the interfacial shear strength decreases due to Poisson contraction of the fibres as the load on the composite is increased. Assuming that the decrease in bond strength is proportional to the stress in the fibre, and the fibre is shorter than l_c, the following is obtained

$$\tau = \tau_0 - \frac{KF}{\pi r^2} \quad [4.82]$$

where F is the load on the fibre, K a constant and τ_0 the maximum shear stress. So

$$F = 2\pi r l \tau = \frac{2\pi r \tau_0 l}{\left(1 + \frac{2Kl}{r}\right)} \quad [4.83]$$

and the composite strength is given by

$$\sigma_{uc} = \frac{V_f \tau_0}{K}\left[1 - \frac{r}{lK}\ln\left(1 + \frac{Kl}{r}\right)\right] \quad [4.84]$$

The work of fracture depends on the amount of pull-out and on V_f. The average work done per fibre, W_p, is

$$W_p = \frac{1}{12}\pi r \tau l^2 \quad (l \leq l_c) \quad [4.85]$$

assuming that all of the fibres pull-out so that the pull-out length varies between 0 and $l/2$. If only a fraction (l_c/l) of the fibres pull-out, the average work done per fibre is

$$W_p = (l_c/l)\frac{1}{12}\pi r \tau l^2 \quad (l \geq l_c) \quad [4.86]$$

Flexible thermoset matrix

This is the case when the failure strain of the thermoset is greater than that of the fibre, i.e. $\epsilon_{um} > \epsilon_{uf}$. Glass fibre and carbon fibre reinforced flexible epoxies belong to this type of composite. The behaviour of this sort of thermoset composite is very similar to that of most thermoplastic composites. The fracture strength for composites with high fibre volume fractions, with fibres longer than l_c, is given by[1]

$$\sigma_{uc} = \sigma_{uf}\sum_{l_i > l_c}\left[V_{fi}\left(1 - \frac{r\sigma_{uf}}{2\tau l_i}\right)\right] + \sigma'_m(1 - V_f) \quad [4.87]$$

where l_i is the individual fibre length with fibre volume fraction V_{fi}, and σ'_m the matrix stress at the failure strain of the composite.

And if all the fibres are shorter than l_c, the strength of the composites is given by

$$\sigma_{uc} = \sum_{l_i \leq l_c}\left[V_{fi}\left(\frac{\tau l_i}{2r}\right)\right] + \sigma'_m(1 - V_f) \quad [4.88]$$

If some fibres are longer than l_c, and some are shorter than l_c, combining (4.87) and (4.88) gives

$$\sigma_{uc} = \sigma_{uf} \sum_{l_i > l_c} \left[V_{fi} \left(1 - \frac{r\sigma_{uf}}{2\tau l_i} \right) \right] + \sum_{l_j \le l_c} \left[V_{fj} \left(\frac{\tau l_j}{2r} \right) \right] + \sigma'_m (1 - V_f) \quad [4.89]$$

where

$$\sum V_{fi} + \sum V_{fj} = V_f$$

Equation (4.89) can be used to predict the fracture strength of short fibre reinforced flexible thermosets. Using a simplified form of (4.89) it has been found that the mechanical strength of an aligned short fibre composite approaches that of an aligned continuous fibre composite when the fibre length is 10 l_c, and 90% when the fibre length is 4 l_c.

In-plane random and three-dimensional random fibres

Wet lay-up of CSM, and some injection mouldings have fibres randomly oriented in plane and/or three-dimensionally. The fibres may not be completely random but show a degree of preferred orientation and may vary from one part of the moulding to another. The fibres in the composite may be held together in bundles or dispersed as individual fibres. When the fibres are in a bundle form, the bundle will act as the reinforcing unit. Thus the reinforcing efficiency and the fracture strength will depend on the degree of dispersion.

When load is applied to the thermoset composite, assuming the strain is uniform over the cross-section of the composites, since the transverse strain of the composite is smaller than the longitudinal strain, failure will first occur at the bundles in which fibres are oriented transverse to the load direction. As the strain increases, cracks will also occur in bundles in which fibres are oriented at other angles to the load direction.

There is no entirely satisfactory theory to express the fracture strength of composites with random orientations. One approach is to use an empirical method to predict the composite strength. For composites with a brittle matrix, the fracture strength is given by

$$\sigma_{uc} = C_1 C_2 \sigma_{uf} V_f \quad [4.90]$$

where C_1 is the fibre reinforcing efficiency factor, and is the ratio of the strength of aligned short fibre reinforced composite over that of continuous aligned fibre reinforced composite. C_2 is the fibre orientation factor. For mouldings with fibres shorter than l_c, $C_2 = 2/\pi$ for complete in-plane random fibre composites and $C_2 = 1/2$ for complete three-dimensional random fibre composites.[2]

In recent years, probabilistic aspects have been introduced to predict the fracture strength of composites.[4] Some numerical work has also been undertaken to study the fibre reinforcement with two-dimensional and three-dimensional finite element analysis. All of these will invariably enrich our understanding of the micromechanics of short fibre reinforced thermoset composites.

Typical properties of discontinuous fibre reinforced thermosets

In general theoretical strength and stiffness are not achieved in real mouldings because of the complexities of load transfer discussed above. However, for longer fibres (e.g. CSM/unsaturated polyester), the low strain properties approach those for continuous fibres with appropriate corrections for fibre orientation and length. Matrix cracking and fibre debonding contribute to non-linear stress–strain curves and the ultimate strength. Typical properties are given in Fig. 97. For short fibres these depend on a number of factors and Table 48[5] summarizes them.

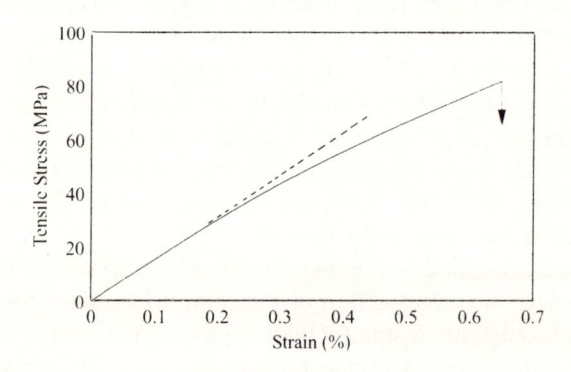

Figure 97 Typical tensile stress–strain curve for injection mouldings of short fibre reinforced phenolic composite. (Average fibre length = 0.52 mm. Fibre volume fraction = 34%)

Table 48. Typical mechanical properties of short fibre reinforced thermosets

	DMC		SMC		Injection Moulding
	Polyester	Phenolic	Polyester	Phenolic	Phenolic
Tensile strength (MPa)	60	30	70–95	70	45–55
Flexural strength (MPa)	126	75	150–185	155	100–150
Flexural modulus (GPa)	11	9	11–13	9	8–12
V_f (%)	15–25	29	25–40	29	20–33

References

1. A Kelly, N H MacMillan *Strong solids*, 3rd edn, Clarendon Press, Oxford, 1986, pp. 213–325.
2. J Aveston, R A Mercer, J M Sillwood, in *Proceedings of Composites – Standards, Testing and Design*, National Physical Laboratory, IPC Science and Technology Press, Guildford, UK, 1974, pp. 93–103.
3. H L Cox, *Br. J. Appl. Phy.*, **3**, 72 (1952).
4. F Hikami, T W Chou, *J. Mater. Sci.*, **19**, 1805 (1984).
5. *Replacement of metals with plastics*, National Economic Development Office (Plastics Processing EDC), London, 1985, p. 19.

CHAPTER 5

Mechanical Properties – macromechanics

5.1 Damage accumulation – fracture mechanics

S L OGIN

In a general composite laminate containing off-axis plies, various types of damage occur under load; these cause load redistribution and can introduce non-linearity into the stress–strain response. The types of damage can be split into three broad categories: matrix cracking, delamination and fibre breakage. These broad categories of damage also occur in components containing a crack or a notch and under cyclic loading.

Matrix cracking in off-axis plies[1] is usually the first significant damage; debonding of fibres lying at an angle to the loading direction occurs at an earlier stage but is much more difficult to detect. As the load is increased, the density of these cracks increases and appears to stabilize at a unique value for a given laminate, the 'characteristic damage state'. Splitting of 0° plies parallel to the fibre direction is also a form of matrix cracking, caused by the Poisson mismatch between plies of different orientations.

Delamination occurs between layers of a laminate and is usually initiated by the interlaminar stresses near a free edge.[2] These out of plane stresses arise because the in-plane transverse and shear stresses in each ply of the laminate, which can be calculated using laminated plate theory, must decay to zero at the free edge. The sign of the interlaminar stresses at the edge, with tensile stresses tending to delaminate the coupon, depends upon the stacking sequence. Delaminations normally initiate at a free edge but can also occur at the intersections of matrix cracks in adjacent plies.

Fibre breakage under static loading before final failure is related to the statistical distribution of flaws along the fibre length; weak fibres break at lower laminate strains. Any stress concentration, e.g. a matrix crack in an adjacent ply can cause additional fibre fractures.[3] The fibre ends produced as a result of fibre fracture are suitable sites for the initiation and growth of fibre/matrix debonds due to the high shear stresses at the interface near the end of a fibre. These debonds are often very difficult to detect with standard microscopical techniques.

Fracture mechanics

The characteristic feature of damage accumulation in composite materials under monotonic and cyclic loading is the concurrent development of the different categories of damage outlined above. In the general case, all forms of damage are present and will interact in complex ways. For example, when loading a crossply laminate containing a notch, splitting of the 0° plies, delamination of the 0° plies from the 90° plies, transverse ply cracking and fibre fracture in the 0° plies occur simultaneously.[4] The damage accumulated in an unnotched composite will contribute in differing degrees to a reduction in laminate properties.

Most engineering materials other than composites are homogeneous and isotropic. In homogeneous isotropic materials, the growth of a crack to cause component failure can be predicted using the theoretical and experimental techniques of fracture mechanics[5] to determine whether it will grow catastrophically to fail the component under the applied load or grow under conditions of cyclic loading or stress corrosion.

The extension of fracture mechanics to composite materials is an area of much theoretical and experimental activity. Agarwal and Broutman give a useful introductory review.[6] Fracture mechanics has been used successfully in developing an understanding of the growth of different categories of damage, e.g. matrix cracking and delamination; the catastrophic growth of macroscopic cracks to failure in certain composites; and the extension of damage under fatigue loading and stress corrosion.

The fundamental equations of linear elastic fracture mechanics were developed for homogeneous isotropic materials. They treat (i) the stress field around a crack tip, from which follows the stress intensity factory, K, and the material property known as the critical stress intensity factor, or fracture toughness; and (ii) the energy balance during crack growth, from which follows the strain energy release rate, G, and the material property known as the critical strain energy release rate, or toughness. The parameters K and G, and hence the material properties, are related in a simple fashion in homogeneous isotropic materials via an expression involving the Young's modulus and the Poisson's ratio of the material. An additional requirement is that self-similar crack growth occurs, i.e. that crack growth occurs by a simple extension of the crack without branching or directional changes.

In composite materials, two factors need to be taken into account when considering the growth of a particular type of damage or the catastrophic growth of a macroscopic crack. Firstly, the expressions used for the strain energy release rate and the stress intensity factor need to be treated with care. For example, the expression for the strain energy release rate based on compliance change with crack length[5] is valid for all linear elastic materials, whether isotropic or anisotropic, homogeneous or heterogeneous. On the other hand, an (isotropic) expression for the stress intensity factor is not necessarily valid for composites, because the anisotropy may introduce a different degree of interaction between the crack tip and the external boundaries of the specimen requiring modifications to the isotropic relationship. For orthotropic composites, which have three mutually perpendicular planes of material symmetry (e.g. a composite lamina), the energy release rate and the stress intensity factor can be related by an expression which depends on the elastic constants[7]. Secondly, crack extension in composite materials is usually not self-similar. For example, an initial through thickness crack in a composite laminate can give rise to fibre breakage, matrix cracking and delamination ahead of the crack tip when loading to failure occurs. An exception to this is the growth of a transverse ply crack in a crossply laminate or crack growth in short carbon fibre reinforced epoxy resin, where very little crack branching occurs at the crack tip.

For composite materials, then, experimental techniques and theoretical analysis related to the strain energy release rate are often a fruitful way to proceed. They can, for example, enable experimental measurements of the material toughness to be made with confidence and, in some cases, enable comparisons between experimental measurements and theoretical predictions of compliance changes[8].

Techniques based on energy changes/compliance changes work well if self-similar crack growth occurs in the composite material, i.e. a simple extension of the crack with very little branching or directional changes. In other cases, it may be possible to take into account the contribution of different categories of damage occurring at the crack tip when trying to make comparisons between theory and experiment. For example, experimental techniques have been developed to quantify increasing resistance to crack growth with increasing crack length, which is often a feature of composite behaviour.[6] However, it is generally very difficult to account satisfactorily for all the contributions to energy changes in a composite when more than one type of damage has to be considered.

In applying fracture mechanics to composite materials, then, the difficulties lie in (i) establishing whether parameters measured are material properties; (ii) including the various types of damage in any theoretical treatment of a problem in an appropriate way; and (iii) identifying the type(s) of damage being measured in experiments.

References

1. J E Bailey, P T Curtis, A Parvizi, *Proc. R. Soc. Lond.* **A366**, 599 (1979).
2. R B Pipes, N J Pagano, *J. Composite Mater.*, **4**, 538 (1970).
3. K L Reifsnider, R Jamison, *Intl. J. Fatigue*, **4**, 187 (1982).
4. P W R Beaumont, *J. Strain Anal.*, **24**(4), 1 (1989).
5. D Broek, *Elementary Engineering Fracture Mechanics*, Martinus Nijhoff, Amsterdam, 1984.
6. B D Agarwal, L J Broutman *Analysis and Performance of Fiber Composites*, John Wiley, New York, 1990.
7. Sih G C, Paris PC, Irwin G R, *Intl. J. Fract. Mech.* **1** (3), 189–203 (1965).
8. L Boniface, S L Ogin, P A Smith, *Proc. R. Soc. Lond.* **A432**, 427 (1991).

5.2 Delamination – fracture mechanics

S L OGIN

Delamination, or the peeling apart of the plies of a laminate, is one of the easiest types of damage to observe but one of the most difficult to deal with quantitatively. It can occur under monotonic or cyclic loading and is a major cause of failure in laminated composite materials.

The driving force for delamination is the mismatch in elastic constants between adjacent plies, or groups of plies, in a laminate. Herakovich[1] provides a very helpful description of the relationship between this mismatch and the interlaminar stresses. For example, in a ($\pm 30°/90°$) CFRP laminate, the Poisson's ratio for the $\pm 30°$ group of plies is much larger than for the 90° plies. When loaded axially, the difference in Poisson's ratio gives rise to non-zero transverse stresses in the layers of the laminate. These stresses must fall to zero at the free edge and equilibrium of stresses requires that non-zero interlaminar stresses are generated. If these stresses are positive at an interface between plies, the plies will tend to peel apart.

The sign of the interlaminar stresses depends on the stacking sequence of the plies and on the sign of the applied load so that reversing the sign of the load from tension to compression simply reverses the sign of the interlaminar stresses. Interestingly, the width of the region within which the three-dimensional stresses exist is approximately equal to the thickness of the laminate.[2]

Having established that a laminate will have a tendency to delaminate, the next question is how to predict the onset and growth of a delamination. Unfortunately, stress analyses suggest that the interlaminar stresses driving delamination become infinite at the laminate free edge. Such predictions are not particularly helpful because they mean that a failure criterion based on a maximum interlaminar stress for delamination to occur is not possible. Hence, an alternative characterization based on fracture mechanics has been developed.

The early work of O'Brien[3] is a useful starting point in understanding the use of fracture mechanics in characterizing delamination onset, and growth. In essence, a general expression for the strain energy release rate with delamination area is derived on the assumption that the only form of damage is delamination and that this produces a decoupling of the laminates into sublaminates. This expression is then used, together with experimental observations of delamination onset, to find the value of the critical strain energy release rate, also called the fracture energy or toughness, for the onset of delamination. The analysis can then be used to predict the onset of delamination in different laminate lay-ups of the same material. The strain energy release rate approach can thus be useful in predicting the onset of delamination if the interfaces likely to delaminate have been identified from a stress analysis and the fracture energy is known. Unfortunately, complications for this simple description of delamination cannot be ignored. For example, the delamination can occur in a crack opening mode, a crack shearing or sliding mode, or a mixture of the two. A rigorous basis for characterizing the material needs to be able to combine the contributions of these modes to the overall cracking behaviour in an appropriate way.[4]

Secondly, a simple theoretical expression for the strain energy release rate, based on delamination as the only type of damage, is independent of delamination size. This means that, once initiated, a delamination should grow catastrophically across the width of the laminate. But delaminations usually grow in a stable manner and an increased strain is required to propagate them in monotonic loading. The stable growth is related to the interaction of the delamination front with other types of damage, particularly matrix cracking. It is possible to characterize experimentally the increasing resistance to delamination with the aid of a delamination resistance or *R-curve*; this quantifies the increase in delamination resistance with delamination area. However, this characterization is no longer general for the material because it depends on the interaction of the delamination with other types of damage; interaction will be affected by structural variables such as ply thickness and stacking sequence.[5]

In addition to growth under monotonic loading, delaminations will grow under fatigue loading. Figure 98 shows an example of a delamination in a $(0_2^\circ/90_2^\circ/\pm 45^\circ)_S$ CFRP laminate. After 550,000 cycles, the delamination is just visible at the laminate edge with the aid of penetrant enhanced X-radiography. After a further 150,000 cycles, the delaminated area is roughly 40% of the laminate area. Fracture mechanics has been used to characterize both the onset of delamination

(a)   (b)

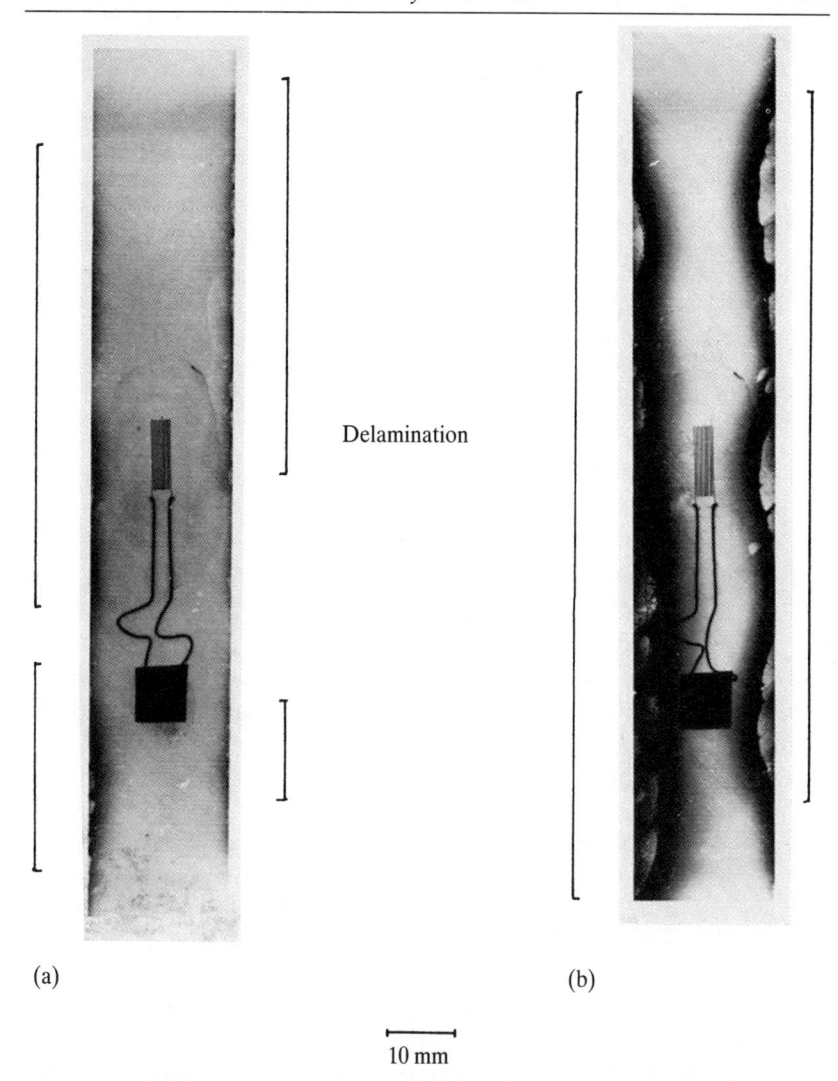

10 mm

Figure 98 Delaminations in a $(0_2°/90_2°/\pm 45°)_5$ CFRP laminate after (a) 550,000 cycles and (b) 700,000 cycles at a peak strain of 0.6%. (Photographs courtesy of L. Boniface)

under cyclic loading and the delamination growth rate. For example, the number of cycles before the onset of delamination for a given cyclic loading has been determined for different fibre/matrix systems and the delamination growth rate has been determined for 'tough' and 'brittle' systems.[5]

References

1. C T Herakovich, *J. Strain Anal.*, **24**, 57 (1989).
2. R B Pipes, N J Pagano, *J. Composite Mater.*, **4**, 538 (1970).
3. T K O'Brien, *Characterisation of Delamination Onset and Growth in a Composite Laminate*, in ASTM STP 775, American Society for Testing and Materials, Philadelphia, 1982, pp. 140–167.
4. S Hashemi, A J Kinloch, J G Williams, *Mixed-mode Fracture in Fiber–Polymer Composite Laminates*, in ASTM STP 1110, American Society for Testing and Materials, Philadelphia, 1991, pp. 143–168.
5. T K O'Brien, *Towards a Damage Tolerance Philosophy for Composite Materials and Structures*, in ASTM STP 1059, American Society for Testing and Materials, Philadelphia, 1990, pp. 7–33.

5.3 *Design considerations for aerospace applications*

G W MEETHAM

Since the early 1930s, aluminium alloys have been the basic materials used in airframe construction for all types of aircraft. With the advent of the jet engine, aircraft speed increased and by 1959 a speed of some 1500 mph was achieved by the F106 Starfighter. This order of speed has become a plateau because the structural heating experienced at higher speeds requires materials with higher temperature capability. For example, the Mach 3 SR71 Blackbird is made essentially of titanium alloy. In many new aircraft designs, however, the dominant position of aluminium alloys is being challenged by polymer composite materials.

The basic requirement of any aircraft, whatever its application, is to deliver the required performance with minimum life-cycle cost. Aircraft weight is a factor of critical importance; its influence ranges from the agility of military combat aircraft to the range and payload of commercial transport aircraft. For example, at current price levels, fuel accounts for some 25% of the total operating cost of a typical passenger aircraft and a weight reduction of 100 kg on such an aircraft could result in a fuel saving of over 60,000 gallons per year. Current designs of fighter aircraft must operate over broader combat envelopes than their predecessors, with supersonic persistance, greater turn rates and higher load factors; structural weight reduction is essential to achieve the necessary performance capability.

It has been shown[1] that the use of new improved materials can make a greater contribution to weight reduction than other advanced design concepts. Reduction in material density can be more effective than increase in strength or stiffness in reducing aircraft weight. Other things being equal, a 10% reduction in density will achieve three times the weight saving that would be possible with a 10% increase in strength or stiffness.[2] This has motivated major activity on polymer composite materials. First generation composites – glass fibre in phenolic or polyester resins – have been used since the 1950s in interior aircraft fittings, such as cargo floors and luggage lockers. The stiffness of GRP is inadequate for structural applications so boron fibre reinforced epoxy resin was used for early components, such as horizontal stabilizers in the F14 Tomcat and F15 Eagle aircraft. Although boron fibre has a stiffness some five times that of glass fibre, its major disadvantage is the high cost inherent in the method of manufacture. Thus the application potential for polymer composite materials increased dramatically with the development in the 1960s of a process for the production of high strength, high modulus carbon fibre. Carbon fibre epoxy composites have excellent specific strength and stiffness (Fig. 99) and the fibre architecture can be tailored to meet anisotropic property requirements. Compared with conventional 2000 and 7000 series aluminium alloys, weight savings of 20–40% are possible in wing, fuselage, elevator, rudder and aileron applications.[1]

First generation carbon fibre composites were susceptible to damage, even from relatively low energy impacts. Significant improvements have subsequently been made. Carbon fibres with higher breaking strain have been developed by attention to precursor quality and processing

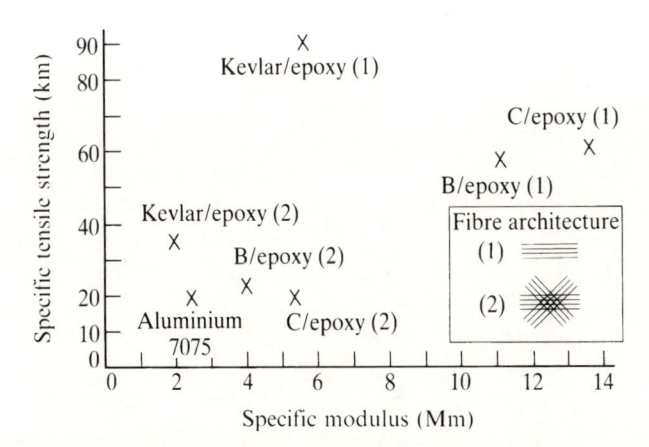

Figure 99 Specific strength and stiffness of fibre composite materials (40–60% volume fraction fibres)

conditions and have been combined with new epoxy resin systems with improved toughness. Impact resistance can be further increased by the use of modified architecture, such as the selective incorporation of woven fabric in the lay-up. Reduction in composite mechanical properties at high temperature, particularly in the presence of moisture has led to the development of resins with improved hydrothermal stability. The manufacturing technology for thermosetting epoxy resin composites allows the production of complex shapes by relatively simple lay-up, moulding and curing processes. Early manufacture was largely by hand techniques; automated processes were progressively introduced to improve quality and reduce costs. Thermoplastic resins suitable for producing structural components with long fibre reinforcement have recently become available. These are potentially more damage tolerant and environment resistant but yet to be solved are manufacturing problems such as prepreg handling, higher processing temperature, etc.

The use of polymer composite material since the mid 1960s has increased with each new type of military aircraft (Fig. 100). Early applications involved the replacement of aluminium in secondary structures not critical to aircraft safety. They have since been applied to primary structures critical to airworthiness and safety. Thus the AV8B Harrier aircraft uses carbon-epoxy composite in the forward fuselage, horizontal stabilizer, wing box structure and skin, overwing fairings, ailerons and flaps – some 25% of the airframe weight. Application to civil transport aircraft has tended to follow military experience. Composites were initially used for items such as leading and trailing edge panels on

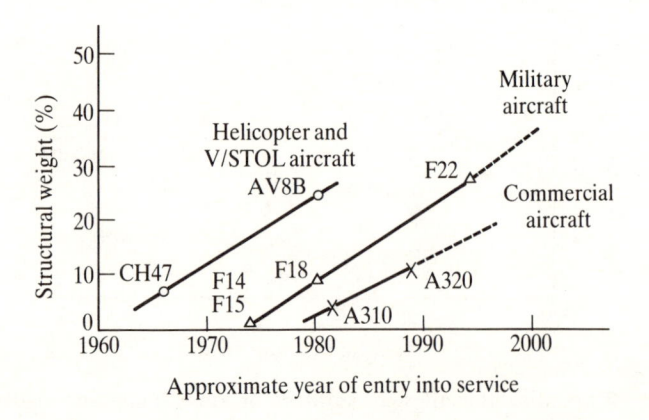

Figure 100 Polymer composite usage in airframes

Airbus A300/A310 wings in the early 1980s. Their use is rather more extensive on the A320 – flaptrack fairings, spoilers, airbrakes, ailerons undercarriage panels and fairings. Some 12% of the airframe weight is in composite materials, giving a claimed weight reduction of more than 800 kg. The new MD 11 aircraft reportedly uses about 3000 kg of composite.

Composite structures have been used in helicopters since the late 1950s and the latest designs use composite extensively for cost and performance considerations. Cost savings have been achieved in situations where complex structures can be fabricated as a single integrated component compared with an assembly of a large number of metal parts. Carbon fibres are used in parts requiring maximum stiffness and compressive strength, whereas the higher breaking strain of aramid fibres has led to their preference where maximum damage tolerance is required. More recently, composite materials have become widely used for helicopter rotor blades.[3] The life of metallic blades is governed by fatigue; composite blades have superior fatigue life and enable more efficient aerofoil designs to be created.

The ability to shape smooth surfaces with a high degree of dimensional accuracy accounts for the almost universal use of polymer composites in high performance gliders where aerodynamic profile accuracy is critical. It is of interest to note that two recent unique aviation achievements would not have been possible without the use of carbon and Kevlar composites. These are the first manpowered flight across the English channel by the Gossamer Albatross and the first non-stop non-refuelled circumnavigation of the world by the Voyager aircraft. In recent years, composites have been used for the primary structure of two executive jet aircraft – the Lear Fan 2100, which contains some 70% composite and the Beech Starship.

Polymer composites were first applied to jet engines during the 1960s; they were used for compressor blades and casings in a lift jet concept where minimum engine weight was a key requirement.[4] Factors such as erosion resistance and temperature capability have prevented their subsequent use in core engines, although the development of high temperature polyimide resins with temperature capability up to 300°C could modify this situation.

The need to minimize weight is critical in space applications so polymer composites are frequently used in spacecraft structures and in associated equipment, such as supports for solar arrays, dish antennae and pressure vessels for propellent gases. Polymer composites also find application because of considerations other than weight saving. For example, their low axial thermal expansion and high thermal conductivity enable high dimensional stability to be achieved in satellite structures, which may experience some degree of thermal cycling during movement in space. Low *dielectric constant* characteristics of glass or aramid fibre composites meet the need for signal transparency in radome applications.

References

1. F H Froes, *Materials and Design*, **10**(3), 110 (1989).
2. C A Stubbington, *Metals and Materials*, **4**, 424 (1988).
3. N G Marks, *Metals and Materials*, **5**, 456 (1989).
4. H E Gresham, C G Hannah *J. R. Aero. Soc.*, **171**, 677 (1967).

5.4 Failure criteria

P A SMITH

The definition of failure of a composite laminate is not straightforward. An obvious starting point is the maximum load the laminate can take; this usually coincides with fibre failure in the principal load-bearing plies. But there are other types of damage, notably matrix cracking, which may develop well before maximum load. Matrix cracking in off-axis plies is often referred to as subcritical damage because it does not lead directly to laminate failure in the same way that critical damage (such as fibre breakage in longitudinal plies) does. This does not mean that matrix cracking is unimportant, as over a period of time, matrix cracks may lead to long-term property degradation. Moreover, there are some applications for composite materials in which it may be desirable to have no cracking at all, in pressure vessels for example.

Consequently for some applications it may be necessary to design with matrix cracking as the critical event while in others overall laminate failure may be appropriate. An alternative critical condition might be one based on residual stiffness. In each case, a failure criterion is used to predict the load at which the various damage events occur then a suitable safety factor applied to evaluate the permissible service loadings. In the present article, we restrict ourselves to considering some of the failure criteria that can be used to predict failure of a lamina under arbitrary in-plane loading. In the final section, the application of these criteria to laminates is considered.

Non-interactive failure criteria

Consider the lamina shown in Fig. 101 (see also Sections 4.1 and 4.4). The applied stresses have been resolved into a component parallel to the fibre direction, σ_1, a component perpendicular to the fibre direction, σ_2, and a shear stress, τ_{12}. There are five distinct characteristic strength properties of a lamina, for each of which there is a corresponding characteristic strain. The properties are a tensile and compressive strength parallel to the fibre direction, X_t and X_c, a tensile and compressive strength perpendicular to the fibre direction, Y_t and Y_c, and a shear strength, S. Methods of measuring some of these properties are described in Section 5.16.

According to the maximum stress failure criterion, the lamina will fail when one of the applied stresses (σ_1, σ_2, τ_{12}) reaches the corresponding lamina strength. The maximum strain failure criterion states that the lamina will fail when one of the strains (ε_1, ε_2, γ_{12}) reaches the critical value associated with one of the five possible modes of failure. Consequently both the maximum stress and maximum strain failure criteria provide predictions of failure mode as well as failure stress. The two criteria are very similar and differ only as a result of Poisson's ratio effects; of the two, engineers tend to prefer the maximum strain failure criterion, as it seems more acceptable physically.

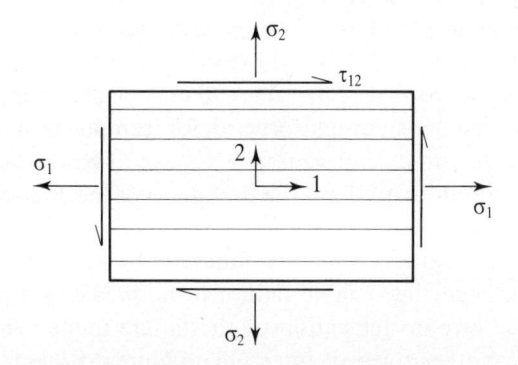

Figure 101 Schematic diagram of unidirectional lamina showing loading applied in the principal material directions 1 and 2.

Interactive failure criteria

The limitation of the maximum stress (or maximum strain) failure criterion is that it does not allow for interaction between stresses. This is particularly important for matrix failure when the combined effect of transverse tension and shear stresses is to lead to failure before either of the stresses reaches their corresponding critical value. The interaction effect was first quantified in the Tasi-Hill criterion[1]

$$(\sigma_1/X)^2 - \sigma_1\sigma_2/X^2 + (\sigma_2/Y)^2 + (\tau_{12}/S)^2 = 1 \qquad [5.1]$$

According to this approach, failure will only occur if the combination of stresses (σ_1, σ_2, τ_{12}) results in the left-hand side of the equation exceeding unity. Equation (5.1) does not distinguish between tensile and compressive strengths; they are assumed to be equal. This problem can be overcome simply by using X_t or X_c (for X) and Y_t or Y_c (for Y) depending on whether σ_1 and σ_2 are positive or negative. As a result of allowing for stress interactions, the Tsai–Hill equation can give good agreement with experimental data. The approach was revised subsequently by Hoffman,[2] who included linear terms in the equation. These enable the problem of the difference in tensile and compressive strengths to be overcome explicitly.

A more generalized form of interactive failure criterion was developed by Tsai and Wu.[3] In this case, additional linear terms result along with a term involving the product $\sigma_1\sigma_2$,

$$A_{11}\sigma_1^2 + A_{12}\sigma_1\sigma_2 + A_{22}\sigma_2^2 + A_{66}\tau_{12}^2 + B_1\sigma_1 + B_2\sigma_2 = 1 \qquad [5.2]$$

The principle of the Tsai–Wu criterion is the same as for the Tsai–Hill criterion; failure will occur if the combination of applied stresses is such that the left-hand side is greater than unity. The six coefficients are regarded as material properties. Five of them, A_{11}, A_{22}, A_{66}, B_1, B_2, can be expressed in terms of the five characteristic lamina strengths, e.g. $A_{11} = 1/X_t X_c$, $A_{66} = 1/S^2$, $B_1 = 1/X_t - 1/X_c$, etc. But there is a slight difficulty with the coefficient A_{12}; this can only be determined by carrying out a biaxial test, not straightforward for composite materials. Since predictions of failure do not generally appear to be too sensitive to the value chosen, this does not limit the usefulness of the Tsai–Wu prediction significantly.

Both the Tsai–Hill and Tsai–Wu failure criteria suffer from the same problem. Although they enable failure to be predicted with reasonable accuracy, they give no indication of the failure mode. This problem is overcome in another form of interactive failure criterion, suggested by Hashin,[4] which distinguishes between fibre and matrix dominated failure. Hart-Smith[5,6] has highlighted other difficulties associated with the criteria presented by Tsai and co-workers. Instead he proposes a criterion based on a modified maximum shear stress (Tresca) for predicting fibre-dominated failures in cross-plied laminates.

Application of failure criteria to laminates

Any of the failure criteria described above can be used to predict progressive ply failures in multi-ply laminates under arbitrary loading and this is often done as part of computer based laminate analysis programmes. The procedure is to calculate the ply stresses in each ply, including any thermal and moisture effects for unit applied load (see Sections 4.12 and 6.3). The stresses in each ply can then be resolved into the principal material directions and the chosen failure criterion applied. This determines the applied load necessary to cause first ply failure, usually matrix cracking in any 90° plies (see Section 4.11). Once damage has occurred, some assumption has to be made regarding the degraded stiffness of a damaged ply so that the redistribution of stress associated with the damage can be deduced. After this has been done, it is necessary to check that the increased stresses in the other plies do not cause any of them to fail. Then the applied load is incremented until another failure occurs and the process repeated until overall laminate failure occurs, usually by fibre fracture in the 0° plies.

References

1. V D Azzi, S W Tsai, *Exptl Mech.*, **5**, 283 (1965).
2. O Hoffman, *J. Composite Mater.*, **1**, 200 (1967).
3. S W Tsai, E M Wu, *J. Composite Mater.*, **5**, 58 (1971).
4. Z Hashin, *J. Appl. Mech.*, **47**, 448 (1980).
5. L J Hart-Smith, in ASTM STP 1120, American Society for Testing and Materials, Philadelphia, 1992, pp. 142–69.
6. L J Hart-Smith, in ASTM STP 1156, American Society for Testing and Materials, Philadelphia, 1993, pp. 363–80.

5.5 *Fatigue – aramid fibre reinforced plastics*

B HARRIS

Introduction

Polymeric fibres broadly designated as aramids or aromatic polyamides were first introduced by Dupont under the tradename Kevlar, although other companies do now market them (see Section 1.3). The majority of research work on fatigue of composite laminates involving aramids specifically relates to the higher performance Kevlar 49 fibre. This is a highly drawn poly(paraphenylene terephthalamide) derivative aimed at the aerospace market but, on account of its characteristic fibrillar nature, it has also found service in many other applications where toughness and damage tolerance are required. The fatigue response of aramid composites referred to in this article therefore relates mainly to Kevlar 49; more recent varieties of aramid have yet to receive the same kind of wide-ranging evaluation.

The other two major reinforcing fibres for reinforced plastics, glass and carbon, are supplied, like Kevlar, as multifilament tows, or cloth, of which the individual filaments are of the order of 10 μm in diameter, but they may both be regarded as essentially monolithic inorganic entities, despite the textile origin of the carbon. The fatigue response of GRP and CFRP is substantially determined by the stiffness of the fibres and by the strength of the interfacial bond. The structure of the fibre, per se, has no bearing on fatigue response, and although glass, by virtue of its moisture sensitivity, shows time-dependent fracture behaviour (stress corrosion), neither glass nor carbon fibres, qua fibres, exhibit any true cyclic fatigue effects. Kevlar fibres, by contrast, have a more complex structure than carbon and glass. As a consequence of the textile manufacturing process, the individual filaments are effectively composed of a multitude of finer fibrils weakly bonded together and there is a distinct skin/core structure. Dobb[1] has shown that Kevlar fibres contain an extensive internal defect structure of microvoids and cracks. It is also observed that moisture weakens the fibre by penetrating this defect structure and by weakening the interfibrillar hydrogen bonding, causing extensive splitting. The core/skin bonding is also weakened, leading to composite failure at this interface. In addition to its effect on moisture sensitivity, the fibrillar structure of the fibres results in relatively poor performance of aramid fibre reinforced composites when subjected to flexural or compression loads. The poor flexural and compression behaviour of aramid fibres is linked to their tendency to form kink bands under compression; this tendency is greater the higher the tensile modulus of the fibre. Kevlar 49 is thus more susceptible to this kind of damage than the lower performance variety, Kevlar 29. A clear indication of the difference between the compression performance of simple unidirectional carbon and glass fibre composites (CFRP and GRP), on the one hand, and Kevlar 49/resin composites (KFRP) is the serious consequence of the composite shear collapse under compression loads that appears to be initiated by the kinking behaviour of the fibre itself within the matrix.

Fatigue of Kevlar fibres

Early work on Kevlar-49 showed that the individual polymer filaments themselves suffered a cyclic fatigue phenomenon. However, fatigue failure only occurred above 80% of their tensile failure load; the fibres are therefore considered to be highly resistant to fatigue.[2] Bunsell also noted that raising the minimum load in a tensile fatigue cycle significantly increased the life for a given maximum load level.[2] Tensile fatigue failure, like monotonic tensile failure, is accompanied by destructive fibrillation of the filament, and it is unclear whether monotonic and fatigue failures are mechanistically distinguishable. The fatigue resistance of Kevlar-29 has also been shown to be sensitive to the maximum load and the load amplitude and it shows a marked deterioration of properties during cycling. The structural damage resulting from the compression kink banding of the fibres, referred to earlier, resulted in serious loss of strength during flexural fatigue cycling of individual filaments; Kevlar-49 is considerably worse in this respect than Kevlar-29.

Fatigue of KFRP

Early reported experimental stress/life results of KFRP composites suggested some diversity of opinion: on the one hand an $S/\log N_f$ curve, S–N plot, that falls linearly with only a slight slope, resembling familiar results for early unidirectional CRFP composites; on the other, a steeply falling $S/\log N_f$ curve, which was not in accord with the high fatigue resistance of the individual Kevlar filaments. In a series of papers on plain Kevlar and Kelvar hybrid composites, Harris and coworkers[4] have shown that the fatigue behaviour of KFRP appears to follow a different pattern from those of either CFRP or GRP. The family of $S/\log N_f$ curves in Fig. 102 shows that for unidirectional, crossply (0°/90°), and $[(\pm 45°/ 0°_2)_2]_S$ laminates, the tensile fatigue curve is very flat at high

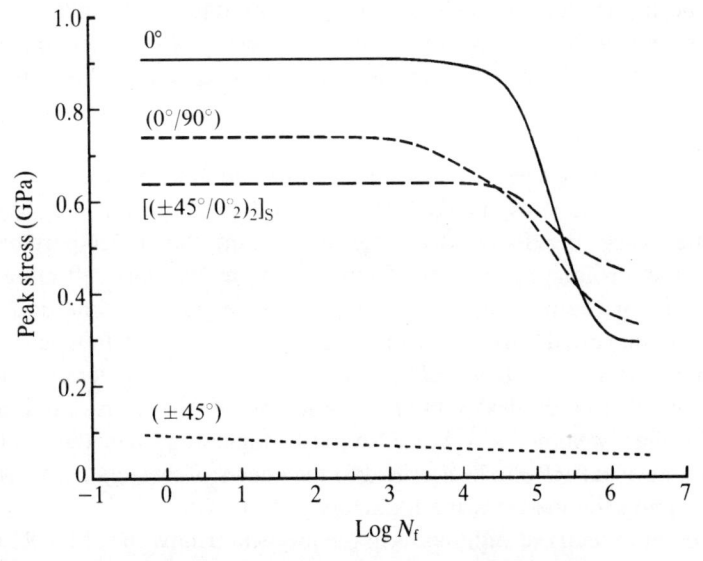

Figure 102 Family of stress/life curves for tensile fatigue (R = +0.1) of Kevlar fibre reinforced composites of various layups

levels (short lives), but that at low to intermediate stress levels this promising high stress behaviour is not sustained; the curve falls rapidly with a downward curvature. This characteristic inverse sigmoidal shape is reminiscent of the response of single Kevlar-29 filaments, has seriously limiting consequences for designers and is shown by all three structures which contain some proportion of unidirectional plies, although the failure mode change that causes it appears to originate earlier and to have somewhat less drastic consequences in the 0°/90° composite. The behaviour of the ±45° laminate reflects, in principle, the low shear fatigue resistance of the matrix resin. An additional effect of the poor shear properties of the fibres themselves is also indicated by the position of this curve. Strain/life data for composites of identical lay-up with the same resin matrix but containing HTS carbon or E-glass fibres fell on a single curve above the Kevlar curve and values were 50% higher at all lives. The $S/\log N_f$ curve for the $[(\pm45°/0°_2)_2]_S$ composite containing 50% 0° and 50% 45° plies, would fall significantly above a notional rule of mixtures line constructed from the separate 0° and ±45° curves, and it appears therefore that the lower working strain levels in a composite containing 0° plies effectively protects the off-axis Kevlar fibres from the deleterious effect of shear forces. Kink banding and fibrillation occur in the Kelvar fibres as a result of cyclic loading; they are illustrated in Fig. 103.

The fatigue response of KFRP composites depends sensitively on the stress ratio, $R = \sigma_{min}/\sigma_{max}$, as shown in Fig. 104. For **R ratios** of 0.1 and 0.01, both pure tension regimes, the stress/life data points overlap to a large extent. As the compression component of the stress cycle increases (increasingly negative R ratio), the peak stress versus life curve is increasingly depressed and the life for a given peak stress is markedly reduced, although all the curves appear to be converging for lives longer than about 10^6 cycles. Raising the R ratio to 0.5, by increasing the minimum tensile stress in the cycle ($\sigma_{min} = 0.5\sigma_{max}$; $\sigma_{mean} = 0.75\sigma_{max}$), has the clearly beneficial effect of extending the flat upper part of the curve and prolonging lives by factors of 5 or 10, thus reflecting the behaviour of the individual aramid fibres under fatigue conditions.[2] It has been suggested[3] that this behaviour in an all-tension fatigue regime and the steeply falling tensile $S/\log N_f$ curves for KFRP materials may be a result of the duplex skin/core structure of the fibres and their interfibrillar weakness. It has also been suggested that the peculiar behaviour of KFRP can be ascribed to the effects of mechanical abrasion in damaging the load-bearing filaments.

Apparently marked differences in the physical behaviour of KFRP and comparable carbon and glass reinforced plastics suggest the operation of radically different fatigue mechanisms in the different composites.

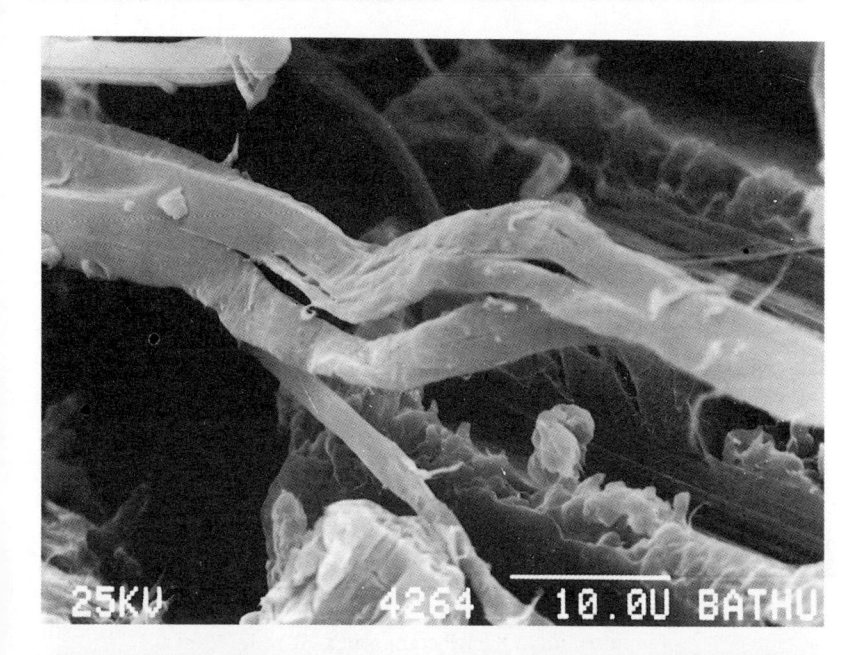

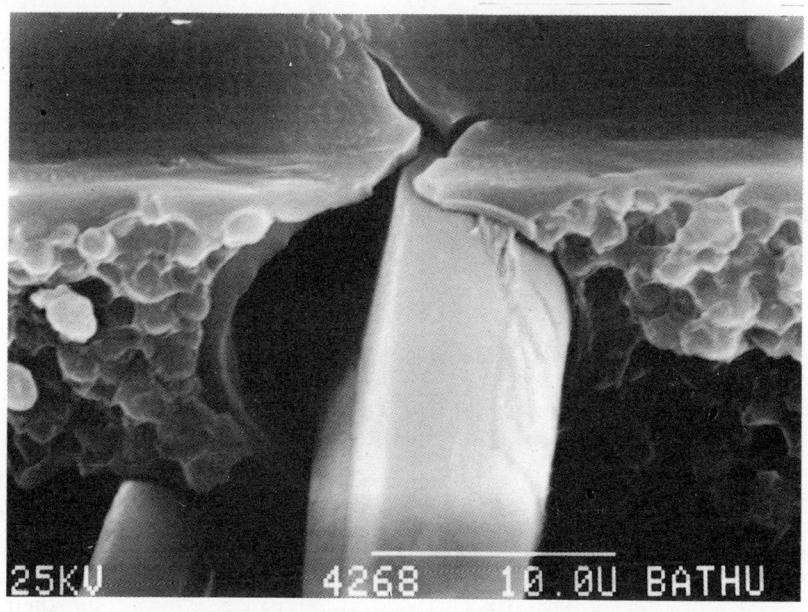

Figure 103 Scanning electron micrographs of kink banding (a) and fibrillation (b) in Kevlar-49 filaments as a result of fatigue of KFRP composites

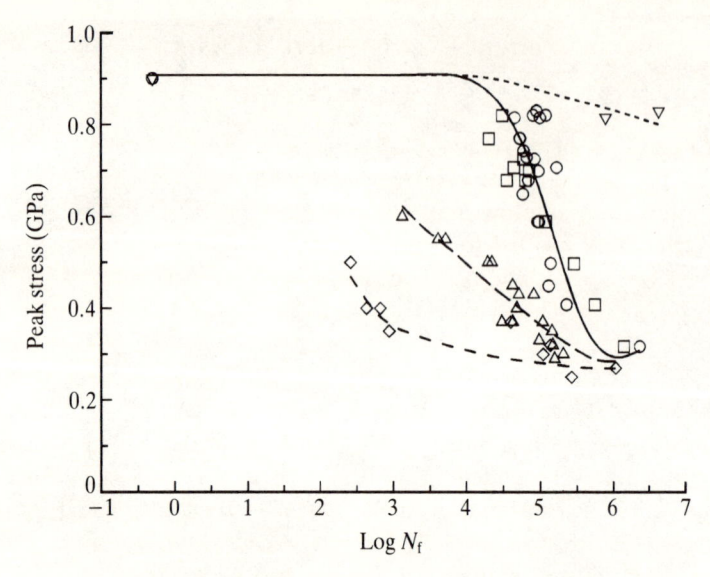

Figure 104 Effect of stress ratio ($R = \sigma_{min}/\sigma_{max}$) on the S/logN$_f$ curves of unidirectional KFRP composites.[7] R ratio: ▽ = 0.5, ○ + 0.1, □ + 0.01, △ −0.3, ◇ −0.6

Nevertheless the pattern of behaviour suggested by Fig. 104, for example, is in line with the general strain-control model of composite fatigue developed by Harris et al.[4] from a proposal of Talreja[5]; it is also in line with the ideas of Curtis.[6]

References

1. M G Dobb in *Handbook of Composites*, Vol. 1, eds W Watt, B V Perov, Elsevier, Amsterdam, 1985, pp. 673–704.
2. A R Bunsell, *J. Mater. Sci.*, **10**, 1300 (1975); **17**, 2391 (1982).
3. C J Jones, R F Dickson, T Adam, H Reiter, B Harris, *Proc. R. Soc. Lond.*, **A396**, 315 (1984).
4. B Harris, H Reiter, T Adam, R F Dickson, G Fernando, *Composites*, **21**, 232 (1990).
5. R Talreja, *Proc. R. Soc. Lond.*, **A378**, 461 (1981).
6. P T Curtis, *Tensile Fatigue Mechanisms in Unidirectional Polymer Matrix Composites*, RAE (Now DRA), Farnborough, Technical Report TR91011, Ministry of Defence, London, 1991.
7. R D Agarwal, L J Broutman, *Analysis and Performance of Fibre Composites*, Wiley Interscience, New York.

5.6 Fatigue – carbon fibre composites

P T CURTIS

Increasing use is being made of fibre reinforced plastics in primary structures because of their high strength and stiffness combined with low density and potentially low unit cost. The aerospace use of carbon fibre composites CFRP has led to significant weight savings, although strain levels have been kept low, of the order of 0.3–0.4%, because of impact damage and notch sensitivity in compression.

Because of these very low design strains, aerospace manufacturers do not perceive fatigue as a major problem and most designs are statically determined. There are exceptions to this, notably helicopter rotor blades. The decision to exclude fatigue behaviour from the primary design process is based on the evidence that, at a design limit strain two-thirds of this ultimate value, materials comfortably withstand large numbers of fatigue cycles, equivalent to many aircraft lifetimes, without failing, and additionally that damage does not grow at these low applied strains.

The simplest form of composite material is one with all the carbon fibres in the principal load direction. Typical plots of peak tensile stress versus cycles to failure (*S–N* plots) are shown in Fig. 105 for a range of carbon fibre reinforced materials.[1] For unidirectional materials under tensile loading, the fibres carry virtually all the loads, so the tensile fatigue behaviour might be expected to depend solely on the fibres. Carbon fibres are not sensitive to fatigue loading, so good fatigue

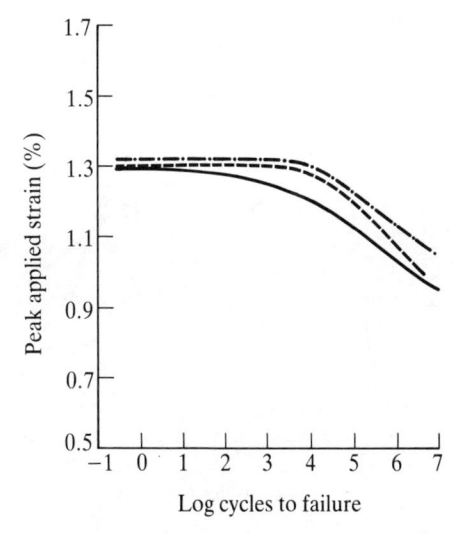

Figure 105 Tensile fatigue – UD CFRP. Comparisons of fibres in the same epoxy matrix: — = standard fibre; — · — = high strain fibre; – – – = intermediate modulus fibre

behaviour should result. However, experimental evidence has shown that the slopes of the *S–N* curves are determined principally by the strain in the matrix.[1,2] Consequently plots of mean strain versus cycles to failure are often more instructive.[2]

In recent years, materials manufacturers have substantially improved the performance of these materials, with carbon fibre strengths doubling and stiffness increasing by up to 40%. In addition, matrix resins have been developed with much improved toughness and environmental resistance. The effect of varying the carbon fibres on the fatigue behaviour is relatively small, although the overall level of the *S–N* curve on the stress axis will be determined by the strength of the fibres (Fig. 105). In general, the stiffer the fibres, the lower the working strain in the matrix material and the flatter the *S–N* curve. The matrix materials do show differing sensitivities to fatigue. For unidirectional materials, many of the newer tougher matrices appear to show increased sensitivity to fatigue (Fig. 106),[1] with steeper *S–N* curves, although this is often not translated into the laminated composites.

The damage development process in these materials is dominated by the resin matrix and interface, although clearly the load-bearing fibres have a role to play. All non-metallic fibres have a statistical distribution of strength determined by flaws, thus some of the weakest fibres will fail even at low stresses early in a fatigue test.[1,3,4] This leads to stress concentrations in the polymeric matrix and at the fibre/resin interface.

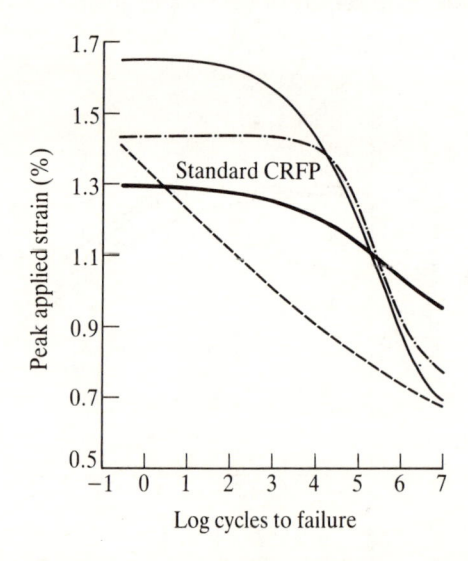

Figure 106 Tensile fatigue – UD CFRP. Comparison of different matrices with similar fibres: —·— = tough epoxy; – – – = thermoplastic; — = tough BMI blend; ▬ = standard epoxy

These stress concentrations cause damage to develop in the form of microcracks. Similar damage may develop at local microdefects such as voids, misaligned fibres or resin-rich regions. They develop in fatigue, especially along the fibres, tending to isolate fibres and rendering them ineffective as load carriers. As a result, fibres become locally overloaded and further fractures occur. Close to failure, the materials may show extensive longitudinal splitting along the fibres, leading to the brush-like failures characteristic of continuous carbon fibre reinforced polymers. The rate of this degradation process in the matrix and at the interface is a function of the bulk strain in the matrix as well as the nature and fatigue sensitivity of the resin material.

Composite materials are usually used in laminated form; layers are arranged so the fibres are oriented in the principal stress directions. On increasing the percentage of off-axis fibres in a laminate, the static properties are reduced as there are fewer fibres available to support the applied stresses. The slope of the S–N fatigue curve increases in relation to the drop in static strength (Fig. 107). This is because the layers with off-axis fibres are more easily damaged in fatigue as their mechanical properties are resin dependent. Transverse plies, with fibres at 90° to the test direction, develop transverse cracks upon the first tensile load cycle or with increasing numbers of cycles.[5] Angled plies, with fibres typically at ±45°, will also develop intraply damage. As these layers support little

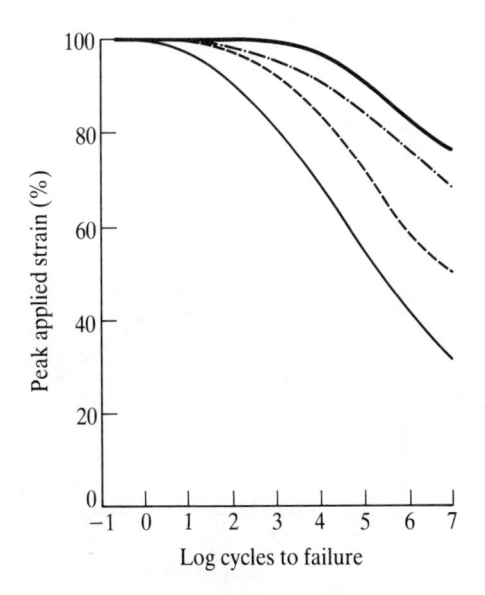

Figure 107 Normalized S–N curves for laminates with varying percentages of UD layers: ▬ = 100% UD; –·– = 50% UD; --- = 25% UD; — = 0% UD

axial load, the damage has a minimal effect on the axial tensile strength or stiffness but can have a noticeable effect in fatigue loading. The stress concentrations at the ends of intraply cracks can lead to the initiation of delamination between the layers and maybe to a loss of integrity, with potential for environmental attack and certainly reductions in compressive strength.[3,5] Alternatively, the cracks may propagate into adjacent primary load-bearing layers and seriously weaken the material.[1] Ultimate tensile failure is still determined by the fibre failure, thus the tensile S–N curves for multidirectional laminated carbon fibre composites are still relatively shallow, but steeper than for fully unidirectional materials.

At present, it is undoubtedly the compressive properties of carbon composites that limits their wider use in aerospace applications. In fatigue loading, carbon composites are much more sensitive to compression than to tension. Ultimately, the worst fatigue case for carbon composites is fully reversed fatigue or tension–compression loading. The poorer behaviour (Fig. 108) is because many of the laminate plies have no fibres in the test direction and develop intraply damage. This causes local layer delamination at relatively short lifetimes. This is less serious in tensile fatigue as the layers containing fibres in the test direction continue to support the majority of the applied load. But in compression, tensile induced damage of this type can cause local layer instability and layer buckling; this leads to macro-instability failure at relatively short lifetimes.

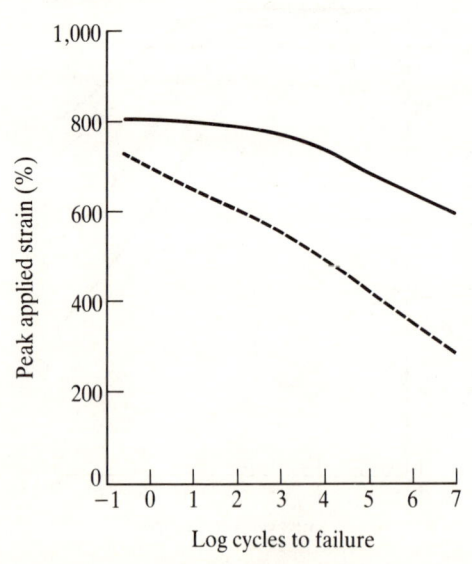

Figure 108 A comparison of the tensile and reversed axial fatigue behaviour of multidirectional CFRP: —— = zero tension; – – – = tension–compression

Improvements to metallic materials and the increasing importance of cost mean that composite design strains must be increased to make more efficient use of materials. Newer tougher composites have an increased fatigue sensitivity,[1] so fatigue behaviour is also increasing in importance and once again has become an issue to be seriously considered in the design process.

References

1. P T Curtis, RAE (Now DRA) Technical Reports TR86021 (1986), TR87031 (1987), TR82031 (1982), Ministry of Defence, London.
2. R Talreja, *Proc. R. Soc. Lond.*, **A378**, 461 (1981).
3. R D Jamison *et al.*, *Effects of Defects in Composite Materials* in ASTM STP 836, American Society for Testing and Materials, Philadelphia, 1983.
4. P T Curtis, *Intl. J. Fatigue*, **9**, 377 (1991).
5. M G Bader, L Boniface, *Fifth International Conference on Composite Materials, ICCM V*, eds W C Harrigan Jr, J Strife, A K Dhingra, AIME, Metallurgical Soc. Inc., Warrendale, Pennsylvania 1985, pp. 221–32.

5.7 Fatigue – glass fibre reinforced plastics

B HARRIS

Introduction

Fatigue is a name given to various forms of deterioration of materials under variable loading conditions. In metallic materials, it is associated with the incremental propagation of a single crack followed by catastrophic failure when the stress intensity associated with the crack reaches some critical value. This occurs at applied load levels well below the normal monotonic failure load and gives rise to two graphical manifestations of results: the stress/life or $S/\log N_f$ (S–N) curve and the Paris law curve. In the S–N curve, the fatigue stress is plotted against the logarithm of the number of cycles to failure and shows a falling tendency; the higher the peak stress, the shorter the fatigue life. The Paris law curve is a log/log plot of crack growth rate and is a linear function of the stress intensity range, ΔK, except at very low stress intensity, where a threshold ΔK is approached, and at very high stress intensity, where behaviour characteristic of monotonic fast fracture is approached.[1] Composite materials exhibit quite different behaviour.

Damage in composites

Composite materials as a whole, unlike metals, are inhomogeneous (on a gross scale) and are often anisotropic. They accumulate damage in a general rather than a localized way, and failure does not always occur by propagation of a single macroscopic crack on a plane normal to the major applied stress. In glass fibre reinforced plastics (GRP), fine fibres with stiffness of the order of 70 GPa and strengths of the order of 2 GPa are embedded in polymeric solids, thermosets or thermoplasts, with instantaneous stiffnesses of 3 GPa or less, with negligible strengths, and with time-dependent responses to stress. The failure strains of these resins may range from 0.5% for a very brittle polymer to 5% or more for a tough thermoplast or a toughened thermoset. The brittle fibres exhibit the usual variability of strength; distributions are often characterized by the two-parameter Weibull model and coefficients of variation (standard deviation/mean) are of the order of 10–20%. The consequences of these physical features of the two quite disparate components of a GRP are important in determining their response to cyclic stress. As for any composite, on first loading, the weaker fibres will begin to break in a random fashion throughout the loaded region. Locally, the loads shed by the broken fibres will be transferred into neighbouring fibres by way of the much weaker matrix. At reasonable working strains, an unbroken fibre reinforcement network carries the majority of the load on the composite and the matrix viscoelasticity is not called into play. But as fibre breakages occur, the overall extensibility of the composite will increase and creep deformations of the resin will occur, first locally, at the scale of the fibre fractures, then on a more macroscopic scale, in the composite as a whole. Some of the viscoelastic deformation will be reversed on unloading and further fibre damage will occur on reloading. This damage is exacerbated in GRP, although not necessarily in stiffer composites such as carbon fibre composites (CFRPs), by the reversals of load involved in fatigue cycling. This is easily demonstrated by monitoring the fibre breakage with acoustic emission methods. It is clear that fatigue is a cycling and not simply a time dependent phenomenon.

The overall deterioration process is manifested as a general loss of both stiffness and residual strength as cycling proceeds. These changes are often monitored as indicators of fatigue damage. Progressive deterioration has given rise to the concept of 'wear-out' during fatigue; the effective life of the material is determined by the point at which the residual strength of the composite has fallen to the level of the peak cyclic stress, as illustrated in Fig. 109. In practice, there is always a great deal of variability in the strengths of fibre composites, and this is again

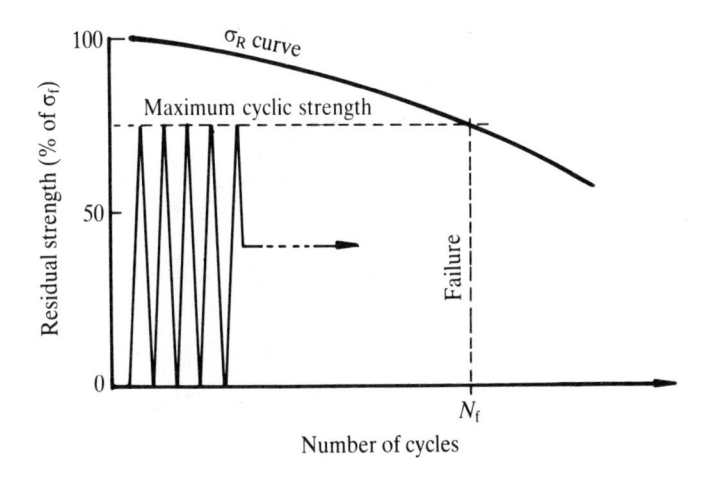

Figure 109 Degradation by 'wear-out' until the residual strength falls to the level of the maximum cyclic stress, at which point failure occurs

exacerbated by the stochastic nature of damage accumulation during fatigue. Thus, the residual strength curve always exhibits scatter and, as fatigue regimens in service are usually variable, it becomes a statistical problem to define the effective life, or a reasonably 'safe' design life for a GRP composite. An additional problem is that in such materials 'failure' may not involve complete separation into two parts so it may be difficult to define what we mean by failure, e.g., the elastic stiffness may be reduced by cycling, and 'failure' may have to be defined in terms of some critical level of loss of rigidity.

The previous discussion implies a unidirectional composite with all the fibres arranged in the direction of the applied stress. In practice, such a design is rare and there are almost always certain proportions of the fibres arranged at different orientations in order to cope with non-axial loads. In any ply where the fibre/resin interfaces are at a significant angle to the major stress axis, other forms of damage will manifest themselves, including resin cracking (in the more brittle varieties of plastic) and fibre/matrix decohesion or debonding. These damage modes will modify the overall composite properties and the cracks they cause may themselves initiate further damage in the main load-bearing plies. For such complicated structures, the derivation of a perfect model of fatigue damage is very difficult. Figure 110 shows the effect of peak stress during tensile fatigue loading on the residual strength of a 0°/90° glass fibre/epoxy laminate. In fatigue, as in resistance to other forms of damage, some composite lay-ups are more resistant to crack growth than others.

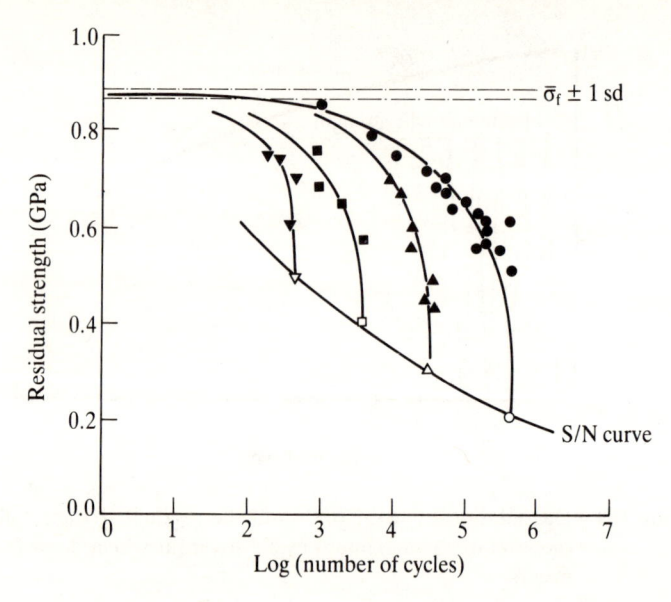

Figure 110 Changes in the residual strength during cycling of 0°/90° glass/epoxy laminate of different stress levels (65% relative humidity).[3] Stress, σ_{max}: ● 200 MPa, ▲ 300 MPa, ■ 400 MPa, ▼ 500 MPa

Various cumulative damage models have been suggested for the fatigue of GRP, some concentrating on global models of damage and others dealing with specific damage mechanisms.[2,3] A useful global power model[3] permits the results of Fig. 110 to be normalized into a single curve of the form

$$S^n + C^m = 1 \qquad [5.3]$$

The quantity S is a stress function, $(\sigma_R - \sigma_{max})/(\sigma_f - \sigma_{max})$, where σ_f is the normal dynamic tensile strength (measured at the same testing rate as that involved in the fatigue cycling), and σ_R is the residual strength of the laminate after cycling for a given time to a peak fatigue stress σ_{max}. C is a cycles ratio, $(\log n + 0.3)/(\log N_f + 0.3)$, where N is the number of cycles sustained at a stress level for which the fatigue life is N_f and the factor 0.3 is $-\log(1/2)$, the notional 'life' associated with the dynamic tensile strength. Equation (5.3) represents a family of curves which, with suitably chosen values of n and m, fit data for a range of composites, including those reinforced with carbon and Kevlar-49 fibres, as shown in Fig. 111.

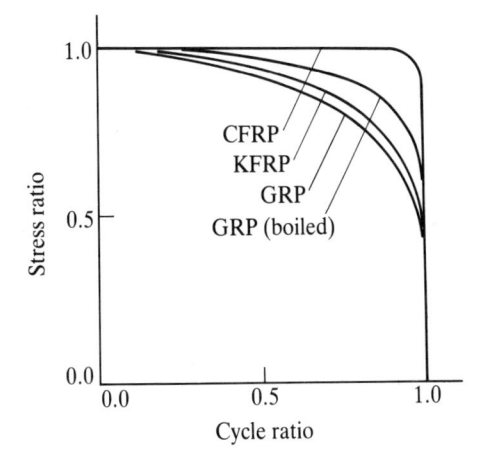

Figure 111 Curves showing normalized residual strength versus normalized cycles ratio for 0°/90° laminates of glass, carbon and Kevlar 49 fibre reinforced epoxy resin[3]

Fatigue phenomena in GRP

The conventional stress/log life curves for composites appear to be of widely varying shapes, depending on the constituent materials. For example, unidirectional composites containing very stiff fibres, such as high modulus carbon, exhibit very flat, almost horizontal, $S/\log N_f$ curves for repeated tensile cycling described by the relationship

$$\sigma = \sigma_f - B \log N_f \qquad [5.4]$$

where σ is the peak tensile stress in the cycle, σ_f is the monotonic tensile strength, and B is a constant. The lower the stiffness of the reinforcing fibres or the higher the proportion of fibres in off-axis plies or the greater the level of non-tensile stress involved in the cycling regimen (i.e. shear or compression), the greater is the slope of the $S/\log N_f$ curve. The steeper the slope of the fatigue curve, the lower the fatigue resistance of the composite. But the fatigue curves for GRP often appear to be curved rather than linear, as illustrated in the summary diagram of typical results shown in Fig. 112. Mandell and coworkers[4] have shown that fatigue damage in a wide variety of GRP materials can be explained simply in terms of the gradual deterioration of the load-bearing fibres. They show that much of the fatigue data for composites with long or short fibres, in any orientation, and with any matrix – thermosets and thermoplasts – fit (5.4) reasonably well and that the ratio σ_f/B, obtained by normalization of the data by dividing all fatigue stresses by the monotonic tensile strength, has a constant value of about 10 with very little dispersion.

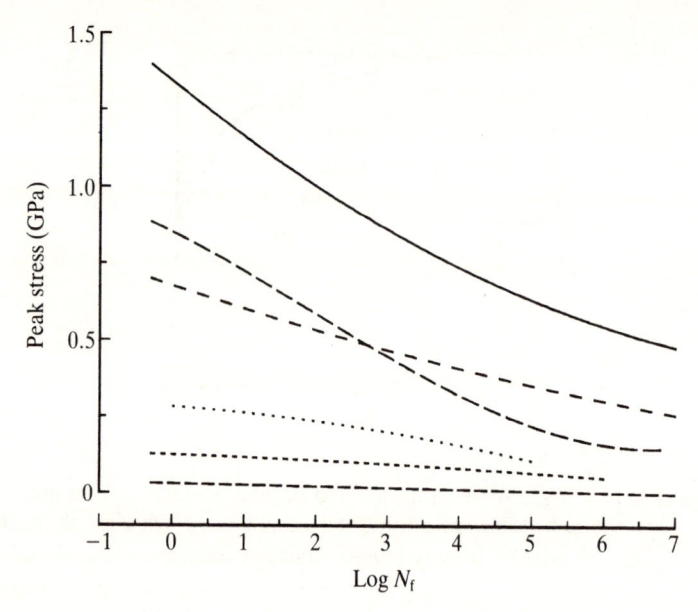

Figure 112 Stress/life curves for various GRP materials. ——— 0°; — — — [(0°,90°)₅]ₛ; – – – – [(±45°,0°₂)₂]ₛ; ⋯⋯ woven cloth laminate; - - - - - CSM/polyester; – – – – DMC

Clearly, the normalization process has the effect of removing variations due to differences in fibre volume fraction. The conclusion that the controlling fatigue mechanism in GRP is the gradual deterioration of the load-bearing fibres is logical and inevitable. But what controls the actual life of any given composite is the manner in which other mechanisms, such as transverse cracking in the 90° plies or local resin cracking in woven cloth composites, modify the rate of accumulation of damage in the load-bearing plies.

Talreja[5] has proposed a general model in which the strain/log life curves for polymer matrix composites may be thought of in terms of three regimens within which separate mechanisms control fatigue failure. At high stress levels, fibre breakage and/or interfacial debonding occurs; this leads to failures within the normal scatter band for the strength of the 0° plies. The fatigue curve in this region is therefore very flat. At somewhat lower cyclic stresses, this statistical fibre breakage still occurs but does not lead to composite destruction in lives so short that no other mechanisms have time to occur. These other mechanisms, such as matrix cracking and interfacial shear failure, can therefore influence the overall damage state, providing the composite working strain level is sufficiently high, and the slope of the fatigue curve begins to increase. But there is a notional fatigue limit for the matrix resin; if working strains do not rise above this level, the composite should not, in principle, fail in fatigue. The strain/log life curve should therefore flatten out again, and something like an endurance limit should be observed. The generic shape of the strain/log life curve therefore resembles the schematic diagram Fig. 115. Whether or not some or all of these stages actually occur depends on the characteristics of the constituents and the lay-up geometry. High modulus CFRP show only the first stage; GRP materials frequently show all three stages. The wider implications of this model have been discussed by Harris et al.[6]

Factors affecting the fatigue of GRP

Fibres

We have already referred to characteristic material parameters such as fibre stiffness, strength, volume fraction and distribution. As the use of short fibres results in a greater level of matrix loading through shear, short fibre composites inevitably exhibit poorer fatigue resistance. Local failures in the matrix are easily initiated, especially in the region of fibre ends, and they can impair the integrity of the composite even when the fibres remain undamaged. There is some evidence that for cycling at a given fraction of the failure stress, the fatigue life increases rapidly with increasing fibre aspect ratio until it levels off at some critical value of l_c/d, above which the fatigue strength is a constant proportion of the monotonic strength.

Moisture

In reinforced plastics, water softens resins leading to higher failure strains. This should tend to increase fatigue resistance by inhibiting local crack growth. But water also weakens the reinforcing glass fibres so the net result is more likely to be a reduction in fatigue resistance.

Frequency

Metallic fatigue is rarely affected by cycling frequency, but because the monotonic strengths of GRP are markedly rate dependent, by virtue of the viscoelastic character of the matrix and because the strength of the glass fibres is time dependent,[7] the fatigue response will vary with test frequency. Unlike metals and CFRP composites, GRP composites cannot readily dissipate heat and they gradually heat up during cycling. As the temperature rises, their strengths fall, thus fatigue failure is hastened. Short fibre composites are particularly prone to this deterioration. Studies on a wide range of GRP materials of various lay-ups and with different moisture constants have shown that the fatigue behaviour of GRP can be rationalized by carrying out fatigue tests at constant rate of stress application (RSA) rather than at constant frequency, which is the usual practice[8]. By monitoring the specimen temperature rises during cycling at different RSA, Sims and Gladman showed that the temperature dependence of the composite tensile strength transposed directly to the fatigue strength. They showed that fatigue tests at different rates could be normalized with respect to the tensile strength, corrected for hysteretic heating and measured at the same RSA as the fatigue, to give a single linear master fatigue curve.

Stress concentrations

In metals, local stress concentrations can be dissipated to some extent during cycling by plastic deformation. Similar effects occur in composites, although the mechanisms of deformation are different. It is sometimes found that a fatigue sample with a hole will break at a site remote from the hole because the local high stress level acts to strengthen the material near the hole. Local hysteretic heating will also reduce the local notch sensitivity of the material.

References

1. M F Ashby, D R H Jones, *Engineering Materials*, Pergamon Press, Oxford, 1980.
2. J T Fong in *Damage in Composite Materials*, ed. K L Reifsnider, ASTM STP 775, American Society for Testing and Materials, Philadelphia, 1982, pp. 243–266.
3. T Adam, R F Dickson, C Jones, H Reiter, B Harris, *Proc. Inst. Mech Eng. Mech. Engng. Sci.* **200**(C3), 155 (1986).
4. J F Mandell in *Developments in Reinforced Plastics 2*, ed. G Pritchard, Applied Science, London, pp. 67–108.
5. R Talreja, *Proc. R. Soc. Lond.*, **A378**, 461 (1981).
6. B Harris, H Reiter, T Adam, R F Dickson, G Fernando, *Composites*, **21**, 232 (1990).
7. C J Jones, R F Dickson, T Adam, H Reiter, B Harris, *Proc. R. Soc. Lond.*, **A396**, 315 (1984).
8. G D Sims, D G Gladman, *Plastics & Rubber: Materials and Applications*, **1**, 41 (1978); **2**, 245 (1980).

5.8 Fatigue – life prediction – role of materials and structure

P IRVING

Introduction

Fatigue damage is the process of progressive failure that occurs when a material is subjected to cyclic loads; the maximum stress is less than the static stress necessary to fail the material in a single load application. Prediction of fatigue performance is essential in engineering because:

(i) the process of material selection to match service conditions must be made quickly and accurately;
(ii) the designer must predict the service durability in fatigue from knowledge of the material and of the service conditions.

This section describes the influence on fatigue performance of composite constituents, the matrix, the fibres and their bonding and arrangement.

Definition of failure

In metallic materials, failure in fatigue occurs via propagation of a single or very few cracks, the later stages of which occur extremely rapidly. Composite fatigue failure is different; it is characterized by diffuse damage progression in which cracks and delaminations spread throughout the many interfaces in the material. The symptoms of this damage growth are:

(i) visual or other non-destructive indications, such as surface splits and delamination;
(ii) a progressive reduction of strength with increasing load cycles;
(iii) a progressive reduction of stiffness with increasing load cycles.

Final failure or separation occurs when the residual strength has reduced to that of the maximum applied cyclic strain level. Because of the gradual nature of the deterioration, it is frequently necessary to impose a definition of failure other than that of specimen or component separation. Figure 113 shows stiffness reduction for a carbon fibre laminate. Figure 114 shows life to separation of unidirectional carbon fibre, glass fibre and aramid fibre composite materials.

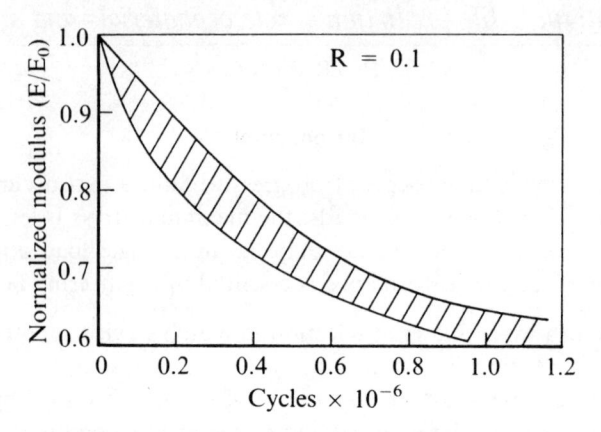

Figure 113 Stiffness degradation during fatigue of $(45°/90°/-45°/0°)_s$ CFRP laminate (adapted from Talreja[1])

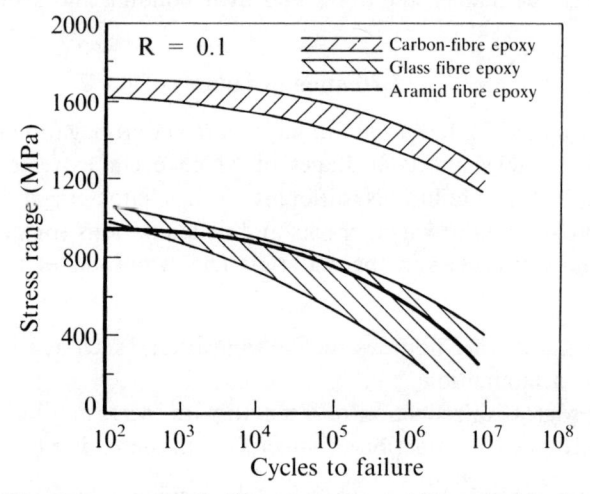

Figure 114 Fatigue strength–life curves for unidirectional carbon, glass and aramid fibre (adapted from Curtis[6])

Material variables determining fatigue performance of unidirectional laminates

The basic failure modes of all composite materials in static and fatigue loading are:

(i) matrix cracking,
(ii) fibre breakage,
(iii) fibre–matrix debonding,
(iv) delamination.

The process of fatigue failure consists of the progressive operation of all these; the manner and speed is dictated by the constituents and the lay-up of the composite. The strength, stiffness and failure strain of the fibres, matrices and interfaces are crucial in determining the resistance to fatigue. Talreja[1] (see Section 5.6) recognized that the different fatigue life behaviour of carbon and glass fibre composites with different elastic moduli may be accounted for by considering the strains which fibres of different stiffness impose on the matrix.[3] In a unidirectional continuous fibre composite material stressed parallel to the fibres, strains in the composite, the fibre and the matrix will be equal, and

$$\epsilon_c = \epsilon_f = \epsilon_m \quad [5.5]$$

where ϵ_c, ϵ_f and ϵ_m are the strains in the composite overall, in the fibres and in the matrix.

The stresses in the individual components are given by

$$\sigma_c = \sigma_f V_f + \sigma_m (1 - V_f) \quad [5.6]$$

where σ_c, σ_m and σ_f are the stresses in the composite, the matrix and the fibre, respectively and V_f is the volume fraction of fibres.

Plots of fatigue life against applied cyclic strain (Fig. 115a) reveal three regimens in which different failure mechanisms are operating:

(i) a region dominated by fibre fracture,
(ii) a region dominated by a mixture of interfacial shear failure and matrix cracking,
(iii) a region dominated by matrix cracking.

The dominant region will be determined by the relative values of the failure strain of the matrix and the fibres under fatigue loading. Fibre fracture will dominate carbon fibre composites where the failure strain of the matrix exceeds that of the fibre; the fatigue curve will consist of a scatter band of data almost parallel to the cycles axis. If fibre failure strain is increased until the fibre failure strain exceeds that of the matrix, fibre fracture will occur only at the highest strains and at low cycles; interfacial shear and matrix cracking will predominate at longer cycles and the stress/life gradient will increase significantly. This is illustrated in Fig. 115 a, b and c where glass fibre (b) and Type I carbon fibre composite (c) behaviour are compared.

Strain life diagrams provide a conceptual framework for selection of resin and fibres to form a laminate. However, there are many other complicating factors controlling fatigue performance. The insights provided by the strain diagram approach imply that fatigue performance will be enhanced by increased resin ductility and that increased fibre fraction, V_f, will maintain constant fatigue strength when assessed in

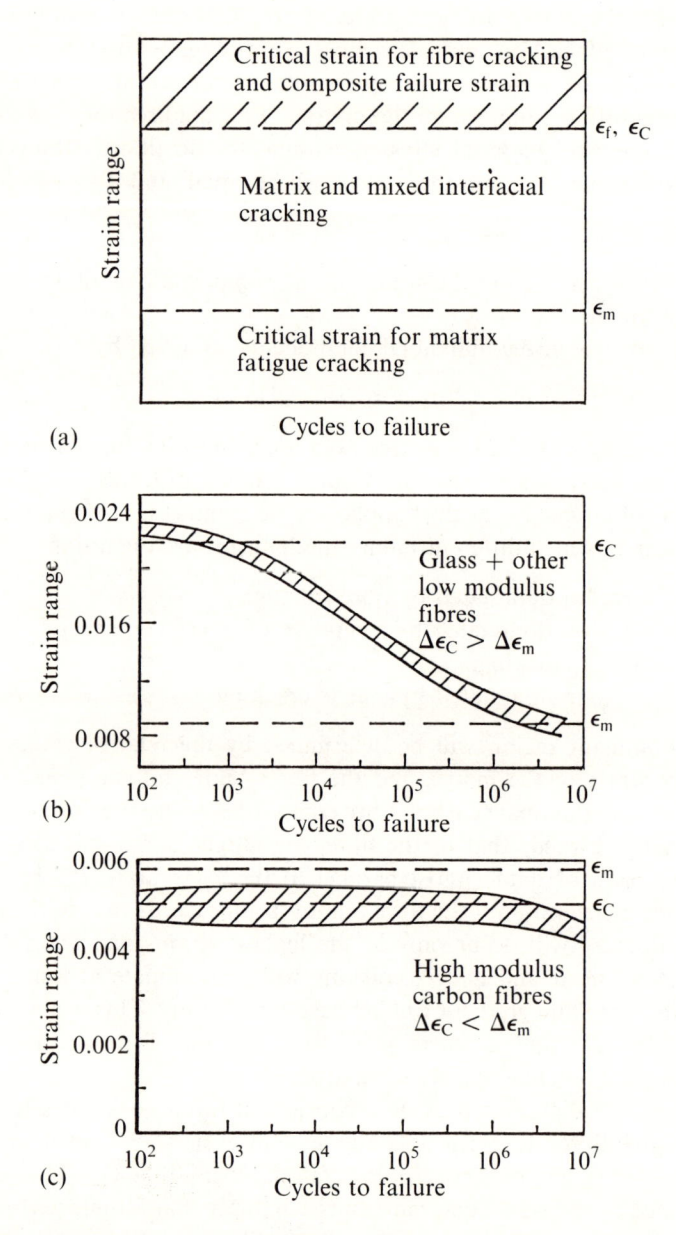

Figure 115 Strain life plot of fatigue mechanisms showing regions of different composite failure (adapted from Talreja[1])

terms of strain amplitude but will increase fatigue performance when assessed in terms of stress, as $E_c = E_f V_f + E_m(1 - V_f)$ and $\sigma_c = E_c \epsilon_c$.

It has been found that the influence of tougher matrix resins is to increase the fatigue performance of the composite at low and intermediate lives ($<10^6$ cycles), but at high cycles, performance is degraded[2] in comparison with standard resin systems. Fatigue experiments conducted on carbon fibre composites with epoxy resins of a range of toughnesses and failure ductilities and on composites with high ductility thermoplastic matrices all demonstrate this effect (see Fig. 106). Fatigue tests on the matrix alone and on 90° unidirectional plies produce the same result. A similar effect occurs during fatigue of metals. High ductility is required for optimum low cycle performance, high strength for the best high cycle fatigue strength. Fatigue strength in unidirectional composites is also influenced by:

(i) fibre strength,
(ii) interfacial bond strength,
(iii) crack resistance of the matrix,
(iv) homogeneity of fibre distribution,
(v) matrix defects and strain concentration effects,
(vi) lack of fibre wetting.

Many of these will also influence the tensile strength of the laminate and Mandell and others[3] have considerable evidence that fatigue strength of unidirectional laminates is closely correlated by the static tensile strength. This is a good, if approximate, approach to optimizing the fatigue strength of unidirectional laminates.

Fatigue of crossply laminates

Use of laminates containing plies other than 0° or use of woven fabrics degrades the fatigue performance from that obtained with uniaxial fibres when considered in terms of stress or strain (Fig. 107). The behaviour of crossply laminates in fatigue is extremely complex and mechanisms of crack initiation and growth are extremely important in determining damage progress. Fatigue damage occurs both at the specimen edges and internally within the individual plies. A sequence of failure for a 0°/90°/90°/0° laminate is as follows.[4,5]

(i) Early in the life, transverse matrix cracks appear in the 90° plies. These extend across the complete width and thickness of the 90° ply. They grow in number and density until a saturation value, termed the 'characteristic state' is achieved (typically one crack per millimetre). As the 90° plies carry little of the laminate load, this damage is subcritical; the cracks are a way for the low failure strain 90° plies to accommodate the strain imposed by the 0° plies; the 0° plies carry the majority of the load. Some reduction of laminate stiffness will result. However, the damage is a necessary precursor to subsequent events, as it results in a redistribution and concentration of stress at the crack tips.

(ii) Longitudinal cracks nucleate and grow in the 0° plies under the influence of cyclic stresses perpendicular to the applied load axis, generated by restraint of Poisson contraction strain in the 0° plies by the 90° plies. Again, these cracks will not cause significant reduction of load bearing capacity.

(iii) Interply delamination cracks will initiate at the specimen edge under the influence of stresses perpendicular to the ply thickness, up to 10% larger at the composite edge than in the interior. This process occurs simultaneously with stage (ii) and will cause significant stiffness reduction.

(iv) Internal delaminations initiate and grow at the intersections of the matrix cracks in the 0° and 90° plies, propagating along the interface.

(v) The 0° ply fibre fracture occurs at the stress concentration at the tip of the 90° ply matrix cracks.

(vi) Final failure occurs when sufficient fibres have broken so as to exceed the load-carrying capacity of the remaining fibres.

Experiments on crossply laminates with toughened resin systems have shown superior performance compared with standard epoxies. It has also been reported that use of toughened systems increases the fatigue cycles required to initiate the transverse 90° ply cracks and so delays the onset of the damage progression to failure. Delamination crack growth is of particular importance in practical application of composites, as it can occur without precursor mechanisms at component edges, holes and other discontinuities. Like crack growth in metals, modifications to material properties have only secondary influence on crack growth rates.

References

1. R. Talreja *Proc. Roy. Soc. Lond.*, **A378**, 461–475 (1981).
2. K Schulte, K Freidrich, G Hostenkamp *J. Mater. Sci.*, **21**, 3561 (1986).
3. J F Mandell in *Developments in Reinforced Plastics 2*, ed. G Pritchard, Applied Science London, 1982, pp. 67–108.
4. K Reifsnider (ed.) in *Fatigue of Composite Materials*, Elsevier, Amsterdam, 1991, pp. 1–77.
5. K Schulte, W W Stinchcombe, in *Application of Fracture Mechanics to Composite Materials*, ed. K Friedrich, Elsevier, Amsterdam, 1989, pp. 273–325.
6. P T Curtis in *Advanced Composites*, ed. I K Partridge, Elsevier, Amsterdam, 1989, pp. 331–368.

5.9 Fatigue – life prediction – damage calculation and assessment

P IRVING

Introduction

In order to design with composite materials, it is necessary to develop models that relate fatigue properties previously measured in the laboratory to service performance in a real structure. Ideally, predictive models should be able to calculate performance incorporating the influence of:

(i) the microstructure of the material created by the component manufacturing process,
(ii) the stress and strain fields generated by component geometry and macroscopic arrangements of fibre and laminates,
(iii) the varying load cycle spectra imposed by the service environment.

The level of understanding necessary for the above has not yet been developed for polymer composite materials. Validation for current applications is performed instead by extensive laboratory service simulation tests; the service loads on the component are recreated, if no or negligible damage growth has occurred in stresses or lives in excess of the service requirement, the component is fit for purpose. The procedure is conservative but lacks predictive capability and has to be repeated for each new component, material and service situation. Current design strains for carbon fibre laminates in aerospace applications are less than 0.3%, this is less than strain amplitudes that cause fatigue failure in undamaged composite laminates. Much fatigue prediction research has aimed to predict the growth of delaminations under cyclic loading. Pressures on designers will inevitably cause component design strains to increase into the fatigue sensitive region.

Models for prediction of damage growth

When polymer composites are subjected to cyclic loading, damage in the form of fibre and matrix cracks and delaminations causes progressive reduction of the stiffness and strength. The composite fractures after a number of cycles is determined by the loading history and the composite material. Models of damage growth seek to predict the rate of reduction of stiffness and strength and the cycles at which failure occurs. Examples of a stress/life curve for unidirectional laminates are shown in Fig. 114 and the degradation of stiffness in Fig. 113.

Over the past 15 years, many damage growth laws have been proposed; Sendeckyi[1] characterizes them into four types:

(i) empirical fatigue models,
(ii) residual strength models,
(iii) stiffness reduction models,
(iv) actual damage mechanisms based models.

In addition, a separate category considers failure as resulting from delamination crack growth alone; expressions have been derived for G_c, the strain energy release rate for delamination crack growth, and crack growth rates, da/dN, have been characterized in terms of ΔG_c.

Empirical fatigue

Empirical fatigue theories have been developed to correlate particular sets of data. There are many different forms and their merit resides in their ability to predict performance. Their disadvantage is that the limits of applicability cannot be determined except by experiment. Typical of this class is the expression used by Mandell[2] to relate fatigue performance to the ultimate static strength of the composite.

$$\sigma_a = \sigma_{uc} - b \log N_f \qquad [5.7]$$

where σ_a is the applied stress amplitude, σ_{uc} is the ultimate tensile strength, N_f is the number of cycles to failure and b is a constant.

Residual strength degradation

Residual strength degradation theories have the unfortunate property that it is not possible to assess their damage state non-destructively, as there is no established correlation between NDT indications in composite and residual strength. Nevertheless, they have the merit of being based on measureable degradation behaviour so they have some use as a predictive technique. Failure occurs when the reducing strength of the composite becomes equal to the applied stress. They are usually based on an expression of the form

$$\sigma_r = \sigma_a[1 + (N_f - 1)f]^s \qquad [5.8]$$

where σ_r is the residual strength and f and s are constants. In some versions of the residual strength model, s and f are functions of the applied load ratio, $R = \sigma_{min}/\sigma_{max}$, where σ_{min} is the minimum stress in the load cycle and σ_{max} is the maximum stress.

Stiffness degradation

Stiffness degradation theories assume that damage in the laminate, whether caused by fibre breakage, matrix cracking or by delamination, all cause stiffness degradation, which can thus be used as an index of the rate and extent of damage growth. When sufficient damage has accumulated, laminate failure results. In order for life predictions to be made, a relation must be established between the damage, D, its rate of accumulation with cycles, dD/dN, and the resultant life. Additionally, the relation between the loading parameters, σ_a and R, and the damage state, D, must be found. Loading variables may then be related to life, via stiffness degradation.

Once again, many models have been proposed. Typical of the genre is the one due to Poursatip et al.[3] Based on measurements on carbon fibre/epoxy laminates $[45°/90°/-45°/0°]_s$ they found that

$$\frac{dD}{dN} = -2.857 \left(\frac{1}{E_o} \frac{dE}{dN} \right) \qquad [5.9]$$

where E_o is the original stiffness and dE/dN the rate of stiffness degradation. Incorporation of R ratio effects and the relation between dD/dN and σ_a followed by integration leads to an expression of the form

$$N_f = A \left(\frac{\sigma_a}{\sigma_{uc}} \right)^B \left(C \left(\frac{1-R}{1+R} \right) \right)^P \left(1 - \frac{\sigma_a}{(1-R)\sigma_{uc}} \right) \qquad [5.10]$$

with $A = 3.108 \times 10^4$, $B = -6.393$, $C = 1.22$, $P = 1.6$ and $\sigma_{uc} = 586$ MPa.

Laminate failure mechanisms

Models based on laminate failure mechanisms are directly based on observable damage. Because of the complexity of the mechanisms, quantitative approaches are difficult to develop and apply. Approaches have been made by Wang.[4] The Reifsnider critical element theory also falls into this category.[5]

Delamination

In real composite structures, edges, holes, bonded joints and ply dropouts are all sites of potential initiation of delamination cracks. In addition, premature delamination initiation can occur via impact damage.

Many workers have examined the influence of number of plies, their orientation and the size of delamination on the strain energy release rate, G_c. It is found that stable crack extension can occur under static conditions in a manner analogous to the *R* curve in metals. Under fatigue cycling, critical values of *G* for crack growth, G_c, are reduced to perhaps 20% of their static value. Values of 50–200 J m^{-2} have been recorded, depending on the lay-up and the matrix material. For edge delamination in a sample subjected to tension loading, G_c is given by[1]

$$G_c = \frac{\epsilon_c^2 t}{2}(E_c - E_c^*) \qquad [5.11]$$

where ϵ_c is the applied strain, *t* the laminate thickness, E_c the original laminate stiffness and E_c^* the delaminated stiffness. ΔG, or G_{max} for a single mean stress, has been related to delamination growth rate in fatigue, analogous to the relation between metallic crack growth rates and the stress intensity factor amplitude ΔK. As in metals, at *G* values less than ΔG_c, delamination growth ceases. Of particular interest is the behaviour in compression: laminate buckling can occur and this increases the operating value of *G* considerably. Hence tension–compression fatigue of laminates, particularly damaged ones containing pre-existing delaminations, produces enhanced values of *G* and high crack growth rates. Delaminations will grow rapidly, even when the laminate experiences entirely in-plane compression. The relation between *G* and the laminate area growth rate, d*a*/d*N*, is of the form

$$\frac{\mathrm{d}a}{\mathrm{d}N} = A(\Delta G)^n \qquad [5.12]$$

where *A* and *n* are constants. Values of *A* and *n* are such that fatigue lives in delamination crack growth will be short; either delaminations will not grow, because ΔG is less than the threshold, or they will be subject to unstable crack extension under static loading, because G_c for static fracture has been exceeded.

Damage accumulation and Miner's law

The prediction of fatigue performance of composite laminates under variable amplitude loading has been extensively investigated. Miner's linear cumulative damage rule[6] is often used for cumulative damage calculations.

$$\sum_{m}^{i=n} \frac{n_i}{N_f} = 1 \qquad [5.13]$$

where n_i is the number of cycles of a given stress range and N_f is the number of cycles to cause failure at that load level, *m* is the number of stress range levels.

It is used extensively to predict the fatigue performance of metals and does give good results. Applications to composite laminates have found it to be non-conservative by up to a factor of 10 on life.[1,3] Such errors are by no means unusual in metals but there is as yet insufficient data from composite tests to state whether there are greater errors in laminates.

Other approaches to cumulative damage have been proposed based on both residual strength and empirical models of damage in composites.[1] In empirical models, notional non-proportional damage curves are empirically developed and used to conduct a Miner summation. Load interaction effects cannot be incorporated. Considerably more work is required before a sound basis for fatigue damage assessment in composites is established.

References

1. G P Sendeckyi in *Fatigue of Composite Materials*, ed. K L Reifsnider, Elsevier, Amsterdam, 1991, pp. 431–483.
2. J F Mandell in *Developments in Reinforced Plastics 2*, ed. G Pritchard, Applied Science, London, 1982, pp. 67–108.
3. A Poursatip, P W R Beaumont, *Composite Sci. Tech.* **25**, 283 (1986).
4. P C Chou, A S D Wang, H Millar *Cumulative Damage Models for Advanced Composite Materials*, Air Force Wright Aeronautical Laboratory Report (AFWAL) TR-82-4083, 1982.
5. K Reifsnider in *Fatigue of Composite Materials* ed. K Reifsnider, Elsevier, Amsterdam, 1991, pp. 1–77
6. M A Miner, *J. Appl. Mech.* **12**, A159 (1945).

5.10 Impact performance – CFRP laminates

G DOREY

The impact performance of a structure is usually associated[1] with the effects of being struck by an impacting object (foreign object impact) or effects of high strain rate loading. The resulting damage depends on the impacting object (density, mass, hardness, velocity, shape, attitude, etc.), on the dynamic response of the target structure (skin thickness, support conditions, etc.) and on the target material properties (modulus, strength, toughness, etc.).[2]

Impact tests are usually designed to simulate particular threats and can vary widely in deformation rates, as shown in Table 49.

In isotropic homogeneous materials, impact damage is determined mainly by the loading; in composite materials, impact damage is influenced significantly by the relatively weak interface between the

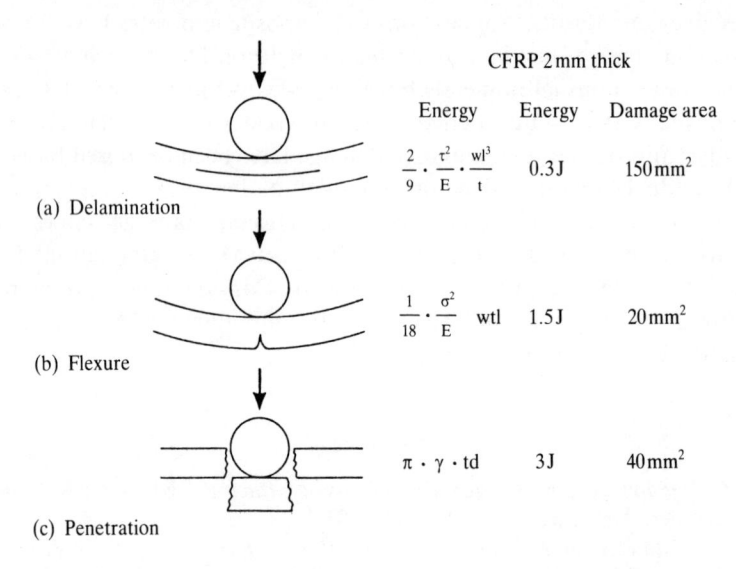

Figure 116 The primary failure modes of composite laminates under impact loading

		CFRP 2 mm thick	
	Energy	Energy	Damage area
(a) Delamination	$\dfrac{2}{9} \cdot \dfrac{\tau^2}{E} \cdot \dfrac{wl^3}{t}$	0.3 J	150 mm^2
(b) Flexure	$\dfrac{1}{18} \cdot \dfrac{\sigma^2}{E}$ wtl	1.5 J	20 mm^2
(c) Penetration	$\pi \cdot \gamma \cdot td$	3 J	40 mm^2

Table 49. Impact tests and associated deformation rates

Quasi-static test, velocity	n mm s^{-1}
Dropweight or pendulum impact	n m s^{-1}
Gas gun projectiles	$n \times 100$ m s^{-1}
Ballistics	$n \times 1000$ m s^{-1}
Hopkinson bar, strain rate	$n \times 10^3$ s^{-1}
Laser induced shock	$n \times 10^6$ s^{-1}

reinforcement and the matrix. In typical carbon fibre reinforced plastic (CFRP) composites, the fracture energy parallel to the fibres (splitting and delamination) is usually less than 1 kJ m^{-2}, whereas the fracture energy perpendicular to the fibres is 50–100 kJ m^{-2}. Composite materials will therefore tend to fracture parallel to the fibres unless constrained not to.

Failure modes associated with impact loading are shown in Fig. 116, together with threshold energies at which damage first occurs. Delamination (Fig. 116a) is caused by shear loads within the laminate and occurs particularly in thick composites, close to supports, or in composites with weak interfaces. In thinner skins and larger spans, the damage is more likely to be a flexural failure on the tensile face (Fig. 116b) or on the compressive face. At high velocities, the target cannot respond quickly enough in flexure; it is effectively rigid and the projectile causes through thickness shear and penetration.

Information on phenomena occurring during impact can be obtained by instrumenting the test facility. Typical load–time traces for dropweight impact are shown in Fig. 117. Similar traces for sandwich panels show separate peaks for each skin.[3]

Penetration of composite materials involves breaking fibres, and the energy to do this is a function of the area under the fibre stress–strain curve. The energies needed to break aramid or glass fibres are much greater than for carbon fibres; this is why aramid and glass fibres are used for lightweight armour and for structural applications where the high specific modulus of carbon fibres is not required. Energy absorption is also required for crashworthiness; Hull[4] has shown that by controlling the failure process, carbon fibre tubes can absorb significantly more energy than metal tubes.

For many structural applications of CFRP, low energy impact can be caused by accidental damage and can result in internal damage such as multiple delaminations, as shown in Fig. 118. This damage causes little effect on the surface and is difficult to detect. It is often referred to as barely visible impact damage or BVID and it can have a significant effect on structural performance, as shown in Section 5.11.

Efforts are being made to modify fibre, matrix and fibre/matrix interface properties to reduce or modify the damage and to make it more obvious from surface effects that impact has occurred so that the impacted area can be inspected in detail.

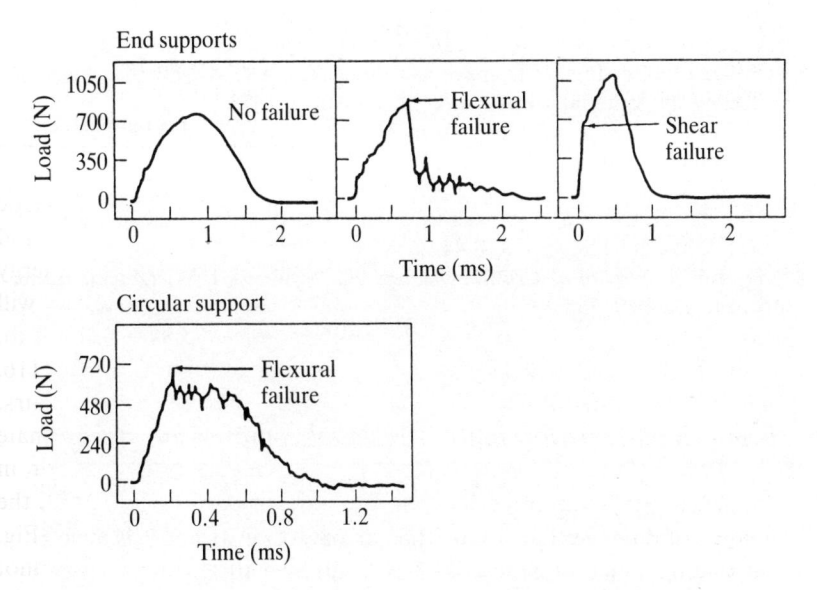

Figure 117 Load-time traces during dropweight impact on CFRP laminates

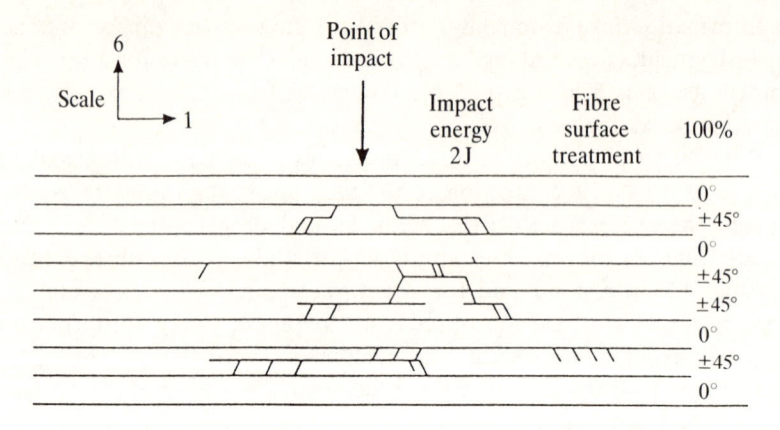

Figure 118 Multiple delaminations caused by dropweight impact on a CFRP laminate

The general conclusions about the impact performance of composite structures are that they can be much better or much worse than structures made from other materials, depending on all the relevant parameters, and that specific designs should take account of and, if possible, should tailor the failure modes.

References

1. Foreign Object Impact Damage to Composites, ASTM STP 568, American Society for Materials and Testing, Philadelphia, 1975.
2. G Dorey in *Composite Materials in Aircraft Structures*, ed. D H Middleton, Longman, London, 1990 pp. 50–68.
3. M W Wardle in *Progress in Science and Engineering of Composites*, eds T Hayashi, K Kawata, S Umekawa, Japan Society for Composite Materials, Vol. I, Tokyo, 1982, pp. 837–44.
4. D Hull in *Structural Crashworthiness*, eds N Jones, T Wierzbicki, Butterworth, London, 1983.

5.11 Impact performance – residual compression strength

G DOREY

The kinds of damage that occur when a composite structure is struck by an impacting object are described in Section 5.10. In most cases the damage is localized and therefore has little effect on global properties such as stiffness. But it can have a significant effect on strength, referred to as residual strength or strength after impact. The different forms of damage can affect the various composite strengths in different ways.[1] For instance, delamination reduces shear strength and compression strength but has little effect on tensile strength, whereas broken fibres have more effect on tensile strength. Typical effects of dropweight impact on residual strengths of CFRP laminates are shown in Fig. 119.

Delaminations occurred at incident energies between 1 and 3 J (1 J is approximately the energy of 100 g dropped through 1m) and reduced the compressive strength by 50% but had little effect on tensile strength. At incident energies above 3 J, fibres were broken and this reduced the tensile strength. These thresholds would be different for different thicknesses and stacking sequences but the shapes of the curves would be similar for the same fibre and matrix. Tougher matrix materials, such as toughened epoxies[2] or thermoplastics,[3] have greater delamination energies or interlaminar fracture toughnesses and this reduces the incidence of delamination. An example is shown in Fig. 120, where residual compressive strengths are compared for laminates of carbon fibre–epoxy and carbon fibre–PEEK (poly ether ether ketone, a semicrystalline thermoplastic).

Specimens for the determination of residual strengths have to be large enough (typically 200 mm long and 100 mm wide) to contain the damage (typically 50 mm in diameter) and be representative of practical

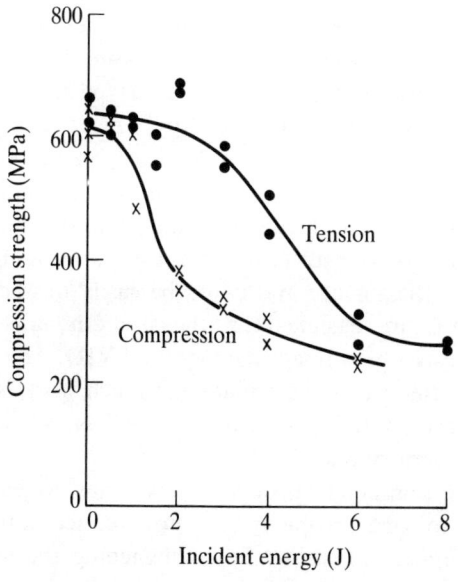

Figure 119 Residual tensile and compressive strengths after dropweight impact on a carbon fibre–epoxy [0°/90°/0° ± 45°/0°]s laminate

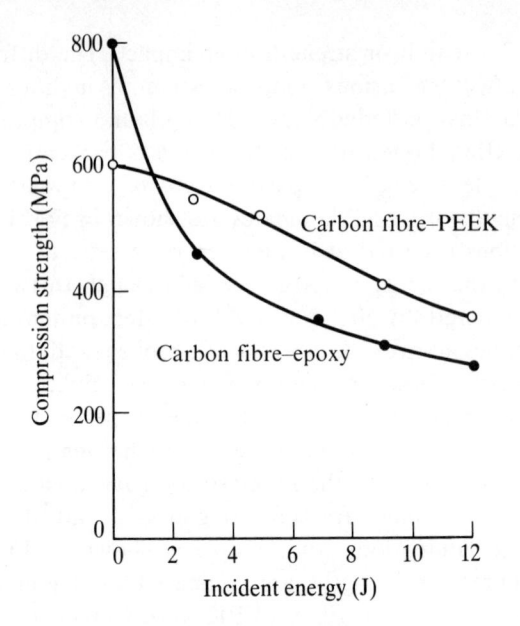

Figure 120 Residual compressive strengths of carbon fibre–epoxy and carbon fibre–PEEK $[\pm 45°/0°_3/\pm 45°/0°_2]_S$ laminates after dropweight impact

laminates; in the case of compression, they have to have antibuckling guides. It takes a number of specimens to generate residual strength curves so this can be expensive in material. Some standard compression after impact tests use a standard laminate (typically 32 plies thick), a standard size and a standard impact energy. This is equivalent to a single point on Figs 119 or 120 and this could clearly be misleading.

The reductions in compression strength occur because the multiple delaminations precipitate local buckling and delamination growth leading to collapse of the structure. Because the delaminations are not clearly visible on the surface and could be easily missed on inspection, aircraft structures are designed on the strength values of laminates containing barely visible impact damage or BVID. This is significantly below the full strength of the laminate and much work is being done on modelling the delamination failure mechanism[4] as well as improving the material impact performance.

When altering composite properties, care has to be taken so that improvements to one property do not reduce another property unacceptably. Figure 121 shows how toughening the matrix resin can reduce the shear modulus so that the matrix cannot support the fibres in compression and a shear buckling failure is precipitated. It has also been shown that CFRP with tougher matrix resins tends to have steeper

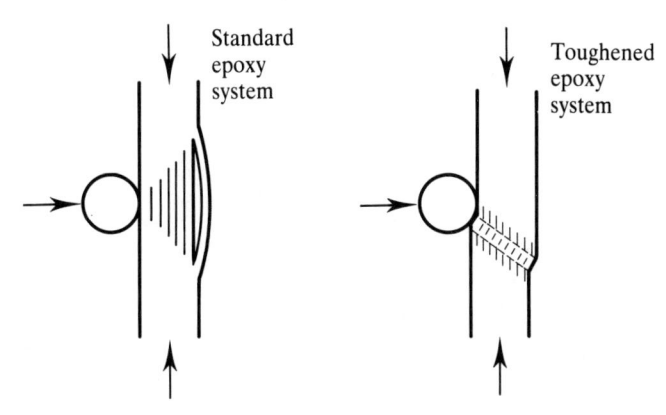

Figure 121 Failure modes associated with compression loading of impact damaged CFRP laminates

fatigue curves so design strains based on impact might have to be replaced with values based on fatigue for some applications.

In general, the tougher resin materials have had a poorer compression strength under hot-wet conditions, where again the shear modulus is reduced (see Section 6.1). There has had to be a compromise between the two design critical properties of compression after impact and hot-wet compression strength. Matrix materials improvements have been directed at improving both of these quantities together, as shown in Fig. 122. Other attempts at improvement have included altering fibre distributions such as in woven, braided and knitted preforms (see Section 5.13) and in designing hybrid laminates combining different composite materials in the various plies. Most of these have resulted in some improvement in impact performance but at the expense of some other structural property; their use depends on the specific design requirements.

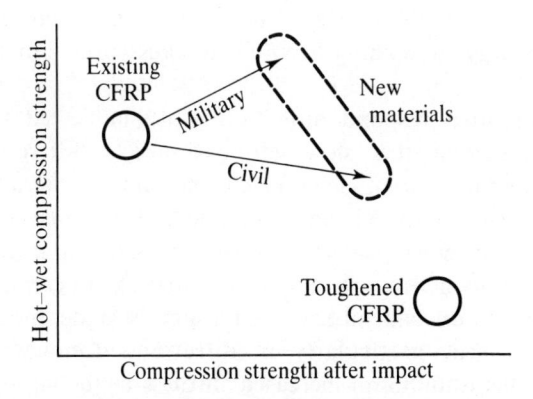

Figure 122 The development of improved CFRP materials combining critical properties

References

1. G Dorey in *Advanced Composites*, ed. I K Partridge, Elsevier Applied Science, London, 1989 pp. 369–98.
2. J G Williams, M D Rhodes in *Composite Materials: Testing and Design*, ed. I M Daniel, ASTM STP 787, American Society for Materials and Testing, Philadelphia, 1982.
3. S M Bishop in *Developments in Science and Technology of Composite Materials*, (*ECCM-1*), Bordeaux, eds A R Bunsell *et al.*, Elsevier, London, 1985.
4. W T Chester, M J Pavier, *Composites*, **21**, 23 (1990).

5.12 Impact resistance – fibre reinforced polymers

P HOGG

The impact response on a composite material is usually taken to cover the behaviour of thin shell composite structures subjected to high rate accidental loading from contact with a foreign object, such as a dropped tool or runway stones. There are a multitude of alternative impact situations (the crushing of structural members of an automobile during a crash is a good example) but the above represents probably the greatest problem in service for composites because the impacts occur out of plane to the reinforcement where the materials are particularly weak. It is instructive to reflect on the relative performance of polypropylene and carbon fibre reinforced epoxy composites. A quasi-isotropic plate fabricated from continuous aligned carbon fibres might exhibit an in-plane modulus of 50–55 GPa, and a tensile strength of 650–750 MPa. Polypropylene in contrast has an in-plane modulus of 1–2 GPa and a tensile strength of only 25 MPa. When subjected to a falling weight impact test, a 2 mm thick sheet of both materials would require impacts of similar energy, something like 40 J, to cause total penetration by the striker.

Structures often respond differently when subjected to high speed deformation compared to slow deformation. The differences can result from a number of factors, including changes in the material properties at high rates of strain and changes in static and dynamic response of the structure to load, which causes an effective change in near and far field stress states. Polymer matrix composite materials, in common with most plastic based materials, tend to exhibit a slightly higher modulus as strain rate is increased, particularly in directions dominated by matrix properties; this leads to an increased stiffness in the structure. Yielding in most epoxy and polyester matrix composites is limited, even at slow strain rates, and strain rate effects do not in general lead to any significant change in the fracture mechanisms. There will be exceptions to this where matrix properties are especially rate sensitive, particularly with some thermoplastic matrix composites.

In the main, differences between impact and slow rate testing in composites results from the increased stiffness of the composite and from the dynamic response of the structure. If impact rates of loading are considered to involve projectiles striking the composite at speeds in the range $1–10\,\mathrm{m\,s^{-1}}$, deformation and fracture are qualitatively similar to slow testing, even if quantitative differences appear. Under ballistic conditions, where the speed exceeds $100\,\mathrm{m\,s^{-1}}$, and where stress waves are significant, different fracture modes can occur and damage is highly localized.[1]

The shape, size and mass of the striker and the composite structure are further variables that can alter the effective impact performance of a composite. The discussion that follows is restricted to fibre composites loaded between 1 and $10\,\mathrm{m\,s^{-1}}$ with test data generated using an instrumented falling weight impact machine where a hemispherical striker (20 mm diameter) strikes a simply supported plate (60 mm square on a 40 mm diameter support ring). The mass of the striker varies but is always significantly greater than the mass of the specimen. The data is specific to the particular test geometry but is indicative of the trends exhibited in falling weight impacts in general.

The impact performance of fibre reinforced composites depends on many factors, including the nature of constituents, fibre, matrix and interface; the construction and geometry of the composite; and the test conditions. It is useful to divide a discussion of impact performance into two categories; impact performance under relatively low energy impacts, where the composite is damaged but still capable of performing its primary function; and impact performance under higher energy impacts, where the composite is completely ruptured or penetrated by the striker. The influence of material variables on performance is found to be different in these two categories.

Low energy impacts

When a composite plate is subjected to a low energy impact, the plate deformation absorbs the energy of the striker elastically at first and subsequently by a combination of microfracture processes and further elastic deformation. If the matrix is a thermoplastic, a degree of plastic deformation may also occur, particularly at the point of contact between striker and specimen, although this is usually restricted to a small indentation. The fracture processes that operate include transverse and intraply shear cracking, delamination and fibre fracture. Tensile fibre fracture is limited during the early stages of impact deformation but can occur on the tensile surface of the plate. Similarly, some evidence of compressive buckling can be found in some cases at or near the surface of the specimen contacting the striker. By far the main fracture modes are delamination and transverse/shear cracking. Delamination is particularly prominent if the composite consists of well-defined layers as in prepreg based laminates. The position and nature of the cracking that occurs in a given plate is very dependent on the geometry of the specimen and the relative interlaminar/in-plane shear and transverse strengths of the composite. Out of plane impact is essentially a bending deformation and this will induce both shear in the centre of the specimen and tensile and compressive stresses at the top and bottom surfaces. The thickness to span ratio of the plate controls the relative magnitudes of tensile and compressive stresses and, coupled with the relative strengths, they determine the position and extent of cracking. In most cases, where the composite plate is relatively compliant, cracking initiates as transverse damage near or at the bottom tensile surface; this induces the subsequent formation of interconnected delaminations. If the interlaminar strength is poor, delaminations at the neutral axis of the plate may be the first form of damage. Alternatively, if the stiffness of the plate is increased, for example due to an increase in thickness, then high contact stresses can develop and damage can initiate near to the striker at the upper surface of the plate. Again, this damage begins as transverse or shear cracks through the width of a layer or ply and subsequently leads to an interconnected series of delaminations and shear/transverse cracks that radiate through the thickness of the plate away from the point of impact, Fig. 118.

The matrix and interfacial strength are key variables in determining the nature and extent of cracking for a given low energy impact blow. The tougher the matrix and the stronger the interface, the less damage is formed. In aligned composites formed from prepregs, the extent of delamination has been linked to the various fracture properties of the laminate itself, such as mode I and mode II interlaminar fracture toughness.[2] Apart from their effect on interfacial strength during impact, fibres determine the local stress distribution through the composite by controlling stiffness. Similarly, fibre arrangement, stacking sequence and volume fraction can have a significant influence for related reasons, although additional effects may arise in some cases because of unfavourable ply sequences, which effectively create weak interfaces.[3]

The respective role of matrix and fibre may be judged from Fig. 123, which shows the width of delaminations formed after low energy impact of two carbon fibre composites, one with a PEEK matrix, another with a comparatively brittle epoxy and a third with an epoxy matrix and glass

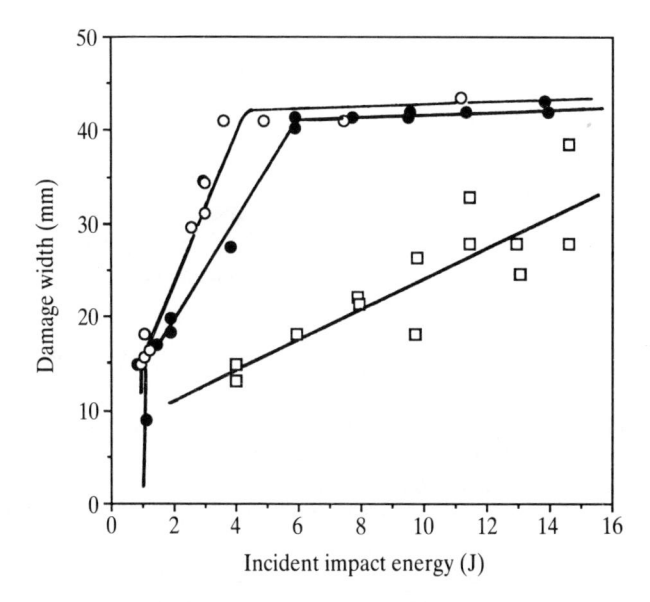

Figure 123 Width of delaminated zone as a function of impact energy for 2 mm thick, quasi-isotropic laminates with different fibres and resins, but identical specimen dimensions; ○ = carbon fibre–epoxy; □ = carbon fibre–PEEK; ● = glass fibre–epoxy

fibres. The extent of delaminations is a critical damage parameter if the composites are to continue to perform a structural role, for example under compression loads in an aircraft.[4]

High energy impacts

Impact tests on plates are component tests rather than tests to measure material properties. The complete penetration of a plate by an impacting striker therefore indicates a structural failure rather than a critical material property. The force–displacement curves generated during such tests reveal a complex series of peaks and troughs, Fig. 124. As the displacement of the specimen increases, the effective stiffness of the plate begins to decrease as various microfracture processes occur. Sudden large load drops are linked to major fracture processes, such as tensile fracture of fibre bundles or complete plies. The plates continue to sustain higher loads even after fracture processes have commenced, partly due to geometric stiffening effects, until the onset of major fibre fracture at or near to the tensile surface of the plate. The peak force itself does not

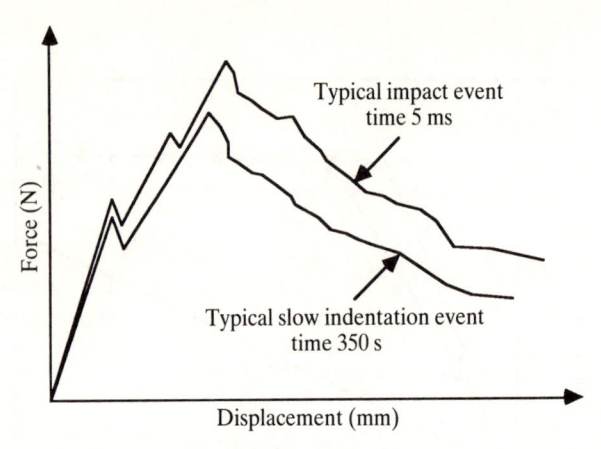

Figure 124 Load–deflection curve for carbon–PEEK composite subjected to falling weight impact compared to equivalent curve for slow indentation by a striker

represent any material parameter directly but is linked to the strength of the reinforcing fibres and strongly influences the energy absorbed by the composite during complete penetration by the striker.

The absolute energy absorbing capacity of composite plates and their ability to withstand a given impact blow without rupturing (i.e. the peak force) are largely functions of the fibre type and are little influenced by the matrix or detailed fibre arrangement. Figure 125 is a master curve for glass fibre composites showing the total energies absorbed as a function of volume fraction. The volume fraction has been multiplied by the thickness of the specimen to allow a comparison of a wide range of different materials. These include thermosets and thermoplastics, quasi-isotropic prepregs, woven laminates and random fibre composites, with volume fractions ranging from 0.2 to 0.6 and specimen thicknesses from 2 to 4 mm. A large scatter can be seen in the data points but this is similar to that typically observed with a single material (fibre–resin arrangement) where thicknesses and volume fraction are allowed to vary. Similar behaviour is shown by carbon fibre composites, Fig. 126. The data shown includes thermoplastic and thermosetting carbon fibre prepreg laminates. The carbon fibre composite results are compared with glass fibre composites both for maximum force generated during the impact and the total energy absorbed during complete penetration. It is clear that glass fibre composites perform better than carbon fibre composites under high energy impact conditions and evidence suggests this is largely a consequence of the higher strain to failure of glass fibres.

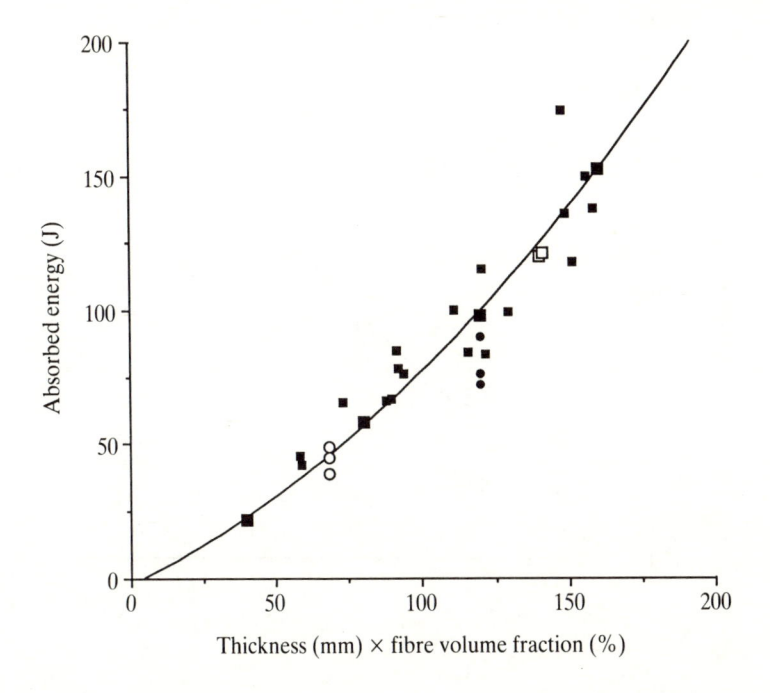

Figure 125 Relationship between energy absorbed during complete penetration versus thickness multiplied by fibre volume fraction for glass fibre composites. Specimens: A (■) are random and woven fibre thermosetting resin laminates; B (■) are SMC laminates, C (○) are random fibre thermoplastic GMTs; D (□) is a quasi-isotropic laminate produced from continuous aligned fibre reinforced polypropylene; and E (●) are continuous quasi-isotropic laminates produced from continuous fibre epoxy prepregs

References

1. G Dorey in *Sixth International Conference on Composite Materials, ICCM VI/ECCM* 2, Vol. 3, ed. F L Matthews *et al.*, Elsevier Applied Science, 1987, London, pp. 3.1–3.26.
2. R E Evans, J E Masters in *Toughened Composites*, ASTM STP 937, ed. N J Johnston, American Society for Materials and Testing, Philadelphia, 1987, pp. 413–436.
3. W J Cantwell, J Morton, *Composites*, **22**, 347 (1991).
4. J C Prichard, P J Hogg, *Composites*, **21**, 503 (1990).

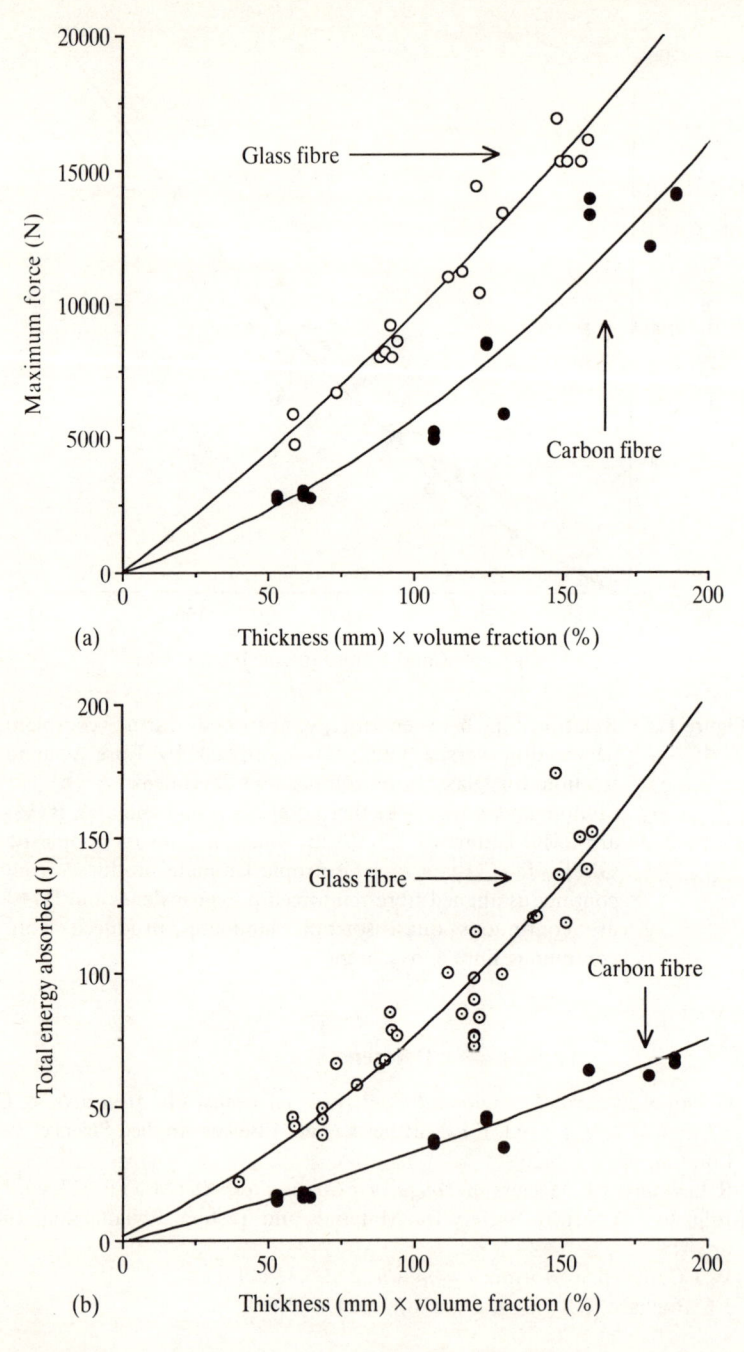

Figure 126 A comparison between a) peak forces and b) total energies absorbed by a wide range of glass fibre composites (data from Fig. 125) and carbon fibre laminates. The carbon fibre data includes both thermosetting and thermoplastic matrix systems

5.13 Knitted fabric composites – properties

F K KO

Introduction

The property of a composite depends on fibre orientation, θ, and fibre volume fraction, V_f, which are functions of the composite's fibre architecture and preforming process. In knitted structures, the key geometric parameter is the stitch length (loop length in a unit cell) which is influenced by yarn diameter and material property. Accordingly, we will first discuss the relationship between the geometric properties and the engineering parameters for a knitted fabric reinforced composite. The effect of the preform fibre architecture on the mechanical properties of knit composites will also be reviewed.

Geometric properties

Yarn diameter, d

The diameter of a yarn determines the gauge (number of needles per unit width) to be used on the knitting machine. Yarn diameter also governs the thickness of a knitted fabric and the dimension of a unit cell. In the knitting process, the knitting yarns are inevitably bent over a sharp radius of curvature equivalent to the size of the needles, which range from 0.5 – 2 mm. For a given fibre of modulus, E_f, and ultimate strength, σ_{uf}, the critical yarn bending diameter, d_c, is related to the needle diameter, D, by the following relationship:

$$d_c = \frac{\sigma_{uf} D}{E_f} \quad [5.14]$$

A plot of critical yarn diameter for various high modulus fibres is given in Fig. 127.

Stitch length, L

Stitch length or loop length is the most fundamental geometric parameter influencing the quality of a knitted structure. Stitch length dictates the stitches per unit area and consequently affects the fibre volume fraction and the level of distortion or segmental fibre orientation of the knitting loop in a knitting unit cell. For a weft knit, the stitch length is approximately 16 times the yarn diameter d. The unit cell dimension is wale spacing × course spacing × fabric thickness or qpt (where $q = 4d$, $p = 3.5d$, $t = 2d$). The theoretical fibre volume fraction, V_f, for Jersey

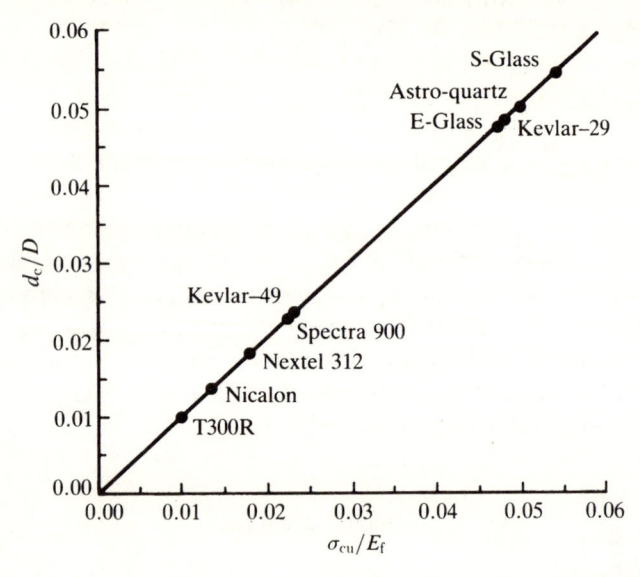

Figure 127 A plot of critical yarn diameter for various high modulus fibres

knit (plain knit) is therefore ~ 0.45 or 0.35 if a closed packing of fibres in a yarn is assumed. To take fibre orientation and the three-dimensional nature of the knitting loop into consideration, for a fibre packing fraction κ, the following geometric relationship can be established for a Jersey knit,

$$V_f = \frac{\pi \kappa}{8(q/d)(p/d)} \left\{ \frac{2 + q/d}{[1 + \tan^2(2d/q)]^{1/2} \cos \theta} + \frac{p/d}{1 + \tan^2(2d/p)} \right\} \quad [5.15]$$

Similarly, in a Tricot stitch for a warp knit structure, the stitch length has been shown to be

$$L = [(p - d)^2 + (nq - 3d)^2]^{1/2} + [(p - 2d)^2 + d^2]^{1/2} + [(p - 2d)^2 + 4d^2]^{1/2} + 15.4d \quad [5.16]$$

The fibre volume fraction for a Tricot knit is ~ 0.11.

For the MWK structure, the unit cell geometry has been idealized by Du and Ko[1]. The insertion yarns are assumed to have a race-track cross-section, whereas the stitch yarn is assumed circular in cross-section. The knit stitch is assumed to have the tightest loop construction and the curved loop is idealized to a rectangular shape. The dimensions of the unit cell are X, Y and Z, respectively, corresponding to the 0° axis, 90°

axis and the thickness axis vertical to the 0°–90° plane. In the following section, the expressions for yarn dimensions, unit cell dimensions, yarn volumes, yarn lengths, fibre volume fractions and criteria for geometric limit are summarized.

The boundary condition that defines the limiting geometry of the MWK fabric can be summarized by the following equations,

$$\phi_x \geq \max\left(1 + \frac{d_s}{w_y} \cdot \frac{w_{\pm\theta} - d_s(1 + \cos\theta)}{w_y \sin\theta}\right) \quad [5.17]$$

$$\phi_y \geq 1 + \frac{2d_s}{w_x} \quad [5.18]$$

where θ = angle of bias yarns to 0° yarn (x-direction), d_s = diameter of stitch yarn, w = width of insertion yarns and ϕ = thread count (number of insertion yarns per unit length, end m^{-1}. Subscripts x and y refer to 0° and 90° orientations, respectively. The degree of fabric tightness is expressed by tightness factor, η,

$$\eta = \frac{1 + d_s/w_y}{\phi_x} \quad (\leq 1) \quad [5.19]$$

Using (5.17)–(5.19), we can relate V_f to key process variables, including bias yarn orientation angle, θ, ratio of stitch to insertion yarn linear density, λ_s/λ_i, and fabric tightness factor, η. The fibre volume fraction relation in Fig. 128 shows that, for a given set of processing parameters, only a limited window exists for MWK fabric construction. The window is bounded by two factors: yarn jamming and the point of 90° bias yarn angle. Fabric constructions corresponding to the curve marked 'jamming' are at their tightest allowable point. As the bias yarn angle increases, it results in the most open structure.

When $\theta < 30°$, jamming occurs in the entire range of yarn linear density ratios from zero to infinity. When θ is in the range 30–40°, the fibre volume fraction decreases with the increase in yarn linear density ratio until jamming occurs. When $\theta = 45°$, the fibre volume fraction decreases with the increase in yarn linear density ratio to a minimum at about $\lambda_s/\lambda_i = 1$, and starts to increase until jamming occurs. When $\theta \geq 60°$, the fibre volume fraction has the same trend as when $\theta = 45°$, but yarn jamming never occurs. The fibre packing in the yarns, taken as 0.75, establishes the upper limit of fibre volume fraction for the fabric.

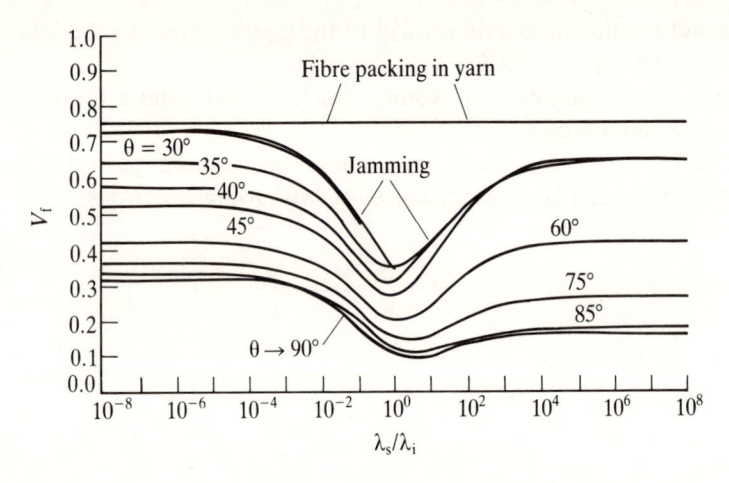

Figure 128 Fibre volume fraction versus ratio of stitch-to-insertion yarn linear density (tricot stitch, $\kappa = 0.75$, $p = 2.5$ kg m^{-3}, $f_i = 5$, and $\eta = 0.5$)

Mechanical properties

There is a limited database for the mechanical properties of knitted composites. The common belief is that knitted structures do not have the linearity nor sufficient fibre volume fraction to meet the requirements of structural applications. The analysis of the geometric properties of knitted structures above demonstrated quite the contrary. Recent experimental evidence presented by Verpoest[2] showed that the lower bound strength for weft knitted composites is equivalent to that of the sheet moulded compounds whereas the knitted composites reached an upper bound strength equal to that of the woven composites when the knitted structure is properly compressed. The strength and stiffness of the MWK composite should be expected to approach that of the angle ply unidirectional composites. In this brief review, the role of inlay yarn in weft knit and warp knit will be illustrated.

Weft knit composites

The tensile properties of 1×1 rib knitted carbon fabric reinforced epoxy composites made by resin transfer moulding were studied by Krishna and Hull.[3] By increasing the tow size of the staple knitting yarn from 1.3 k to 3.9 k it was found that the fibre volume fraction increases from 10.5% to 31.5% resulting in a two-fold increase in modulus, from 10 to 20 GPa and a close to two-fold increase in strength, from 57 to 91 MPa. The warpwise

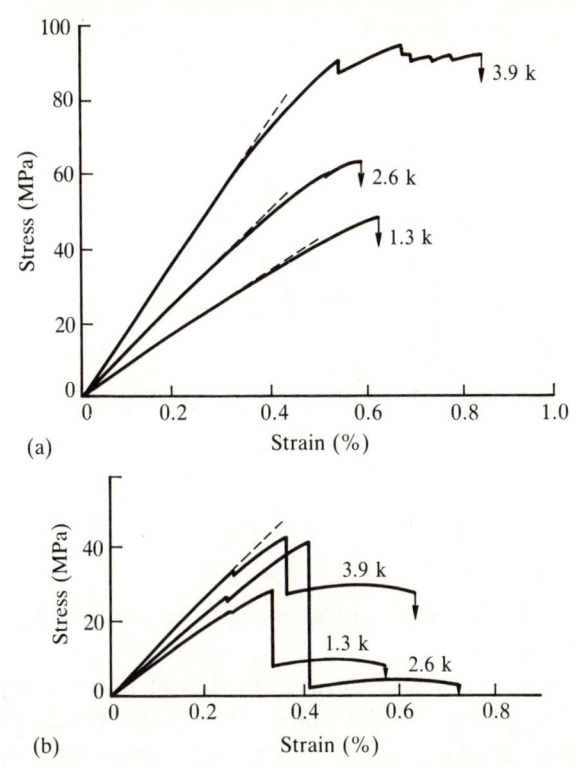

Figure 129 Stress–strain curves for knitted composities (a) warpwise (b) weftwise

(wale) strength and modulus were found to be higher than the weftwise (course) direction. The typical warpwise and weftwise stress-strain curves for the knitted composites are shown in Fig. 129.

The weftwise strength and modulus of the composites were shown to be increased significantly by inserting 6 k and 12 k continuous carbon fibres. No significant effect on the tensile modulus and only a small reduction in strength was observed in the warp direction. Figure 130 shows the typical stress–strain curves for the inlay composites tested in the weft direction.

The authors also noted that failure of the composites propagates in a brittle manner through the resin-rich region; this was attributed to the sharp bending curvature of the knitting yarn in the fabric.

Warp knit composites

The precise automatic placement of multiple layers of linear fibres in a wide range of fibre orientations makes multiaxial warp knit (MWK) preforms an attractive candidate for structural composites. One of the

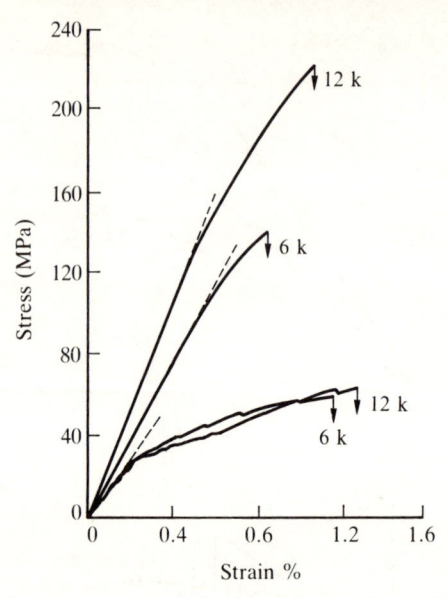

Figure 130 Stress–strain curves for inlay composites tested warpwise

earliest studies of the mechanical properties of MWK composites was carried out by Ko et al.[4] In this study, MWK graphite preforms with a 0°/90°/±45° fibre stacking sequence were impregnated with Epon 828 to produce composites having a fibre volume fraction of 60%. The MWK composites have linear stress–strain behaviour at strength and modulus levels of 500 MPa and 68 GPa, respectively.

More recently, a comprehensive study of carbon composites reinforced by MWK fabrics produced by the Liba and Mayer type machines was carried out by Dexter et al.[5] These preforms were stacked together to form a 16 layer laminate having a fibre orientation of $[-45°/0°/45°/90°]_{2S}$ and $[(-45°/45°/0°/90°)_2; (\pm 45°/-45°/0°/90°/0_2]$ for the Liba and Mayer MWK fabrics respectively. The Liba preforms were stitched together with carbon yarns, while the Mayer preforms were stitched together with Kevlar 29 yarns. The preforms were vacuum impregnated with Hercules 3501-6 epoxy resin to produce a composite having a fibre volume fraction of 59%. Table 50 provides a comparison of the mechanical properties of the MWK composites with quasi-isotropic laminates.

It can be seen that, except for tensile and compressive strength, the MWK composites compare favourably or are the equal of the prepreg laminates in performance. Damage resistance of the MWK composites, as shown in the CAI tests, is significantly higher than that of the laminated composites. Combining these interesting mechanical properties with high levels of producibility, multiaxial warp knits are serious contenders for composite structural composites.

Table 50. Mechanical properties of MWK and laminated composites

	Liba	Mayer	Prepreg laminate
Tensile strength (MPa)	350	385	476
Compressive strength (MPa)	385	490	560
Tensile modulus (GPa)	43	45	49
Compressive modulus (GPa)	42	42	48
CAI strength (MPa)	203	154	126
OH tensile strength (MPa)	364	–	–
OH compressive strength (MPa)	315	350	308

CAI = compression after impact, OH = open hole.

References

1. G W Du, F K Ko in *Proceedings of Textile Composites in Construction*, Lyons, Jun. 1992, Part 2, eds P Hamelin, G Verchery, Editions Pluralis.
2. I Verpoest in *Proceedings of Fiber-Tex*, Philadelphia, Oct. 1992 Drexel University, NASA CP-3211, Washington DC 1993.
3. S R Krishna, D Hull in *Composites Design, Manufacture and Application*, (*ICCM VIII*), Vol. 1, Ch. 6, paper A, eds S W Tsai, G Springer, SAMPE, Covina, California, USA 1991.
4. F K Ko, C M Pastore, J M Yang, T W Chou in *Composites 86: Recent Advances in Japan and the United States*, eds K Kawata, S Umekawa, A Kobayashi, Japan Society for Composite Materials, Tokyo 1986.
5. B Dexter in *Proceedings of Fiber-Tex*, Philadelphia, Oct. 92 Drexel University, NASA CP-3211, Washington DC 1993.

5.14 Laminate theory

P A SMITH

Mechanics of a lamina

To understand the mechanics of a laminate based on plies consisting of unidirectionally reinforced material it is necessary first to consider the behaviour of a single unidirectional lamina. In doing this we ignore the heterogeneity of the composite, i.e. the fact that it consists of fibres and matrix, and treat it as having overall elastic constants in any direction which will in detail depend on fibre and matrix properties and the fibre volume fraction. For in-plane loading (the only type of loading appropriate to consider in a brief introductory article), see Fig. 131, there are five elastic constants which appear in the stress/strain relations. The terms E_1 and E_2 are used to denote the Young's moduli parallel and

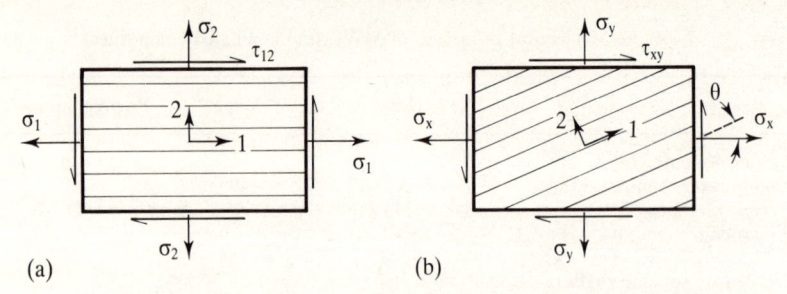

Figure 131 Schematic diagrams of a unidirectional lamina showing loading in (a) principal material directions 1 and 2, (b) coordinate system x–y

perpendicular to the fibre direction, respectively, the so-called principal moduli of the lamina. The quantity G_{12} is the shear modulus appropriate to shear loading in the 1–2 coordinate system. There are also two Poisson's ratio for a lamina: ν_{12} gives the strain in the 2-direction as a result of a strain in the 1-direction, while ν_{21} gives the strain in the 1-direction as a result of a strain in the 2-direction. Although there are five elastic constants, they are not independent of each other. The principal moduli and Poisson's ratios are related by the expression

$$\nu_{12} E_2 = \nu_{21} E_1 \quad [5.20]$$

so that four independent elastic constants (G_{12} and three from E_1, E_2, ν_{12}, ν_{21}) characterize the response of a lamina to in-plane loading. This may be contrasted with the two independent constants (two from Young's modulus, E, shear modulus, G, and Poisson's ratio, ν) needed to characterize the behaviour of an isotropic material under similar loading conditions.

If the lamina is loaded by stresses σ_1, σ_2 and τ_{12} (Fig. 131a), then the resulting strains ϵ_1, ϵ_2, γ_{12} can be found from

$$\epsilon_1 = (\sigma_1/E_1) - \nu_{21}(\sigma_2/E_2)$$
$$\epsilon_2 = (\sigma_2/E_2) - \nu_{12}(\sigma_1/E_1) \quad [5.21]$$
$$\gamma_{12} = \tau_{12}/G_{12}$$

This can be rewritten in matrix form as

$$\begin{pmatrix} \sigma_1 \\ \sigma_2 \\ \tau_{12} \end{pmatrix} = \begin{pmatrix} Q_{11} & Q_{12} & 0 \\ Q_{12} & Q_{22} & 0 \\ 0 & 0 & Q_{66} \end{pmatrix} \begin{pmatrix} \epsilon_1 \\ \epsilon_2 \\ \gamma_{12} \end{pmatrix} \quad [5.22]$$

where the Q_{ij} terms (the so-called lamina principal reduced stiffnesses) are expressed in terms of the elastic constants E_1, E_2, ν_{12}, ν_{21}, G_{12}, for example $Q_{11} = E_1/(1 - \nu_{12}\nu_{21})$.

Laminate theory

If the lamina is loaded in an x–y coordinate system, defined by the angle θ made by the x-direction with the fibres (Fig. 131b) then simple stress analysis gives the stress–strain relation in the x–y coordinate system as

$$\begin{pmatrix} \sigma_x \\ \sigma_y \\ \tau_{xy} \end{pmatrix} = \begin{pmatrix} \bar{Q}_{11} & \bar{Q}_{12} & \bar{Q}_{16} \\ \bar{Q}_{12} & \bar{Q}_{22} & \bar{Q}_{26} \\ \bar{Q}_{16} & \bar{Q}_{26} & \bar{Q}_{66} \end{pmatrix} \begin{pmatrix} \epsilon_x \\ \epsilon_y \\ \gamma_{xy} \end{pmatrix} \qquad [5.23]$$

where the $\bar{Q}_{ij}$ terms (the transformed reduced stiffnesses) are expressed in terms of the principal reduced stiffnesses, Q_{ij}, for example $\bar{Q}_{11} = Q_{11} \cos^4\theta + 2(Q_{12} + 2Q_{66})\sin^2\theta \cos^2\theta + Q_{22}\sin^4\theta$. The presence of the $\bar{Q}_{16}$, $\bar{Q}_{26}$ terms means there is coupling between extensional and shear deformation except when $\theta = 0°$ or $90°$. Equation (5.23) is the fundamental equation describing the deformation of a lamina under in-plane loading and is the basis of laminated plate theory analysis.

Laminate theory – loading and deformation

Laminated plate theory in its simplest form is able to calculate the overall response (midplane strains, curvatures) of an arbitrary lay-up of laminate under applied loading (forces, moments) (Fig. 132). It is assumed that the laminate deforms under conditions of plane stress, i.e. through-thickness stresses are neglected, and that normals to the laminate remain normal after deformation, i.e. there is no slippage between the layers.

The equation that describes the loading deformation behaviour of a laminate is obtained from strain compatibility and overall equilibrium.[1–3] It may be expressed in matrix form as

$$\begin{pmatrix} \mathbf{N} \\ \mathbf{M} \end{pmatrix} = \begin{pmatrix} A & B \\ B & D \end{pmatrix} \begin{pmatrix} \epsilon \\ \kappa \end{pmatrix} \qquad (5.24)$$

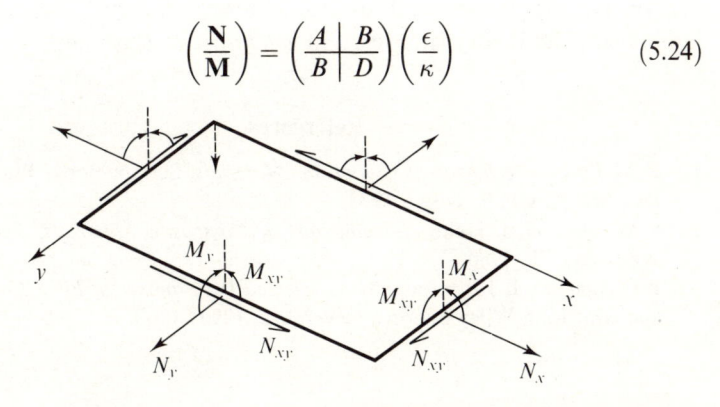

Figure 132 Applied loading on a laminate expressed as a system of forces per unit width (N_x, N_y, N_{xy}) and moments per unit width (M_x, M_y, M_{xy})

where **N** and **M** are column vectors of the applied loading, forces (N_x, N_y, N_{xy}) and moments (M_x, M_y, M_{xy}), and ϵ and κ are column vectors of the resulting deformation, midplane strains (ϵ_x, ϵ_y, γ_{xy}) and curvatures (κ_x, κ_y, κ_{xy}). The terms A, B and D are 3×3 matrices whose coefficients depend on the lamina elastic properties (i.e. the Q_{ij} terms for each ply), the lay-up of the laminate and the stacking sequence.

On calculating the overall deformation of a laminate from (5.24) it is then a simple matter to recover the stresses in the individual plies by substitution in (5.23).

At first sight, (5.24) appears complicated and cumbersome. Fortunately, for many practical situations it is simplified considerably, as many of the terms vanish. For example, if the laminate is midplane symmetric, i.e. the stacking sequence either side of the laminate midplane is identical, there is no coupling between extension and bending; all the terms in the B matrix are zero.

Furthermore, in balanced laminates, where for every ply oriented at $+\theta$ there is a corresponding ply oriented at $-\theta$, no coupling exists between extension and shear, therefore some of the terms in the A matrix are zero.[1-3]

Concluding remarks

In practice, laminated plate theory can be carried out most conveniently using a microcomputer based package, a number of which are available commercially. Software packages enable additional factors such as temperature and moisture effects to be incorporated and may additionally use the calculated stresses in each ply in conjunction with some of the standard lamina failure criteria (see Section 5.4) to predict progressive ply failure and overall laminate failure. The subject of laminated plate theory is covered in a number of textbooks.[1-3]

References

1. R M Jones, *Mechanics of Composite Materials*, Scripta Book, Washington, DC, McGraw-Hill, Tokyo 1975.
2. S W Tsai, H T Hahn, *Introduction to Composite Materials*, Technomic, Westport, CT, 1980.
3. B D Agarwal, L J Broutman, *Analysis and Performance of Fiber Composites*, 2nd edn, John Wiley & Sons, New York, 1990.

5.15 Non-destructive evaluation of composites

B HARRIS

In a brief review of this kind, it is impossible to cover thoroughly all non-destructive evaluation (NDE) procedures currently used for assessing the quality of composites and the reader is referred to the extensive literature for details of practical experience. Reviews of the subject have been published earlier.[1-4]

Optical inspection

In translucent GRP, inspection by transmitted light can give a good indication of the presence of pores, poor wetting out, delaminations and gross inclusions. Light transmission may also be correlated with the fibre content, V_f. Loss of transparency is often associated with the development of fibre–resin debonding and resin cracking (stress whitening). In non-translucent composites, the only feasible visual inspection techniques relate to observation of surface damage. The detection of surface cracking may not necessarily reflect the true state of deterioration of a composite. Detection of surface cracks and deeper cracks open to the surface may be enhanced by dye penetrant methods.

For critical components and areas previously identified as being susceptible to local damage, moiré methods may be suitable. Photographic grids are printed on the surface of the composite and irradiated by coherent light. The resulting interference fringes give clear indications of local stress concentrations and deformation, including those arising from subsurface damage. Laser holographic methods have also been used for the inspection of composite vessels and were shown to be capable of detecting bond failures and early fatigue damage in filament–wound structures. Limitations on the resolution of the early experimental techniques were related to stability and cumbersome photographic processes; recent television hologram interferometry has recently been able to make use of electronic image processing and the technique of electronic speckle pattern interferometry (ESPI) has considerably changed the potential of holographic methods for real-time investigation of damage and defects.[4]

Microwave methods

At microwave frequencies, the dielectric constant of glass is much greater than the dielectric constants of resins; any technique for measuring the average dielectric constant of a GRP composite can therefore be correlated reasonably satisfactorily with the glass content of the material.[4] A typical instrument consists of an open-ended coaxial resonator; the resonant cavity is closed when the instrument is applied to a GRP surface. As the instrument is moved over the surface, changes in the local average glass content are indicated by changes in the resonant frequency. Since any defect that affects the microwave penetration will also cause a resonance shift, the method can also monitor deterioration of a structure under load, once its dielectric response in the undamaged state is known.

Radiographic methods

A good deal of information about composite quality can be obtained by X-ray inspection. Contact radiography with sources of 50 kV or less yields high contrast photographs from low density materials, such as GRP, because of their low inherent filtration. The linear absorption coefficient of glass is about 20 times that of most resins; film density measurements can therefore be correlated with glass fibre content per unit surface area, provided the material is unpigmented. Fibre distribution, quality of weave and the presence of large laminating defects can be easily investigated; there is evidence that failures often occur at manufacturing defects, which are revealed by X-rays even when the defects do not appear to reduce the composite strength. The sensitivity of contact radiography can be improved by impregnating the composite with a radio-opaque material such as zinc iodide. Resolution of cracks of the order a few millimetres long and 0.1 mm deep is feasible in GRP and CFRP, and delaminations are easily resolved. Recent developments in low energy radiographic methods, with specific reference to image quality, image processing and real time analysis, have been described elsewhere.[4]

Thermal imaging

When a uniform heat flux is supplied to a plate, any anomalous variations in the resulting temperature distribution in the plate are indications of structural flaws in the material. Similarly, in an otherwise uniform material working under variable loads, local damage will give rise to changes in the hysteretic thermal losses in the body of the material. Thermal imaging, with limiting detection of temperature differences of about 0.2°C, is easily carried out by means of infrared television photography; colour differentiation analysis gives better resolution than black and white photography. The method has been used to monitor fatigue and other damage in a variety of composite materials (see Sections 5.5–5.7) and has successfully indicated changes in stress patterns near stress concentrators and other damage that results in heat dissipation. Surface temperature has been found to be a sensitive indicator of dispersed and localized damage. Damage has been detected in glass/epoxy laminates at low stresses and frequencies; detection is more difficult in materials of higher thermal conductivity. Testing frequencies greater than 5 Hz are needed to detect fatigue damage in CFRP because of their high thermal conductivity.

In recent years, thermography has been extended by the use of transient heat sources and rapid scanning of the induced thermal fields by TV/video compatible infrared imagers, such as form the basis of the pulse video thermography (PVT) technique.[4] This is a convenient and versatile extension of conventional thermography capable of rapid detection of near surface defects.

An associated thermal NDE method is that of thermoelastic stress analysis. The familiar thermoelastic effect, where a temperature change accompanies adiabatic elastic deformation of a body, has been embodied in commercial equipment of the kind known as SPATE (stress pattern analysis by measurement of thermal emission). This system is said to offer spatial resolution down to 1 mm and temperature discrimination of about 0.002°C for stress changes of about 2 MPa in steel. The application of the method to composites has recently been investigated by Potter and Greaves,[5] who draw attention to potential problems of interpretation resulting from the anisotropic nature of composite laminates.

Dynamic mechanical analysis

A cracked bell or railway wheel sounds dull instead of ringing when tapped; the railway wheel tapper has long used this simple concept as an easy NDE method. By extension, if any structure can be forced to resonate in a reproducible manner, its frequency spectrum can be rapidly analysed to establish its characteristic 'sound'. Any damage to the structure could then be expected to change this characteristic spectrum in a recognizable way. The use of a simple 'coin tap' test of this kind to monitor composite fatigue test pieces has been found to be a poor indicator of early fatigue damage because significant indications are often observed only where damage is already visible. More elaborate low frequency vibration techniques for non-destructive evaluation of the integrity of structures have been developed into a commercial system based on the coin tap idea. The difference between good and defective structures can be detected by tapping the structure with a small hammer with a force transducer incorporated into its head.

Ultrasonic techniques

Ultrasonic inspection techniques are perhaps the most widely used of common NDE methods for quality assessment of composites. The velocity and attenuation of an ultrasonic pulse passing through a material provide information about its general physical properties (i.e. its stiffness) and the structure (i.e. the defect/damage state). Certain basic features of the material need the anisotropy to be taken into account, e.g. a composite. Ultrasonic (US) waves in composites are highly attenuated and their velocity and attenuation are strongly dependent on frequency. This frequency dependence is affected by the detailed geometric construction of the laminate. Most of the common US techniques are used with composites, including pulse transit time measurements, pulse echo methods, goniometry and ultrasonic interferometry.

The most highly developed ultrasonic NDE method, based on measurements of through-thickness attenuation, is the C-scan technique, used routinely for inspecting large panels. Synchronized raster scanning motions of the transmitter and receiver (either using irrigated water-jet probes or in a water bath) on opposite sides of the plate allow measurements of the intensity of the transmitted wave to be made as a function of position. The transmitted intensity is used to modulate the brightness of a visual display spot or the density of ink on a diagram at the appropriate x, y coordinates, so as to build up a picture of plate quality. The resolution of the technique is limited by dispersion in the composite and by the beam dimensions. Something like a focused 4 MHz probe is needed to detect defects of the order of 1 mm in size, consequently C-scanning is more reliable as an indicator of general quality than as a detector of specific defects. Large voids, distributions of fine porosity, areas of variation in fibre content and delaminated regions will usually be revealed by a C-scan, but for assessment of the severity of specific defects it is necessary to make standard test plates containing flat-bottomed drilled holes of various depths and diameters for calibration purposes. The resolution of C-scans can be considerably improved by modern colour-differentiation techniques.

One of the better-established NDE methods based on ultrasonics is the principle embodied in commercial equipment of the kind exemplified by the Fokker debond tester manufactured by Wells-Krautkramer. This machine uses swept frequency probes generating shear waves of constant amplitude and compares the response of the test material with that of a standard. The resonant frequency and amplitude of the transducer response are recorded in contact with the standard, and any shifts that occur when the probe is subsequently coupled to the test sample are noted. These shifts must be calibrated against some known or measurable property such as the strength of a lap joint, and they can then be used as a predictor of lap strength. By appropriate calibration the technique is said to be capable of giving more sensitive indications of the presence of flaws, voids, weak bonds, porosity, delaminations, incorrect cure conditions, poor wetting, and poor gap-filling in joints than any other ultrasonic method. This may be an optimistic view.

Acoustic emission methods

Any sudden structural change within a composite, such as resin cracking, fibre fracture, rapid debonding or interlaminar cracking, causes dissipation of energy as elastic stress waves; these acoustic emissions (AE) spread in all directions from the source and are detected by suitable transducer/amplifier systems now well established; triangulation methods may be used to locate flaws in large structures and to assess their severity.

Each stress wave reaches the piezoelectric transducer (usually coupled to a free surface) as a complex wave packet. If the signal is amplified and fed to a counter that identifies all positive crossings of a given threshold, the number of counts recorded will clearly be much greater than the absolute number of microfailure events that have occurred. This type of counting is known as ring-down counting. A better technique is to convert the ring-down signal into a single envelope and identify each envelope as a distinct 'event' by using a counter that recognizes a dead time of, say, $100\,\mu s$ between events. In this event counting, there should be nearly a 1:1 ratio between the number of microfailure events and the number of counts recorded by the equipment. Both types of counting can be used with a time reset system to give an indication of the rate of occurrence of emissions.

For more detailed information than can be gained from merely the numbers of AE events recorded, it is necessary to analyse either the amplitudes of the stress wave pulses or their frequency spectra. In principle, such analysis gives information about the individual mechanisms occurring within the material to give rise to the recorded event signals.

There has been a great deal of research into AE methods, both from the point of view of their use as NDE tools and as a means of investigating damage mechanisms and damage accumulation in composites under stress, but there is currently much disagreement about the value of these techniques. It seems clear that each type of composite investigated must be treated as a specific case; generalization of any kind seems unwise. A brief review of the subject has been given elsewhere[2] but there is a vast literature on the subject.

Conclusions

By comparison with the existing detailed understanding of the working of standard NDE methods for metallic materials, the value of non-destructive inspection methods for composite materials appears to be much less certain. No single method has yet emerged as being universally useful, although it appears that ultrasonic C-scanning is one of the most used and most useful. The use of two or more techniques, especially where the information they give overlaps, is safer than relying on a single method. A vital preliminary to using any technique is extensive calibration in relation to the particular type of composite to be monitored; that way the limitations of the method are completely understood.

The field of application of NDE methods to composite materials is a complex one; potential users of these techniques may well feel unable to select tools appropriate to their specific quality assurance problem. In a recent review, Adams and Cawley[4] have surveyed modern developments in the field and have attempted to ease the problem of selection by presenting a matrix of NDE methods and defect types with indications of usefulness or otherwise.

References

1. B Harris, *Practical Considerations of Design, Fabrication and Tests for Composite Materials*, AGARD Lecture Series, Paper 124, NATO, Neuilly-sur-Seine, 1982.
2. B Harris, M G Phillips in *Developments in GRP Technology 1*, ed. B Harris, Applied Science, London, 1983, pp. 191–247.
3. A Mahoon, *NDT Intl*, **19**, 229 (1988).
4. Various authors, *NDT Intl*, **21**, (1988), special issue on NDE of composite materials.
5. R T Potter, L J Greaves, *Thermoelastic Analysis of Anisotropic Materials*, RAE (now DRA) Technical Report TR 90012, Ministry of Defence, London, 1990.

5.16 Standard test methods for composite materials

E W GODWIN

Mechanical test methods for determining the strength and stiffness of 'conventional' (i.e. homogeneous, isotropic) materials have been used for many decades. The work hardening and yielding behaviour of the material under test simplifies testing and the results are accepted without criticism. Composite materials, especially high strength, long fibre composites, are not homogeneous and may often be highly anisotropic; this presents problems when conventional test methods are used.

A set of techniques for testing composites is evolving[1-3], often radically different from those used with other materials. For instance, the conventional method of ensuring failure at a given point in a test piece is to reduce its local cross-section; the weakness of composites in shear makes this impractical so the ends of test pieces are built up with 'end tabs', which protect the ends of the material whilst providing a softer surface easy to grip.[4]

Standards organizations such as the American Society for Testing and Materials (ASTM) and the British Standards Institute (BSI) publish recommended standard test methods including test piece dimensions and recommendations on testing techniques, methods of strain measurement, etc.

Composite test specimens are generally machined from flat sheet produced to the required thickness. It is unusual to machine composites to thickness, as removal of the resin-rich outer surface of the material and damage to the underlying fibres may give a different result from 'as moulded' material. In any case, through thickness machining would only be admissible in the case of unidirectional or in-plane random fibres, otherwise the effective lay-up of the material would be altered. The edge should be cleanly cut; final abrading or grinding will remove machining marks.

Test pieces should contain a representative sample of the material under test, i.e. correct proportions of fibres in each direction. If the material contains, say, a coarse woven reinforcement this may mean that a large test piece is required. The length must be sufficient to allow a non-axial fibre to run the full width of the specimen plus typically a half-specimen width clearance before each end. A typical tensile test piece might be 200 mm long by 25 mm wide and 1–2 mm thick. The ends might be reinforced with tabs of softer composite or aluminium alloy some 50 mm long. The width of the test piece depends to some extent on the material under test and a requirement to minimize edge effects. Test pieces containing only 0° fibres are less susceptible to edge effects and could be as narrow as 10 mm. Typical test pieces are shown in Fig. 133.

Testing speed should be selected to avoid the influences of creep and viscoelastic effects; failure should occur within 60–180 s. Fatigue testing is generally carried out at frequencies of 5–15 Hz to prevent heating in the material due to hysterisis or friction between the matrix and pulled-out fibres.

Strains can be measured using strain gauges or extensometers, as well as methods such as photoelasticity, holography, and moiré. Some extensometers will measure both longitudinal and transverse strains, for Poisson's ratio determination. Care should be taken that the knife edges of the extensometer neither slip on the test piece nor damage the surface of the material.

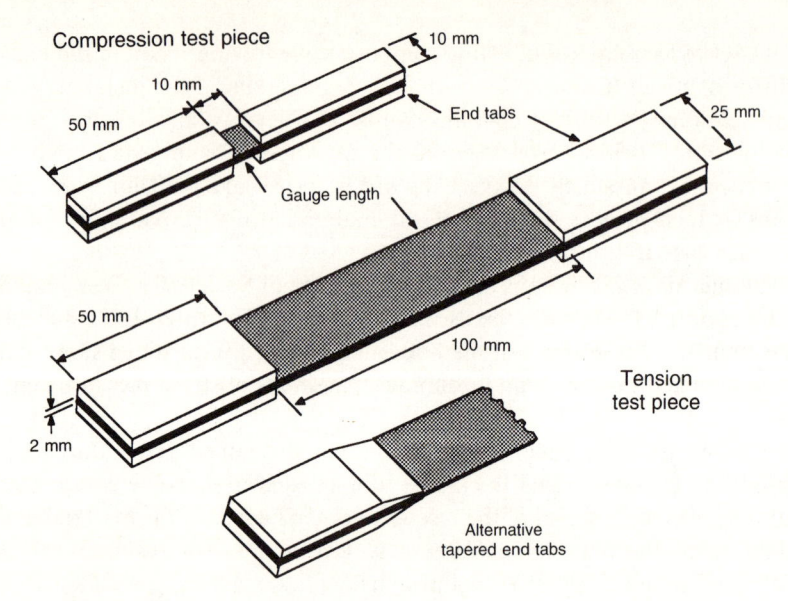

Figure 133 Typical forms and dimensions of composite test pieces

Strain gauges are often more convenient to use and enable strain measurements to be made over a length of 1 mm or less, but they are intended for use on metals so it may be necessary to make corrections for Poisson's ratio effects and mismatch of thermal expansion between the gauge and the test piece. Problems arising from resistive heating of the gauge can be minimized by using high resistance gauges with a large surface area over which heat can be dissipated.

Tensile loads are applied to the test piece through serrated jaws, often designed to wedge together under load to reduce the risk of the test piece sliding within the jaws. This is assisted by yielding of the gripped portion of the test piece and the yielding surface is provided by the end tabs. Hydraulically closed grips allow more accurate control of the gripping force.

A fine crosscut pattern on the jaw faces is recommended for gripping composites: some operators place a layer of emery cloth between the jaws and the composite to reduce the coarseness of the gripping surface. There are currently moves towards replacing traditional bonded end tabs completely with unbonded abrasive cloth.

Composites containing 0° fibres in the outer layers are likely to suffer premature failure in the region of the end tabs due to damage to the outermost load-bearing fibres. Some standards specify that a result should only be regarded as valid if failure has occurred at least some (specified) distance away from the jaws, otherwise the test result is only valid as a lower bound value for the material's strength. Care should be taken with test piece alignment because no yielding can occur to alleviate high local stresses.

Compression testing is complicated by the need to avoid the test piece failing by buckling whilst providing a gauge length sufficiently long to incorporate a representative proportion of fibres and minimize end effects. Often some compromise is necessary to achieve this and it is common to use a gauge length as small as 10 mm for unidirectional material. The test piece is mounted in a special loading jig, designed to be rigid enough to prevent out of plane bending and to allow free axial movement. A typical jig, developed at Imperial College, is shown in Fig. 134.

Multidirectional or woven material, which requires a larger test piece in order to contain a representative sample of the material, is clamped between supports which stabilize the material against out of plane bending. The objection to such antibuckling guides is that they may overconstrain the gauge length. Another compression test is occasionally used but is inconvenient; the material to be tested forms one face of a sandwich beam loaded in bending.

Flexure tests are convenient to carry out and produce results suitable

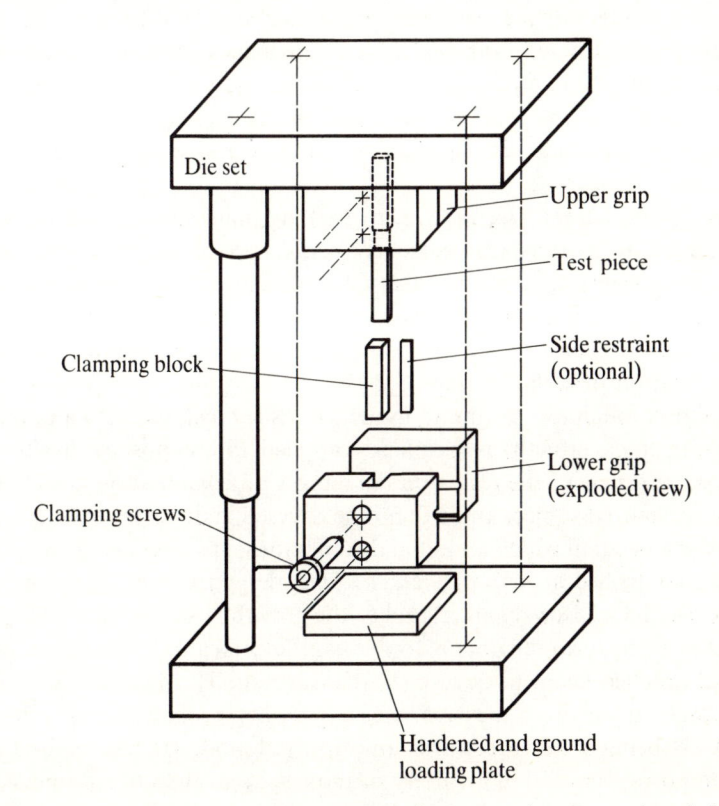

Figure 134 Imperial College compression test rig. Reproduced by courtesy of Dr J. G. Häberle

for many purposes. The test is particularly suited to quality control as simple parallel-sided test pieces are used and the test is quickly carried out. Three- or four-point loading can be used; four-point loading gives a pure bending moment in the central part of the beam. A wide range of beam sizes is used; a span/depth ratio of at least 16/1 is recommended: up to 60/1 may be used. The results are calculated using engineering beam theory, which may require correction due to the large deflections encountered and the fact that the tensile and compressive moduli of many composites are not the same. Moduli and strengths determined in flexure are generally lower than those measured in tension. Objections that the test is invalid as neither pure tensile nor pure compressive stresses are produced can be countered by the fact that flexure represents a mode of loading that is common in practice.

A small span to depth ratio (usually 5/1) allows shear stresses to predominate and provides a simple method for determining interlaminar shear strength. The method is preferable to an alternative one using a double-notched tensile test piece which can fail as a result of through thickness tensile stresses. Flat panels loaded through rigid bars attached to opposite edges of the test piece, 'rail-shear' tests and torsion of a thin walled, filament wound tube each have drawbacks. Rail shear requires a large amount of material and may fail prematurely at the fixing between the material and the loading rails; torsion gives almost pure shear but requires a torsional testing machine and a test piece expensive to fabricate.

The Ioscipescu test (see Fig. 73, p. 273), originally devised for use with 'conventional' materials, is currently being applied to composites with promising results and may be adapted to measure shear properties in all three principal directions. In-plane shear stress and strain data can be calculated by measuring longitudinal and transverse strains during a tensile test on material laid up at ±45°.

Fracture toughness testing in mode I (tensile crack growth) and mode II (shear crack growth) is frequently applied to composites due to the relative weakness of the resin and the danger of delamination growth but it is a specialized subject and will not be covered in detail here. Briefly the tests use a beam in which an initial delamination provides a starting point for the controlled growth of a crack. The energy required to extend the crack may be measured and related to the growth in surface area. The test piece takes the form of either a double cantilever beam (DCB) for mode I, or end notched beam in flexure (ENF) for mode II. Mode III, in which the crack grows in transverse shear, presents problems and work is currently being directed towards an improved mode III test method.

Composites can fail in a variety of modes, even when tested under the same conditions, and a full test report should include a description of the

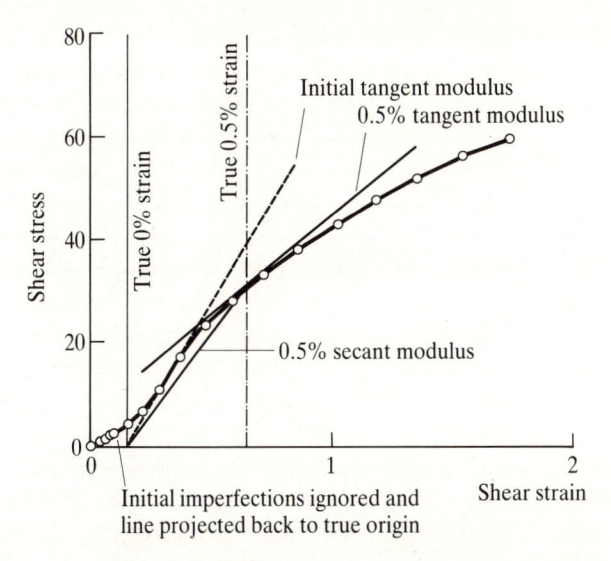

Figure 135 Typical stress/strain curve

failure mode. There is quite a large inherent variability in material properties so reported results will generally be based on the mean of, say, five tests; the standard deviation should also be quoted; it is not uncommon to present all individual test results.

Stiffness can be calculated as an initial tangent modulus or as a tangent or secant modulus at a defined strain, generally 0.25%. It should be made clear what basis was used and whether the measurements were based on the linear portion of the loading curve, Fig. 135.

It has been found that small changes in testing methods can have a significant effect on the results obtained for composites. A great deal of effort is currently being applied to attempting to devise tests and techniques which minimize the possibility of operator, testing machine or other influences affecting results and enable the values obtained to be applied with confidence.

References

1. P T Curtis (ed), *CRAG Test Methods for the Measurement of the Engineering Properties of Fibre Reinforced Plastics*, RAE (now DRA) Technical Report TR 85099, Minsitry of Defence, London, 1985.
2. P T Curtis, B B Moore, *A Comparison of Plain and Double-waisted Coupons for Static and Fatigue Tensile Testing of Unidirectional GRP and CFRP*, RAE (now DRA) Technical Report TR 82031, Ministry of Defence, London, 1982.
3. *Composite materials: testing and design*, ASTM STP 787 American Society for Materials and Testing, Philadelphia, 1982.
4. L J Hart-Smith, *Aerospace Composites and Materials*, **3** (3), 26; **3** (4), 13 (1991).

CHAPTER 6

Environmental aspects

6.1 Mechanical properties under hygrothermal conditions
E G WOLFF

Introduction

Moisture can cause a wide variety of mechanical and thermophysical property changes in polymeric matrix composite materials. Due to generally greater matrix than fibre sensitivity to moisture, properties in matrix dominated directions will be most affected. For example, a unidirectionally reinforced composite subject to moisture content changes will hardly exhibit strength changes in the fibre (0°) direction, but major changes occur in the transverse (90°) direction. Similarly, shear properties are highly affected. Most composites employ laminate construction, where the in-plane properties are nearly isotropic due to fibre orientations in all of the 0°, 90° and ±45° directions. Our discussion assumes such lay-ups unless otherwise indicated. The mechanisms of property degradation are reviewed, to provide general guidelines, followed by static, dynamic and thermophysical properties. The additional effects of thermal and/or stress cycling are treated in Section 6.6.

Mechanism of moisture induced property degradation

Moisture affects all components of the composite, principally the matrix and the fibre matrix interface but also the fibre itself. The plasticization process involves interruption of the van der Waals' bonds between ethers, secondary amines and hydroxyl groups. Polymers with ketones and imides are more resistant to hydrolysis; they have fewer polar groups and this reduces their moisture sensitivity. Crosslink density may increase with increased concentration of secondary amino groups. This allows more freedom of motion and reduces the glass transition temperature. Plasticization reduces residual stresses and increases viscoelasticity. For many matrix materials, the effect of increasing moisture is very similar to an increase in temperature.

There may be a destruction of the polymer/fibre interface due to breaking of polar bonds by chemical reaction, such as the leaching of K_2O and Na_2O, which also promotes pitting in glass fibres. Organic fibres, such as Kevlar and wood fibres (hemicellulose), also absorb moisture, tend to exhibit softening and may even crack internally.

Property changes may be due to combinations of plasticization and mechanical damage from moisture induced swelling (see Section 4.12). Mechanical damage may cause surface crazing/cracking and matrix microcracking. Moisture during curing causes porosity, hence it weakens the mechanical properties. Blistering of the composite surface can occur if it is rapidly heated above the boiling point of water. Water ingression occurs in face sheets of honeycomb laminates via surface microcracks generated by thermal and/or moisture cycling. Mechanisms that promote damage, e.g. high pressure seawater, also promote the ingress of moisture.

Coupled or synergistic effects between moisture induced swelling, plasticization, and damage are difficult to analyse and are often described in terms of altered non-linear viscoelastic properties. Another mode of analysis, known as hygrothermoelasticity, treats coupled sorption, temperature and stress effects (see Fig. 136).

Static mechanical properties

Strength properties such as ultimate tensile (UTS), ultimate compressive (UCS) strength, bifurcation loads, fracture strain and stiffness are the primary static mechanical properties that normally decrease with increased moisture content. While moisture absorption has been known to cause slight increase in the longitudinal (0°) stiffness, the general rule is to expect slight decreases in 0° tensile strength and greater decreases in transverse tensile, flexural and shear strengths. Moisture absorption alone can cause flat composite plates to buckle. Moisture combined with high temperature significantly reduces matrix dominated properties, such as the interlaminar shear strength. Pritchard[1] showed that in many cases, such as with glass reinforced polyesters, it is possible to predict the degradation of mechanical properties purely on the basis of moisture content.

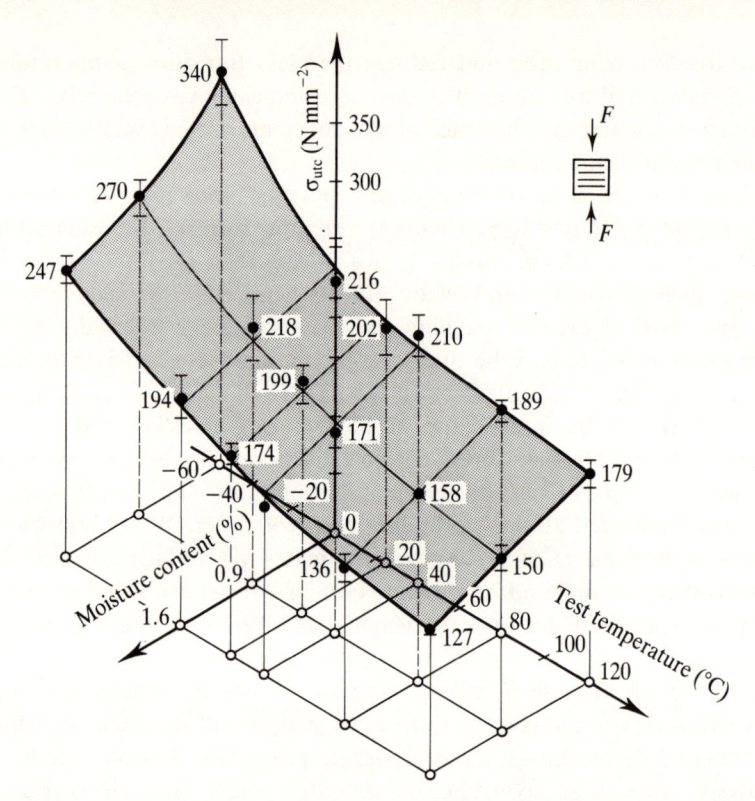

Figure 136 Transverse compressive strength (σ_{utc}) of CFRP-laminate (914C/T300) as a function of test temperature and moisture content. Reproduced from reference 7 with permission.

Time dependent mechanical properties

Fatigue life decreases with percentage relative humidity for typical epoxy adhesives and most composites. Damping capacity tends to increase with moisture content, especially when the fibre also absorbs water, (e.g. Kevlar). Moisture alters the viscoelastic response of composites, such as creep. Wood has long been known to exhibit a 'mechanosorptive' effect, whereby creep is accelerated during changes in moisture content. Wang[2] has recently shown that the same holds true for Kevlar and natural polymeric based composites, but not for nylon or graphite/epoxy. The creep deformation accompanying transient moisture conditions is mainly recoverable if the humidity cycling is continued after unloading. This effect is used for stress relaxation in paper products.

Moisture has a varied role in fracture phenomena. It may lead to spontaneous crack propagation but also slows crack growth under certain conditions. Since moisture reduces residual stresses, it may reduce the tendency for matrix microcracking with continued cooling below the stress-free or curing temperature. Moisture thus has a beneficial role in delaying transverse cracking and cracked samples exhibit lower moisture induced expansion than uncracked samples. Moisture may also have a significant effect on delamination extension at low temperatures.

Thermophysical properties

The coefficient of moisture expansion (CME) is analytically analogous to the coefficient of thermal expansion (CTE). The CTE increases slightly with moisture content, while the CME increases slightly with temperature, depending on the exact mechanism of the moisture absorption. In many applications, however, temperature and moisture may cause comparable (swelling) deformations and the two effects may be difficult to separate. Although analogous to thermal expansion behaviour, the vast difference between moisture (mass) and temperature (thermal) equilibration rates (diffusivities) means that most composites in use contain moisture concentration gradients. Therefore the accompanying changes in properties may vary throughout a composite structure. Wolff[3] has reviewed moisture expansion data and their effects on the dimensional stability of composites.

Conclusions

Potential problems with moisture absorption can be alleviated by a judicious choice of new or modified polymer matrix. For example, many new thermoplastics, which may be highly crystalline, exhibit lower moisture absorption than many epoxy based formulations. For some polymers, such as poly(aryl ether ether ketones), or PEEK, some mechanical properties, such as fracture toughness, may become relatively insensitive to increasing moisture content. Another example is spir-oorthocarbonate monomers, copolymerized with epoxies, to increase impact strength and improve fatigue life.[4] While newer polymers may absorb less moisture, property degradation may still be substantial. Hygrothermal effects may also be controlled by improved composite design, such as with laminae orientations for a zero CME or optimum extension–shear coupling of hybrid laminates.[5] For further information on hygrothermal effects on mechanical properties, the reader is referred to a compilation edited by Springer[6] and a review by Wolff.[7]

References

1. G Pritchard, S D Speake, *Composites*, **18**(3), 227 (1987).
2. J Z Wang, D A Dillard, M P Woolcott, F A Kamke, G L Wilkes, *J. Composite Mater.*, **24**, 994 (1990).

3. E G Wolff in *International Encyclopaedia of Composites*, Vol. 4, ed. S M Lee, VCH Publishers, New York, 1991, pp. 279–323.
4. M R Piggott, P W K Lam, J T Lim, M S Woo, *Composites Sci. Tech.*, **23**, 247 (1985).
5. S J Winckler, S C Hill in *Composites: Design, Manufacture and Application, (ICCM VIII)*, eds S W Tsai, G S Springer, Vol. 1, Ch. 2, Paper I, SAMPE, Covina, California, USA, 1991.
6. G S Springer (ed.), *Environmental Effects on Composite Materials*, Vols 1–3, Technomic, CT, Westport, 1981, 1984, 1988.
7. E G Wolff *SAMPE J.*, **29** (3), 11 (1993).

6.2 Moisture absorption – Fickian diffusion kinetics and moisture profiles

T A COLLINGS

Introduction

It is now well established that fibre reinforced plastic (FRP) structures with resin matrices, when exposed to humid conditions, readily absorb moisture (water) from the environment by a diffusion controlled process. The effect of this absorbed moisture is to degrade the matrix dependent properties and thereby directly influence the load-bearing properties of a laminate especially at high ambient temperatures.

The mechanism of moisture diffusion through a solid is based on the transportation of matter from one part of a system to another as a result of random molecular motions. The motion of a single molecule has no preferred direction. Nevertheless, it is observed that there is a transfer of molecules from regions of high concentration to regions of low concentration. Consider any division between regions of high and low concentrations of moisture. All of the molecules will have a random motion but in the region of high concentration there will be more molecules and hence more movement taking place. Thus there will be a greater number of molecules moving from a region of high to low concentration than will move from low to high. Hence the result is a net transfer of molecules from high to low regions of concentration. Furthermore, since the molecular activity increases with the absolute temperature then the rate of diffusion will also increase with temperature.

Transfer of heat by conduction is also due to random molecular motions and there is an analogy between the diffusion of heat and that of water in a solid. This was first recognized and put on a quantitative basis by Fick[1] in 1885 by adopting the heat equations derived by Fourier in 1822. His theory of moisture diffusion is based on the reasonable hypothesis that the moisture flux is proportional to the concentration gradient.

In the past, it has been usual for other investigators to define moisture degradation of FRP strength properties by an average percentage moisture content, whereas in practice steep moisture gradients exist through the thickness of an FRP laminate and therefore an average value is not very meaningful. It is of more importance that the local moisture concentration at the position of maximum stress is considered; this is usually at the outer surfaces of a laminate.

Although the characteristics of the resin matrix mainly govern the absorption process, fibre stacking and orientation can influence moisture diffusion.[2] This is of importance when calculating the moisture absorbed in complex structures containing many different laminate lay-ups. Clearly, a knowledge of the state of moisture in a laminate by direct measurement or by prediction is central to a fuller understanding of the effects of moisture on the strength properties and the failure mechanisms of FRP laminates.

Moisture constants

The absorption of moisture by an FRP material is dependent upon many factors, e.g. temperature, relative humidity (RH), area of exposed surfaces, resin content, diffusivity and surface protection. All these factors need to be known before any sensible attempt at calculating absorbed moisture can begin. Moisture content, M, is defined as the moisture per unit volume of a laminate expressed as a percentage of the dry weight as

$$M = \left(\frac{W_w - W_d}{W_d}\right) 100 \qquad [6.1]$$

where W_w and W_d are the wet and dry weights of the laminate respectively. To predict mathematically the moisture content and the way it is distributed through a laminate's thickness requires two items of data not usually readily available. These are the equilibrium moisture level and the diffusion coefficient.

Equilibrium level

The moisture equilibrium level, M_∞ is the maximum moisture content that can be absorbed by an FRP laminate for a given relative humidity. A method for determining this in a relatively short time has been described by Collings and Copley.[3]

Diffusion coefficient

The diffusion coefficient, D, determines the rate of moisture uptake and is normally assumed to depend upon absolute temperature T, as

$$\log_e D \propto \frac{1}{T} \qquad [6.2]$$

although the effects of stacking and ply orientation, etc. are also known to have a significant influence. The diffusion coefficient is calculated from

$$D = \frac{\pi h^2}{16 M_\infty^2} \left(\frac{M_2 - M_1}{\sqrt{t_2} - \sqrt{t_1}} \right)^2 \qquad [6.3]$$

where h is the laminate thickness and, M_1 and M_2 are the moisture contents by weight after times t_1 and t_2, respectively, and are taken from the linear portion of the M versus $t^{1/2}$ plot, see Fig. 137. The diffusion coefficient given in [6.3] is in error, as most moisture diffusion measurements include moisture diffusion through six surfaces. For the true one-dimensional coefficient, D_∞, a correction factor given by Shen and Springer[4] can be used,

$$D_\infty = D[1 + (h/b) + (h/l)]^{-2} \qquad [6.4]$$

where b and l are the laminate breadth and length respectively.

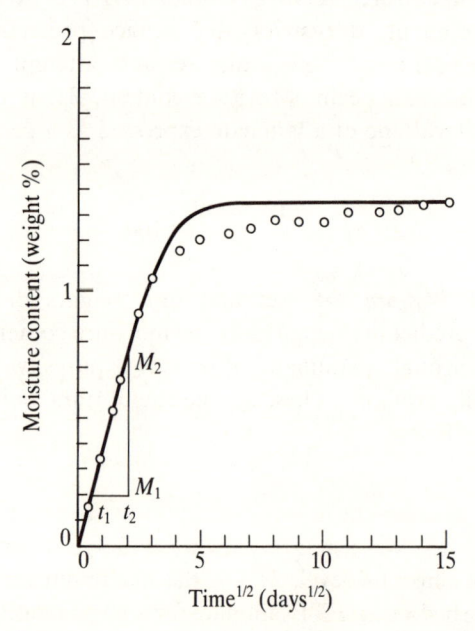

Figure 137 Comparison of predicted (solid line) and measured moisture diffusion (open circles) into CFRP

Moisture prediction

FRP laminates are not homogeneous and are highly anisotropic with three main diffusion dimensions. However, for practical purposes, most FRP structures, away from the laminate edges, can be adequately described using one-dimensional Fickian diffusion, i.e. the through thickness dimension, as it is in this dimension that a knowledge of the moisture distribution is usually required.

Theory

Moisture prediction can be based on the classical theory described by Fick's second law; at time t,

$$\frac{\partial c}{\partial t} = \frac{\partial(D\partial c)}{\partial x(\partial x)} \qquad [6.5]$$

where c is the moisture concentration.

For composites using epoxy matrices, the diffusion coefficient appears to be independent of concentration hence independent of through thickness location, so [6.5] can be written in the form

$$\frac{\partial c}{\partial t} = D\frac{\partial^2 c}{\partial x^2} \qquad [6.6]$$

It is more practical to work in terms of a moisture content, thus a more useful form of [6.6] is

$$\frac{\partial m}{\partial t} = D\frac{\partial^2 c}{\partial x^2} \qquad [6.7]$$

This equation may be solved using a finite difference technique.[3,5]

Theoretical and experimental comparisons

The through thickness moisture distributions for a carbon fibre reinforced plastic (CFRP) are given in Fig. 138. Included in Fig. 138 are the experimental points taken from three specimens fitted with a best fit quartic and the predicted Fickian distribution using a finite difference method to solve Fick's second law.

Figure 137 compares the moisture uptake with root time for both experiment and prediction using Fick's second law. The fit of theory with experiment is not perfect, nevertheless moisture diffusion prediction using Fickian theory is adequate for most practical purposes. It must be remembered that the resin must itself exhibit Fickian type absorption behaviour in order to apply Fickian prediction.

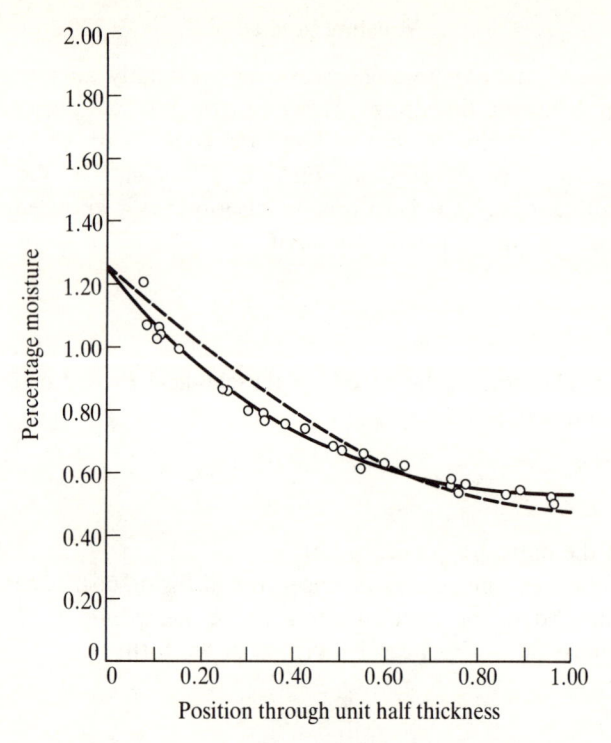

Figure 138 Through the thickness distribution of predicted (solid line) and measured moisture diffusion (open circles) into CFRP; broken line in the Fickian prediction

Diffusion limitations

Most resin systems in current use in the aerospace industry exhibit Fickian type absorption behaviour for most of the required in-service operating temperatures. Unlike heat conduction, moisture diffusion is slow. Thus, in order to provide degraded structural test specimens in a sensible time period, it is sometimes necessary to use elevated temperature to speed up the process. In applying elevated temperature, care must be taken because some resins[3,6] only behave in a Fickian fashion within tight temperature limits. If these limits are exceeded in accelerating the absorption process, diffusion and moisture absorption can become unrealistic, atypical of a resin's natural in-service environment.

References

1. J Crank, *The Mathematics of Diffusion*. 2nd edn, Clarendon Press, Oxford, 1975.
2. T A Collings, Moisture Management and Artificial Ageing of Fibre Reinforced Epoxy Resins, in *Composite Structures 5*, ed. I H Marshall, Elsevier Applied Science, London, 1989.
3. T A Collings, S M Copley, *Composites*, **14**(3), 180–8 (1983).
4. C H Shen, G S Springer, *J. Composite Mater.* **10**, 2–20 (1976).
5. S M Copley, *A Computer Program to Model Moisture Diffusion and its Application to Accelerated Ageing of Composites*, RAE (now DRA) Technical Report TR 82010, Ministry of Defence, London, 1982.
6. J M Whitney, C E Browning, Some Anomalies Associated with Moisture Diffusion in Epoxy Matrix Composite Materials, in *Advanced Composite Materials – Environmental Effects* ASTM STP 658, American Society for Testing and Materials, Philadelphia, 1978.

6.3 Moisture absorption – anomalous effects

F R JONES

Moisture diffuses into polymers to differing degrees depending upon a number of molecular and microstructural aspects:

(i)   polarity of the molecular structure,
(ii) degree of crosslinking,
(iii) degree of crystallinity of a thermoplastic,
(iv) presence of residual 'monomers' and/or hardeners and/or other water attracting species, e.g. glass surfaces.

For composite materials, the most common polymer matrices can be defined as epoxy, unsaturated polyester and related styrenated resins, advanced thermosets and thermoplastics. These materials can have a range of sensitivity to moisture absorption depending on the individual contributions of the above factors. For example, cured epoxy resins can absorb from 2–8% w/w depending upon the base resin and the curing system.[1] Translated into a composite with $V_m = 0.4$, the equilibrium moisture M_∞ content can still be $>3\%$.

Providing the fibre–matrix interface remains bonded, a moisture sensitive interphase is not formed and the water can be distributed throughout the polymer at polar sites or within the free volume that is present. This can be examined by studying the dependence of relative humidity on M_∞. The degree of clustering of the water molecules, presumably within the free volume, is a function of molecular polarity. For typical matrices, the unsaturated polyesters exhibit higher degrees of clustering than that which occurs in the more advanced epoxy resins.[2]

Residual strains

The heterogeneous distribution of moisture can lead to complex effects. As shown in Fig. 139, the temperature dependence of the thermal expansion coefficient becomes highly non-linear as a result of the moisture absorption. In this case, the cured resin has a heterogeneous crosslink density which enhances the differential moisture absorption.[3] For a composite, this enhancement of α_m can lead to a greater mismatch between the ply expansion coefficients $(\alpha_t - \alpha_l)$ and the generation of higher thermal residual strains on cooling a wet laminate. The residual water content of the uncured resin and water absorption before post-

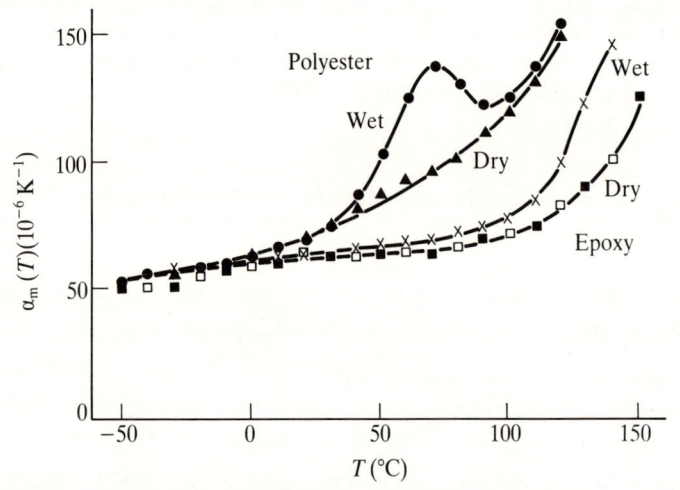

Figure 139 The effect of immersion in distilled water on the linear expansion coefficient α_m for an unsaturated isophthalic polyester resin and a DEGBA/NMA epoxy resin (redrawn from Jones & Mulheron[3] by permission of Butterworth-Heinemann Ltd ©)

Table 51. Experimental and calculated thermal strains at 20°C in (0°/90°)s laminates from isophthalic polyester and DGEBA/NMA epoxy resins

	Polyester			Epoxy
ϵ_{tl}^{th} Experimental	0.34	0.40	0.45	0.28
ϵ_{tl}^{th} Calculated* (%)	0.15	0.22	0.26	0.28
ϵ_{tl}^{th} Calculated† (%)	–	–	0.43	–
Post-cure temperature (°C)	50	80	130	150
Strain free temperature, T_1 (°C)	50	80	90	140
Volume fraction, V_f	0.33	0.33	0.33	0.55

*Calculated from α_m for dry resin.
†Calculated from α_l, α_t for a wet laminate ($M = 0.15\%$).
Experimental values were obtained from the curvature of 0°/90° beams (see Section 4.12).

curing can lead to higher thermal strains in as-fabricated laminates (see Table 51).

Absorption of moisture is accompanied by swelling of the resin (see Section 4.12). This opposes the tendency to contract on cooling. Therefore, moisture ingress by laminates occurs with a reduction in thermal strain. But if the temperature dependence of the expansion coefficient is affected in the manner shown in Fig. 139, a thermal excursion will tend to increase the residual strain within the individual plies (see Section 6.6). The combination of moisture absorption followed by thermal cycling during service leads to fluctuating internal stresses.

Glass transition and modulus

The most important consequence of moisture absorption is a reduction in the glass transition temperature, T_g, of the resin as a consequence of water plasticization. This occurs most efficiently when the water can interact with the resin through hydrogen bonding, causing an increase in free volume. Thus the reduction in T_g on moisture absorption is greater for polar polymers. As a consequence, the effect tends to be greatest for the epoxy resin matrices, especially those which employ amine hardeners since they generate hydroxyl groups in a cured structure. The average decrease in T_g, ΔT_g, is approximately 20 K for each 1% of moisture absorption.[1] Taken together with a range of values of M_∞, it becomes apparent that ΔT_g may be as high as 140 K. For most epoxy resins for aerospace applications, high degrees of crosslinking limit the water absorption and ΔT_g is $\simeq 50$ K. Therefore, a 150°C cured epoxy resin matrix has a maximum useful temperature of $\simeq 100$°C. More highly functional epoxy resins and other resins, such as polyimides and thermoplastics, with higher glass transition temperatures have therefore been examined for application as composite matrices. For the bismaleimide PMR-15, $\Delta T_g = -1.3 \text{ K \%}^{-1}$ with an equilibrium moisture content of 1.3% at 96% RH.[4] In the case of the bismaleimide PMR-15 system, volatiles are generated during cure; these remain occluded in the laminate even after post-curing (see Section 2.5). These volatiles behave analogously to water in Fig. 140, however the reduction in T_g (and hence T_1) appears to offset the enhancement in α_m so that the residual stresses in the as-fabricated laminate are lower than expected. This means that the as-fabricated laminates have an inherent tendency to microcrack during thermal cycling to > 300°C as the plasticizing effect is reduced.[4]

As well as a reduction in T_g, water absorption results in a reduction in the modulus. This is illustrated schematically in Fig. 140 and demonstrates why high performance resins, either of high crosslink

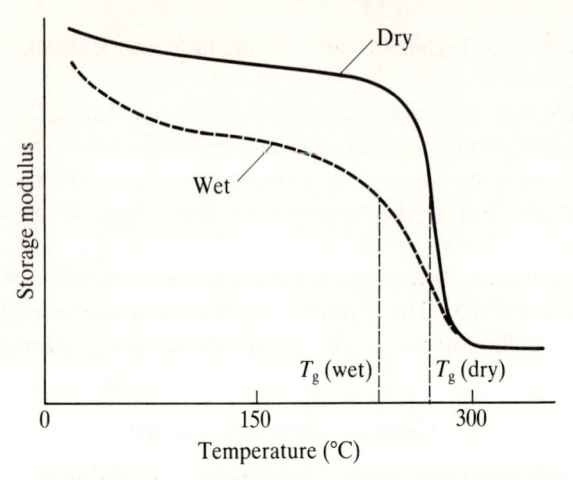

Figure 140 Schematic representation of the effect of water on the storage modulus and glass transition temperature of a thermosetting resin

density or thermoplastic (e.g. PES or PEEK) with lower moisture sensitivity, are of crucial importance to the composites industry.

Adventitious impurities

Examples of residual components from resin synthesis are orthophthalic acid/anhydride in polyesters, NaCl in epoxy resins, and residual curing agents such as dicyandiamide DICY[5] in epoxy resins. These residues can enhance moisture absorption by an osmotic mechanism and can lead to blister formation as discussed in Section 6.4.

The choice of glass fibre and or sizing composition can also affect the moisture absorption and leaching during immersion of the composites. This is illustrated in Table 52, for a zirconia alkali resistant glass fibre,

Table 52. The influence of fibre type on the water absorption of their vinyl ester resin composites

Silane coupling agent	AR-glass A1100*	AR-glass A172†	AR-glass Without silane	Commercial E-glass with silane
Equilibrium moisture absorption (%)	1.03	1	0.86	0.38
Leaching (%)	0.13	0.03	0.12	0.03

* Aminosilane   † Vinylsilane

with differing coupling agents in a common size and a commercially sized E glass roving. For the AR glass a reduction in interlaminar shear strength has been observed which can be attributed to varying degrees of sodium diffusion from the glass substrate into the interphase region.[6]

Modified epoxy resin systems

Most epoxy resins used for advanced composites are polymer modified to provide either flow control, matrix fracture toughness and/or enhanced temperature performance. Under these circumstances, the presence of a second phase or skeletal structure will modify the response of the resin to humid environments, especially when the two components independently have significantly differing responses. Under these circumstances, analagous effects to those described in Fig. 139 may well operate with similar consequences (see Section 6.6).

Conclusions

Many matrix resins absorb moisture reversibly by Fickian diffusion, as discussed in Section 6.2. But resin chemical structure and microstructure are complex, both in crosslink density and polarity; adventitious impurities may be present; these factors cause non-Fickian processes to occur, which may or may not lead to irreversible effects. The presence of water reduces the T_g of the matrix and hence the maximum useful temperature of the composite. Residual stresses may also be enhanced if moisture diffusion leads to a non-linear thermal expansion coefficient/temperature profile.

References

1. W W Wright, *Composites*, **12**, 201 (1981).
2. P M Jacobs, F R Jones, *J. Mater. Sci.*, **25**, 2471 (1990).
3. F R Jones, M J Mulheron, *Composites*, **14**, 281 (1983).
4. M Simpson, P M Jacobs, F R Jones, *Composites*, **22**, 89 (1991).
5. F R Jones, M A Shah, M G Bader, L Boniface in *Sixth International Conference on Composite Materials (ICCM VI/ECCM 2)* vol. 4, eds F L Matthews et al., Elsevier Applied Science, London, 1987, pp. 443–456.
6. D Pawson, F R Jones in *Controlled Interphases in Composite Materials*, ed. H Ishida, Elsevier, New York, 1990, pp. 407–415.

6.4 Osmosis and blistering in GRP

F CHEN and A W BIRLEY

Background

Boats and swimming-pools have been built with glass fibre reinforced polyester (GRP) composites for over 40 years. One drawback of these materials is that they may develop blistering after prolonged contact with water. It has been estimated that as many as 48% of the boats might be affected by blistering, with higher figures for swimming-pools.

Mechanism of blister formation

Since the early 1970s, work has been carried out to determine the cause, effect and cure of blisters in GRP laminates. There are two theories to explain blister formation: osmosis and the residual stress in the laminate.

Osmosis route to blistering

When GRP laminates are immersed in water, the gelcoat acts as a semipermeable membrane. The water permeates through the gelcoat and collects in the microvoids or other 'holes' within the laminates. Chemical reactions take place between water and some of the constituents of GRP, such as the polyesters, glass fibre finish and binder and the curing agents. The hydrolysis products are water soluble, low molecular weight substances which are very active osmotically. More water diffuses in under the osmotic driving force and, when the pressure is high enough, the gelcoat will deform and eventually crack.

Internal stress theory

When a GRP laminate is made, the backing resin and the glass fibre mat are invariably applied to the gelcoat after it has partially cured. The styrene in the backing resin will penetrate into the partially cured gelcoat and cause it to swell at the boundary of gelcoat and backing resin. The styrene then polymerizes and leaves the boundary layer in a swollen stressed state. There is a shrinkage difference between the gelcoat and backing resin as they cure at different rates; this also contributes to the stress formation at the boundary layer. Swelling by the ingress of water will amplify the stress at the boundary layer remaining after manufacture, caused by the different volume changes, and a crack is formed when the stress reaches a critical value.

Characterization of blistering

When blisters are formed on laminates, their locations are randomly distributed and the shapes and sizes are variable. They can be classified in various ways. The British Plastics Federation placed them into six categories according to size and shape. It is also possible to classify them according to the size and frequency of blistering using a paint industry test defined in ASTM D714-87.

The alternative approach is to understand the reasons for blistering; blisters can be classified into four groups according to their origins.

(i) Contaminants: some contaminants in the gelcoat can cause blistering by osmosis.
(ii) Bubbles: a very few bubbles in the gelcoat may develop into blisters.
(iii) Precracks: cracks can be formed before water immersion, due to boundary layer stress. They usually develop into blisters after water immersion.
(iv) Fibre bundles: blisters can be formed along the glass fibre bundles, immediately behind the gelcoat. The main cause of this type of blistering is a combination of internal stress and, at a later stage, osmosis.

A laminate with fibre bundle blisters and few precrack blisters is shown in Fig. 141.

Factors affecting blistering formation

There are many factors which affect blister formation. Experiments show that the chemical composition of gelcoats and backing resin are most important to the blister resistance. Isophthalate based gelcoats are generally superior to those made from orthophthalate. Isophthalic acid/ neopentyl glycol based gelcoats have even better blister resistance than those based on isophthalic acid/polypropylene glycol. It is reported that poly(propylene glycol maleate) resin as both gelcoat and backing resin has shown a remarkable blister resistance.

The thickness of gelcoat also plays a major role on the blister resistance. A very thin gelcoat will dramatically reduce the performance of the laminates.

In most cases, laminates containing powder bound chopped strand mat (CSM) display a lower water uptake and higher blister resistance than emulsion bound CSM.* The use of surface tissue immediately behind gelcoat can significantly reduce the chance of blistering.

*Some of the backing resin may react with other constituents of the mat to provide good contact between fibre and matrix.

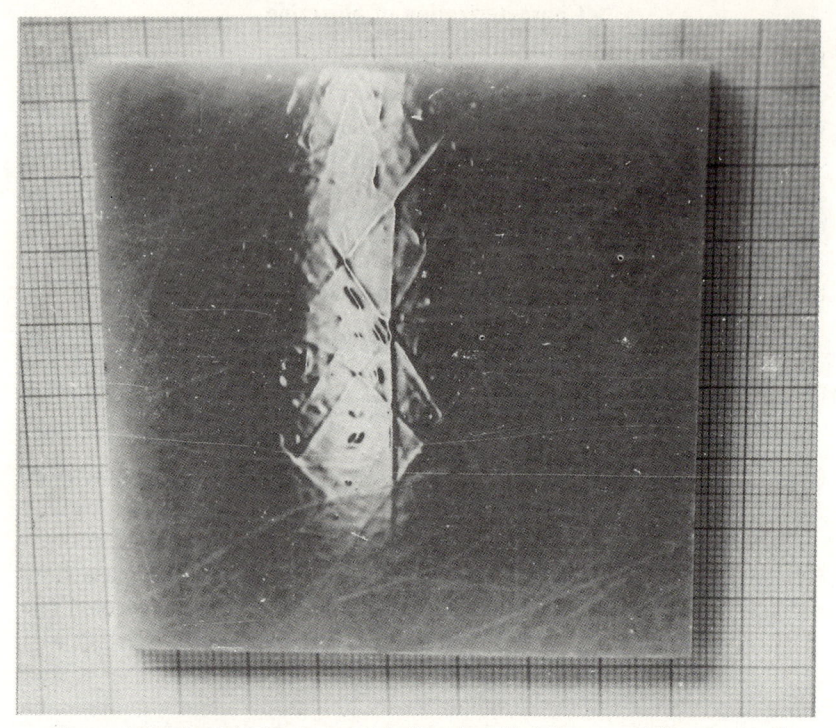

Figure 141 A GRP laminate with fibre bundle blisters and few pre-crack blisters (under fluorescent light)

The levels of initiator, accelerator and diluent added to the polyester resin can have some effects on the blister resistance of the laminates, so can the pigment and filler mixed to the resin. Poor manufacturing conditions and fabrication skills will seriously reduce the blistering resistance. Once the GRP laminate has been made, the time to blister will be affected by the water quality and the temperature. Blisters are more likely to be formed in fresh water and distilled water than in seawater. The higher the temperature of water, the more rapid the blister forms, although it has now been established that the ranking is not necessarily the same under mild conditions compared to the same treatment at 40°C in distilled water.

Conclusions

Blistering in GRP is caused by the stress at the boundary layer of gelcoat and backing resin generated during curing and post-curing stages; this then promotes osmosis, which is a powerful driving force towards self-destruction.

Blister formation is a complex process. It is affected by the compositions of gelcoat, backing resin and glass fibre reinforcement, the curing system of the polyester resin, pigment and filler, manufacturing process and the service environments.

Further reading

1. K G H Ashbee, *Fundamental Principles of Fibre Reinforced Composites*, Technomic Publishing Co. Inc., Lancaster PA, 1989, pp. 267–310.
2. F Chen, A W Birley, *Plastics Rubber Composites Proc. Appl.*, **15**, 161 (1991).
3. F Chen, A W Birley, *Plastics Rubber Composites Proc. Appl.*, **15**, 169 (1991).
4. A W Birley, J V Dawkins, H E Strauss in *Proceedings of the 14th Reinforced Plastics Congress*, Nov. 1984, The British Plastics Federation, London, 1984, pp. 37–42.
5. J S Ghotra, G Pritchard in *Developments in Reinforced Plastics 3*, ed. G Pritchard, Elsevier Applied Science, London, 1984, pp. 63–95.

6.5 Stress corrosion cracking of GRP

F R JONES

Introduction

Environmental stress corrosion cracking (ESCC) is the delayed brittle fracture of a stressed material under the influence of an environmental agency. Both metallic and polymeric materials are susceptible under certain conditions. For polymers, it usually occurs in thermoplastics rather than thermosets. For glassy polymers, such as polystyrene, plasticization by the environment leads to weakening of the craze structure which normally arises as a result of the orientation of polymer molecules into fine fibrils, which span a crack. For semicrystalline polymers, such as polyethene. interfacial regions between the crystallites are weakened. It is difficult to see how a three-dimensionally crosslinked thermosetting polymer can suffer from ESCC without the involvement of chemical degradative mechanisms of the network. There is no evidence for ESCC of thermoset polymers.

For composite materials, ESCC has mainly been reported for E-glass reinforced plastics based on unsaturated polyester (see Section 2.10) vinyl ester (see Section 2.12) and epoxy resin (see Section 2.4) matrices, because of applications in chemical plant and related anticorrosion applications. Some studies on short fibre reinforced thermoplastics have been reported.

While the cured matrix resins are generally resistant to ESCC, E-glass fibres are not resistant to ESCC. In the presence of water this is termed 'static fatigue', and is generally responsible for the weakening of the fibres, during their lifetime. However, under acidic conditions ESCC can occur rapidly. Typically at 0.5% strain in aqueous sulphuric acid, this occurs in 5×10^3 to 30×10^3 s (1.4–8.3 h). Other glass fibres such as S, R and ECR (this glass was developed to counteract this problem) are much less susceptible. AR glass fibres are also highly resistant, presumably because of their high alkali content, and translation from cement matrices to polymer matrices was initiated as a result of this observation.

In alkaline environments, there is no evidence for the synergism between stress and the environment which leads to ESCC, only the general weakening of the fibres as a result of chemical leaching.

The static fatigue of most glasses and the rapid ESCC of E-glass in acidic environments are believed to have the same mechanism, which is a stress assisted hydrolysis of the silica network

$$\equiv Si-O^- \ Na^+ + H_2O \rightarrow \ \equiv Si-OH + Na^+ + OH^-$$
$$\equiv Si-O-Si \equiv + \ OH^- \rightarrow \ \equiv Si-OH + \ \equiv SiO^-$$
$$\equiv Si-O^- + H_2O \rightarrow \ \equiv Si-OH + OH^-$$

For water, this can be attributed to the presence of a high alkaline concentration at the tip of a surface or flaw or crack. Charles[1] explains that ESCC occurs when the rates of hydrolysis of the silica network and crack propagation are similar (i.e. stress-assisted corrosion). When hydrolysis is more rapid than crack propagation then corrosion leads to a rounding of the crack tip and reduction in potential to propagate according to the well-known equation

$$\sigma_{max} = 2\sigma_a (x/\rho)^{1/2}$$

where σ_{max} = stress at the crack tip, σ_a = applied stress, x, ρ are the flaw depth and radius respectively.

If the rate of crack growth is much larger than the rate of corrosion then the environment will have little influence on fibre strength and any time dependence. It is believed that since Na^+ is directly involved then its diffusion to the crack tip is rate determining.

Since the ESCC is clearly pH dependent, the network modifiers and alkali ions must play a significant role. However, it is well established that leaching of Ca^{2+}, Al^{3+}, Fe^{3+}, Na^+, K^+ and other residuals occurs to leave a weakened sheaf around a strong core. This leads to spiral cracking of this sheaf, after the unstressed fibres are stored in aqueous acid for a short period. It is not clear whether the crack forms prior to optical or SEM examination or after removal from the environment but the latter is favoured. There are two confusing observations: (i) that the spiral cracks appear to form when the hydrated sheaf dries out, (ii) that not all of the fibres in a bundle suffer from either spiral cracking or multiple fracture.

It would appear therefore that the leaching of the network modifier ions is also stress dependent and that variable residual stresses are built into the fibres as a result of the water cooling regimens employed in manufacture. The leaching may be further complicated by the complexation of the leachable ions such as Fe (III) within the aqueous solution which can promote the process. This may be the mechanism by which ultraviolet illumination can accelerate ESCC of E-glass filaments.

Composite materials

Fibre composites exhibit tough behaviour because of the large surface area which results from mainly fibre/matrix debonding (and matrix cracking). As a consequence, failure generally results in a fibrous fracture surface with extensive fibre pull-out. However, under ESCC conditions, failure of GRP (principally from E-glass) results in a planar fracture surface typical of brittle behaviour. This can be readily recognized in the SEM microscrope by the presence of mirror-fractured fibres devoid of hackle and river lines. However, since ESCC of the fibres leads to growth of the failure crack through the adjacent resin and fibres, as the stress increases under a constant load with decrease in cross-sectional area the fibres exhibit hackle and the fracture surface becomes less planar (and can exhibit steps) away from the crack initiation zone.

It is clear from the foregoing that the resin matrix has to protect the fibres from both corrosion through leaching of metallic ions or hydrolysis of the network under non-stress applications or from ESCC under stress in aqueous acidic environments. For the former, the presence of resin rich surface layers or gelcoats adequately protect the fibres from diffusion of the environment (see Section 6.2) to the fibres during the lifetime of the component and the prime concern should be the chemical resistance of the matrix resin (see Chapter 2) and the interphase region between fibre and matrix (see Section 1.11). However, for the latter, different principles operate because the presence of a sharp crack can promote a brittle fracture of the composite at service strains as low as 0.05%, which is determined by the susceptibility of the E-glass fibres to the synergism of the stress at the crack tip and the diffusion of the environment along the crack to provide the perfect conditions for ESCC failure. This is illustrated in Fig. 142 where the time dependent failure is shown to exhibit three main mechanisms. In region I at high stress, failure occurs in relatively short times and durability is determined by the greater rate of crack growth compared to rate of diffusion of the environment to the crack tip. This mechanism

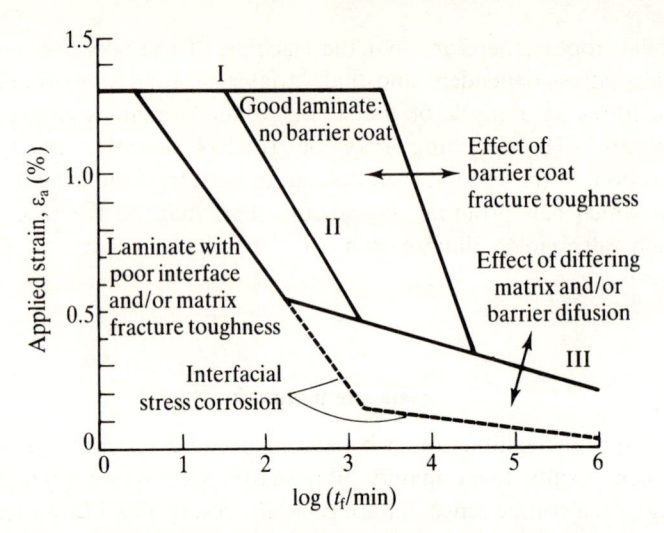

Figure 142 Schematic failure map for a unidirectional E-glass composite under acidic environmental stress corrosion cracking conditions, illustrating the material variables. t_f is the failure time in minutes under constant tensile load

is demonstrated by the nearly horizontal section of the stress corrosion curve and failure times are similar to those in the absence of an active environment. In region II, stress corrosion cracking results from the synergism between the two agencies. At lower stress in region III, the failure is less stress sensitive and occurs over relatively long times. Therefore, the long-term durability at low stress is controlled by diffusion processes and the slow reduction in strength of the corroded glass fibres which occurs.

Types of ESCC failure

Three differing failure modes have been observed by Jones et al.[2] which involve a synergism between the applied (and/or internal) stress and environment. These were designated Type I, II or III which described the nature of the failure:

Type I Brittle fracture within the environment.
Type II Brittle fracture of the unexposed part of the laminate in the atmosphere about 2 cm above the environment.
Type III Damage accumulation within the unexposed part of a non-externally (or low externally) stressed crossply laminate, which is initiated at the surface of the environment.

Type I failure

This is the type of failure which is generally observed in service since it occurs in the submerged composite at applied strains in excess 0.15%. It is initiated by the stress corrosion failure of a glass fibre near the surface of the laminate, as a consequence of environmental diffusion. Attempts to confirm diffusion of the acidic environment through the resin to the fibre have proved difficult but it is likely that the applied stress increases the rate of diffusion. The statistical nature of fibre strength means that a premature fibre break can lead to an enhancement of the local stress in the adjacent fibres by stress transfer (see Section 4.3). Thus a series of stress corrosion nuclei will form in the surface of a $0°$ composite under tension. Brittle failure will occur once the flaw reaches critical dimensions. However in flexure, the stress in the tensile face of the beam may not be uniform and this can lead to the initiation of *one* stress corrosion crack.

In angle ply laminates, such as a $0°/90°/0°$ composite, the presence of a transverse crack generally provides sufficient enhancement of the stress in the $0°$ plies adjacent to the crack for a single ESCC to propagate through the laminate in tension. In flexure, cracks have also been observed to propagate from the inside to the outside of the material.

Similar stress situations can arise if surface cracking of a barrier or gelcoat layer occurs. Thus it has been shown that a cracked gelcoat is more damaging than its absence, under ESCC conditions.[3]

Resin matrix effects

Hogg and Hull[3] have provided the basic understanding of the role of the matrix in ESCC. Thus with a brittle matrix, the resin fillet between fibres will also immediately fracture as a result of fibre ESCC fracture leading to further crack growth as the environment is transported to the crack tip. As a consequence the fracture toughness of the resin will determine the rate of crack growth under stress corrosion conditions. For ductile resins, diffusion of the corrodent through the matrix fillet needs to occur before the next fibre can fail and throw an additional load onto it. Therefore resin properties which compromise fracture toughness and environmental diffusion are required for optional durability. Thus, since chemical resistance is generally obtained with resins of high crosslink density and aromaticity (see Sections 2.10 and 2.11), those resins (e.g. bisphenol polyester) which provide the best chemical resistance do not necessarily provide optimum stress corrosion resistance because of their low fracture toughness. It has been argued that the poorer stress corrosion resistance of a ductile polyester resin

Table 53. Fracture toughness (K_{Ic}) of chemically resistant polyester and vinyl ester resins

Resin	Type	Environment	K_{Ic} (MN m$^{-3/2}$)
Crystic 272	Isophthalic/propylene glycol	Dry	0.79
		Moist (~0.2%H$_2$O)	0.61
		Wet (1.5% H$_2$O) Tested dry	0.55
		Wet (1.5% H$_2$O) Tested wet	0.89
		Dry Tested Wet	0.85
Crystic 272/ 30% Crystic 586	Flexibilized isophthalic/propylene glycol	Laboratory environment, moisture present	0.77
Derakane 411–45	Vinyl ester	Laboratory environment, moisture present	0.75
Crystic 600 PA	Bisphenol polyester	Laboratory environment, mositure present	0.49
Beetle 870	Chemical resistant	Laboratory environment, moisture present	0.46
Atlac 382-05A	Urethane, bisphenol, vinyl ester	Laboratory environment, moisture present	0.45

arises because of a larger moisture diffusion coefficient but it should be remembered that, with lower crosslink density, the chemical resistance is also generally lower, and even partial hydrolysis can reduce the fracture toughness of the resin.

A further complication is the effect of moisture absorption and plasticization on the fracture toughness of the matrix. As shown in Table 53, the moisture content of the resin can influence its fracture toughness as determined by K_{Ic}, the critical stress intensity factor under mode I loading conditions, of an isophthalic unsaturated polyester resin. Since this effect is a consequence of moisture plasticization it will probably have a bigger effect on K_{Ic} for the more ductile resins, so that the chemically resistant resins will provide less protection to ESCC of the fibres, even in the presence of absorbed moisture.[4]

As a consequence, a limiting strain for stress corrosion failure may exist for a particular glass fibre composite, which is governed by the fracture toughness of the resin but the effects of fibre/matrix interfacial strength and/or matrix failure strain especially for angle ply laminates should be ignored. For example, a limiting applied strain for an isophthalic unsaturated polyester resin-based composite of $\approx 0.4\%$ was identified. In general therefore the maximum recommended external applied strain for good quality GRP to provide resistance to ESCC is 0.3%.[5]

Type II

Providing the interface between fibre and matrix remains intact, then the fracture toughness of the wet matrix appears to be the most important factor which determines the resistance of composite to ESCC and failure occurs by Type I fracture as described above. However, the interface or interphasal region is also subject to synergistic effects and stress-assisted corrosion of the interface can also occur. This effect can lead to a time dependent reduction in transverse strength under a constant load in the presence of the environment. For crossply laminates, transverse cracks which will form in the 90° ply at lower strains than normal in the absence of the environment, will promote a Type I ESCC adjacent to the first 90° ply crack. Similar arguments can be applied to angle ply laminates, more typical of the configuration of filament wound pipes used in service (see Sections 4.4 and 4.10).

However, at very low applied strains <0.15%, for a particular epoxy resin/glass system, Type II failure of the unexposed part of the laminate occurred before transverse cracks could completely propagate across the 90° ply. This type of failure was also observed to occur in 0° epoxy composites at similar strains (<0.1%) and in 0° polyester laminates at much higher strains (0.6%). Whereas for the former, catastrophic fracture occurred after 18 days at 0.09% strain and 243 days at 0.07% strain, for the latter 359 days at 0.6% strain were required.

Extensive studies demonstrated that Type II ESCC fracture is a consequence of a time dependent stress-assisted degradation of the interface, which leads to wicking of the environment along the debonded fibres, drawing with it an acidic solution of corrosion products. On reaching the "dry" part of the material, the salts such as calcium sulphate, potassium, sodium aluminium sulphate (from sulphuric acid) or phosphates (from phosphoric acid) precipitate at the fibre/matrix interface providing an additional 'crystallization' pressure which, together with a small applied stress, initiates an ESCC crack which is environmentally fed by wicking of the environment. It should be pointed out that the growth of a single crack will also be a consequence of the fact that the rate of stress corrosion of the interface is larger than the rate of diffusion of the environment through the resin from the exposed face of the laminate. Thus Type II failure occurs under conditions that encourage crystallization of the corrosion products and is inhibited when the non-submerged composite is exposed to a high humidity (e.g. a part filled closed container) or in hydrochloric acid where the calcium or aluminium salts are very soluble. Similarly with highly insoluble salts, such as the phosphates, or in the presence of a more concentrated acid, e.g. sulphuric, where the glass corrosion products are less soluble, the precipitation can occur within the environment and the premature quasi-Type I crack can be initiated.

Type III

Type III is a damage accumulation process in the unexposed laminate. It has identical origins as the Type II mechanism except that in the absence of an external load a failure crack does not form. This is because in a crossply laminate, the 0° fibres are put into compression by the thermal strains which develop during fabrication (see Section 4.12). A consequence of corrosion product precipitation in the 'dry' half of the laminate, therefore, is the transverse cracking of the 90° and 0° plies near the surface of the environment, which progresses up the unexposed part of the composite. Whereas the Type I/II mechanisms operate only in acidic environments, Type III stress-assisted environmental damage accumulation can also occur in alkaline environments. For example, with aqeuous sodium hydroxide this can occur more rapidly since the rate of corrosion of the glass is larger than in the presence of an aqueous acid.

The effect of acid concentration on ESCC

ESCC of glass fibres occurs predominantly in acidic environments because the leaching of the network modifiers occurs simultaneously with the stress-assisted environmental rupture of the silica network. Alkaline environments hydrolyse the silica network causing a general weakening of the fibres.

Thus the pH of the environment has a profound effect on the ESCC of E-glass fibres, with a maximum rate in environments of pH ≈ 0. This occurs in aqueous solutions with $[H_3O^+] = 1$ M i.e. in ≈ 0.5 M H_2SO_4 or 1 M HCl. Thus the highest probability of failure of GRP occurs in dilute aqueous acids of similar pH.

Laminate construction for ESCC resistance[5]

ESCC of E-glass fibre composites can be effectively inhibited by careful choice of resins, fibres and laminate construction. These factors are discussed below and the theoretical aspects are illustrated in Fig. 142. Providing a resin is chosen according to the rules provided above and in Tables 53 and 54, additional benefits can be achieved by careful laminate construction which is intended to prevent the formation of sharp cracks:

Table 54. Generalized trend in corrosion and stress corrosion resistance of E-glass laminates with different matrices

Unsaturated polyester resin		Vinyl ester resin	Chemical resistance	ESCC resistance
Orthophthalic	diethylene glycol ethylene glycol propylene glycol			
Isophthalic	diethylene glycol ethylene glycol propylene glycol neopentyl glycol	Vinyl esters	↓	↑
Fumarate	bisphenol A			
		Urethane modified bisphenol A		

flexible resin gelcoats, random mat reinforcement of gelcoats, or the use of stress corrosion or more chemically resistant glass or other fibres in the barrier coat. In these cases, the structural wall of the composite can still be provided by the cheaper E-glass fibres.

For hand lay-up composites (see Section 3.7) the barrier layer is generally achieved with the application of an appropriate gelcoat to the mould surface followed by a C-glass veil and the structural composite provided by powder bound CSM (see Sections 1.10 and 1.11). It is not usual to recommend differing resins for the gelcoat and structural layers, since interlaminar failure has been observed when this technology is used.

Woven rovings and CSM in optimum combination can provide excellent chemical resistance to barrier layers of laminates used for chemical plant. For filament wound vessels (see Section 3.6) the barrier layer is a resin-rich layer of laminating resin provided by employing chopped E-glass, C-glass or polyester fibre veil overwound with layers of hoop windings and 90° woven tape or with helical windings of continuous fibres at ±53°. The RPM (Stanton plc) system employs the former fibre structure with alternate layers of graded sand fillers.

Centrifugal casting (Johnston Pipes Ltd – Armaflo) (see Section 3.3) has the advantage that a highly ductile surface resin ($\epsilon_{um} \approx 25\%$) can be graded into the less ductile structural resin. In this case, there is no need to employ chemical resistant fibre veils since the volume fraction of chopped fibre can be graded from zero at the surface through to 50% in the structure and simultaneously from random to circumferential by varying the speed at which the mould is spun. Furthermore, alternate layers of compacted sand-filled resin can be incorporated in the structural laminate.

ECR glass flake can also be used to provide reinforcement to the barrier layers. Resistance to highly corrosive environments can be further improved by employing ECR, Chemglass, S, R, glass fibres instead of E-glass for the structural wall.

Conclusions

Environmental stress corrosion cracking (ESCC) of glass fibre reinforced plastics occurs in acidic environments in a brittle manner. This is because the commonly used E-glass fibre reinforcement is highly susceptible. However, careful choice of resin and laminate construction can provide the composite with ESCC resistance at acceptable applied strains for industrial applications such as chemical plant and sewage treatment.

References

1. R J Charles, *J. Appl. Phys.* **29**, 1549 (1958).
2. Environmental effects on fibre reinforced plastics, special edition of *Composites*, **14**, 169–305 (1983).
3. P J Hogg, *Progress in Rubber and Plastic Technology*, **5**, 112–55 (1989)
4. F R Jones, *J. Strain Analysis*, **24**, 223 (1989).
5. *Proceedings of Plastics for Pipeline Renovation and Corrosion Protection*, Plastics and Rubber Institute, London, 1985. (Various papers).

6.6 Thermal spiking/thermal cycling effects

F R JONES

Introduction

In service, composite materials are often subjected to a thermal excursion. For example, in an advanced aircraft, a supersonic dash can increase the wing skin temperatures by a minimum of 100°C. Furthermore, for engine housings, the material is also subjected to a thermal cycle, less rapid but to a higher temperature. In the former, the aircraft is only required to fly fast intermittently so that moisture absorption between thermal excursions can be significant. To simulate this service requirement, composites are subjected to thermal spikes during the monitoring moisture absorption.

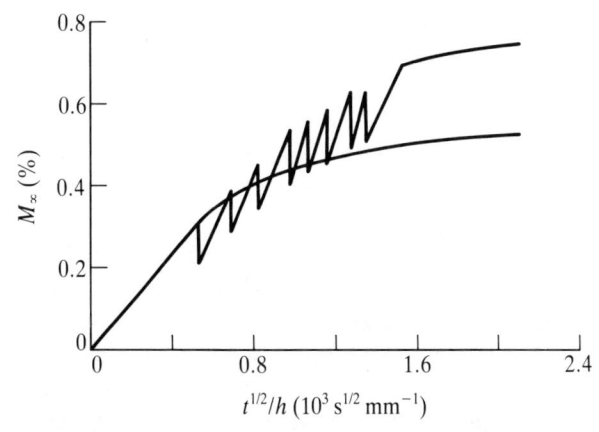

Figure 143 Comparison of moisture absorption by a 0° carbon fibre bismaleimide modified epoxy laminate (Narmco 5245C) under isothermal conditions (continuous line) and subject to intermittent thermal spikes to 150°C (saw-tooth). The environment was 96% RH at 50°C. The saw-tooth illustrates partial drying during thermal cycling and the subsequent enhanced water gain

The effect of thermal spiking on moisture absorption

The timescale for moisture equilibration is long; the effect of a series of thermal excursions on moisture diffusion kinetics is important. As shown in Fig. 143, this can lead to an enhanced moisture absorption by both crossply and unidirectional laminates. The saw-tooth shape arises because the laminate loses water during the thermal spike but on re-immersion in a humid environment, absorbs moisture at an enhanced rate to a higher level. The exact mechanism of this phenomenon is still not clear and is the subject of continuing research. One suggestion is re-equilibration of the network molecular structure of the matrix with the redistribution of free volume. It is also demonstrated that damage in the form of microcracks can occur in some resin castings.[1] The damage is believed to arise through environmentally induced stresses, which result from a moisture concentration gradient. This leads to a constraint on the outer surface which is trying to expand by swelling against the higher stiffness of the 'dry' core and the generation of compressive surface stresses. However, after a thermal cycle tensile stresses can be induced in the surface under conditions (i.e. rapid cooling) when time-dependent viscoelastic equilibration cannot occur.

For the composites, the exact mechanism of enhanced moisture absorption may differ from system to system. For example, at high spiking temperatures, a reduction in overall weight has been observed.[2,3] This loss in weight occurs simultaneously with the generation of damage in the form of transverse cracks in both 0° and 90° plies of a 0°/90°/0° laminate.[2] Careful examination of the coefficients of linear expansion of the dry transverse ply and the wet transverse ply has led to the identification of a critical temperature T_{crit}. Above this temperature, microcracking parallel to the fibres will occur in the plies of the laminate during thermal spiking. The arguments are based on the induction of higher thermal strains during thermal spiking as a result of the larger average expansion coefficient of the wet 90° ply, as described in Sections 6.3 and 4.12. This effect is illustrated in Fig. 144, where the expansion coefficient of the 0° ply is taken to be approximately zero and temperature independent because it is dominated by the longitudinal expansion coefficient of the fibres. Thus the area under the curves represents $(\alpha_t - \alpha_l)(T_{crit} - T_2)$; this determines the magnitude of the induced thermal strain on cooling from T_{crit}; when this exceeds the ply failure strain, transverse cracking occurs (see Section 4.11; for detailed equations see Section 4.12). The example in Fig. 144 shows that in the presence of absorbed moisture, $T_{crit\ (wet)}$ is below the strain-free temperature, which is related to T_g, so that the combination of water uptake and thermal cycling can induce microcracking. For the dry composite, $T_{crit\ (dry)}$ is greater than T_g and thermal cracking cannot be induced. Resistance to this damage mechanism will also be determined by the maintenance of the transverse strength after moisture absorption. In this respect, the contribution to the mechanism of the interface or interphase region between fibre and matrix needs to be resolved.

Below $T_{crit(wet)}$, thermally induced enhanced moisture absorption still occurs at spiking temperatures as low as 120°C, which appears to coincide with the onset of the glass relaxation mechanism for the wet resin, as shown in Fig. 144. Further confirmation is needed to establish the generality that the additional free volume of the hot-wet matrix is responsible for the higher moisture absorption.[1-4]

Conclusions

Thermal spiking can result in an enhanced moisture absorption by fibre reinforced composites subjected to a humid environment. The phenomenon has been observed to occur in a range of epoxy resin systems, which provide the matrices for composite materials where these effects could be critical. The mechanisms that operate are complex; at particular temperatures and laminate configurations, particular mechanisms may dominate.

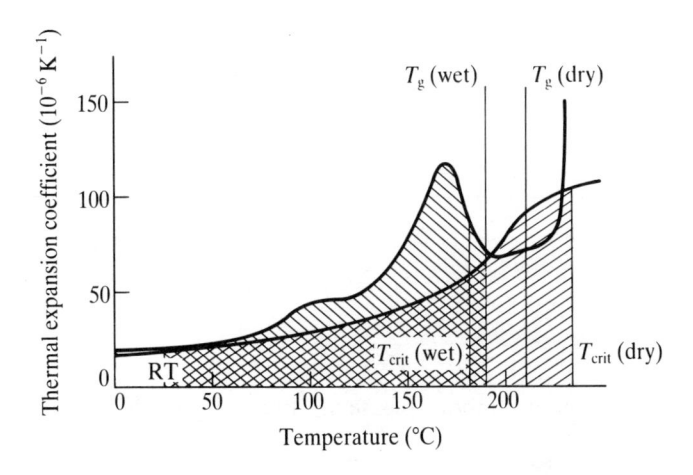

Figure 144 The temperature dependence of the transverse thermal expansion coefficients of the bismaleimide modified epoxy resin based carbon fibre composite (Narmco 5245C) dry (lower curve) and wet (upper curve). The cross hatching indicates the relative thermal strain induced on cooling from T_{crit} (dry) and T_{crit} (wet) (lower and upper curves respectively). T_{crit} is the temperature to which a crossply laminate needs to be heated and then cooled to induce a value of ϵ_{tl}^{th} to exceed ϵ_{ut} and cause thermal cracking. Since T_g (dry) is below T_{crit} (dry) thermally induced cracking does not occur but the reverse situation operates for the wet laminate and thermally induced cracking occurs.

References

1. C E Browning, *Polym. Eng. Sci.*, **18**, 16 (1978).
2. P M Jacobs, F R Jones in Composites, Design, Manufacture and Application, eds S W Tsai, G S Springer, *Proceedings of ICCM/VIII, Jul. 1991, Honolulu*, SAMPE, Covina, 1991, Vol. 2, Ch. 16, Paper G.
3. T Collings, S M Copley, *Composites*, **14**, (1981).
4. G Clark, D S Saunders, T J van Blaricum, M Richmond, *Composites Science & Technology*, **39**, 355 (1990).

Glossary

For specific symbols, please see Index on p. 400.

General terms

α = expansion coefficient
β = moisture swelling coefficient
γ = shear strain
γ = surface free energy
ϵ = strain
η = Halpin–Tsai geometry function
θ = angle of fibres to applied stress
v = Poisson ratio
ρ = density
σ = stress
τ = shear stress
d = diameter
D = diffusion coefficient
E = tensile or Young's modulus
G = shear modulus
G = strain energy release rate
K = stress intensity factor
L = fibre length
M = moisture content
P = load
r = radius
R = fibre aspect ratio in short fibre composites
R = stress ratio in fatigue
RH = relative humidity
T = temperature
t = time
V = volume fraction

Subscripts

∞ = property at infinite time
1,2 = in-plane property
1 = component designation (*see* p. 227)
2 = component designation (*see* p. 227)
2 = transverse direction, in plane
3 = transverse direction, through thickness
I = mode I crack opening (*see* G_I and K_I)
II = mode II crack opening (*see* G_{II} and G_{II})
a = applied
av = average
c = composite
c = critical
crit = critical
ct = composite in transverse direction
deb = debonding (*see* τ_{deb})
e = embedded
f = fibre
fr = fragment
g = glass (*see* T_g)
H = hybrid (laminate, *see* E_H)
i = interface
l = longitudinal direction
l = longitudinal direction of a lamina, ply or unidirectional composite
l = property of longitudinal ply
ll = property of longitudinal ply in longitudinal direction
lt = property of longitudinal ply in transverse direction
lts = property for transverse splitting of longitudinal ply
m = matrix
m = maximum property
max = maximum property
min = minimum property
t = property of transverse ply
t = transverse direction of a lamina, ply or unidirectional composite
tl = property of transverse ply in longitudinal direction
tt = property of transverse ply in transverse direction
u = ultimate
uc = ultimate property for composite (e.g. σ_{uc} = failure stress or strength of composite)
uf = ultimate property for fibre (e.g. ϵ_{uf} = failure strain of fibre)

ul = ultimate property of a longitudinal composite
um = ultimate property for matrix (e.g. ϵ_{um} = failure strain of matrix)
ut = ultimate property of transverse ply
utl = ultimate property of transverse ply in a laminate (e.g. ϵ_{utl} = transverse cracking strain of a crossply laminate)
x = in-plane off-axis property of unidirectional lamina
y = through thickness off-axis property of unidirectional lamina

Superscripts

θ = angle of fibres to applied load
1 = longitudinal (or fibre) direction of a lamina or unidirectional composite
2 = transverse of 90° direction of a lamina or unidirectional composite, in plane
3 = transverse or 90° direction of a lamina or unidirectional composite, through thickness
gel = property of gel
min = minimum property
P = Poisson
th = thermal
$'$ = stress or strain on one component when other fails (e.g. σ'_m = stress on matrix a fibre fracture strain)

Bibliography

General texts on composites

1. B D Agarwal, L J Broutman, *Analysis and Performance of Fibre Composites*, 2nd Edn, Wiley, New York, 1990.
2. K H G Ashbee, *Fundamental Principles of Fibre Reinforced Composites*, Technomic, Basel, 1989.
3. L J Broutman and R H Krock (eds), *Composite Materials*, 6 Vols, Academic, London, 1974.
4. K K Chawla, *Composite Materials – Science and Engineering*, Springer-Verlag, New York, 1987.
5. T-W Chou, *Microstructural Design of Fibre Composites*, Cambridge University Press, Cambridge, 1992.
6. T-W Chou (ed), *Structure and Properties of Composites*, VCH, Weinheim, 1993.
7. M J Folkes, *Short Fibre Reinforced Thermoplastics*, Research Studies Press, Letchworth, UK, 1982.
8. J C Halpin, *Primer on Composite Materials*, 2nd Edn Technomic, Stamford CT/Basel 1992.
9. G S Hollister and C Thomas, *Fibre Reinforced Materials*, Elsevier, London, 1966.
10. D Hull, *An Introduction to Composite Materials*, Cambridge University Press, Cambridge, England, 1981.
11. R M Jones, *Mechanics of Composite Materials*, McGraw-Hill, Washington, DC, 1973.
12. A Kelly, *Strong Solids*, 2nd Edn, Clarendon Press, Oxford, England, 1973, 3rd Edn with N H MacMillan 1990.
13. A Kelly and Yu N Rabotnov, *Handbook of Composites*, Vols 1–4, Elsevier Science, Amsterdam, The Netherlands, 1985.
14. A Kelly, *Concise Encyclopaedia of Composite Materials*, Advances in Materials Science and Engineering, Vol 3, Pergamon, Oxford, 1989.

15. G Lubin, *Handbook of Composites*, van Nostrand Reinhold, New York, 1982.
16. J G Morley, *High-performance Fibre Composites*, Academic Press, London, 1987.
17. I K Partridge (ed.), *Advanced Composites*, Elsevier, London, 1989.
18. R Talreja (ed.), *Damage Mechanics of Composite Materials*, Elsevier, Barking, 1992.
19. S W Tsai, *Composites Design 1985*, Think Composites, Dayton, Ohio, 1985.
20. S W Tsai and H T Hahn, *Introduction to Composite Materials*, Technomic, Westport, Connecticut, 1980.
21. J R Vinson and T-W Chou, *Composite Materials and their Structures*, Applied Science, Barking, 1975.
22. W Watt (ed.), *New Fibres and their Composites*, Royal Society, London, 1980 (Special Issue of Proc Roy Soc, London, A294 1980).
23. F W Wendt, H Leibowitz, N Perrone (eds), *Mechanics of Composite Materials*, Pergamon, Oxford, 1970.

Specialist texts
Matrices

1. A V Boenig, *Unsaturated Polyesters*, Elsevier, Amsterdam, 1964.
2. J A Brydson, *Plastics Materials*, 5th Edn, Butterworth, London, 1988.
3. F N Cogswell, *Thermoplastic Aromatic Polymer Composites*, Butterworth-Heinemann, Oxford, 1992.
4. B Ellis (ed.), *The Chemistry and Technology of Epoxy Resins*, Blackie, Glasgow, 1992.
5. A Knop and L A Pilato, *Phenolic Resins*, Springer-Verlag, Berlin, 1985.
6. H Lee and K Neville, *Handbook of Epoxy Resins*, McGraw-Hill, New York, 1967.
7. J A Manson and L H Sperling, *Polymer Blends and Composites*, Heyden/Plenum, New York, 1976.
8. C A May, *Epoxy Resins Chemistry and Technology*, Marcel Dekker, New York, 1988.
9. D Wilson, H D Stenzenberger, P M Hergenrother (eds), *Polyimides*, Blackie, Glasgow, 1990.

Reinforcements

1. J-B Donnet and R C Bansal, *Carbon Fibres*, 2nd Edn, Marcel Dekker, New York, 1989.
2. K L Loewenstein, *The Manufacturing Technology of Continuous Glass Fibres*, Elsevier, Amsterdam and London, 1973.

3. D R Lovell, *Carbon and High Performance Fibres Directory*, 5th Edn, Chapman and Hall, London, 1991.
4. E P Pleuddemann, *Silane Coupling Agents*, Plenum, New York, 1991.
5. W Watt and B V Perov (eds), *Strong Fibres*, Handbook of Polymer Composites, Vol. 1, North Holland, Elsevier, Amsterdam, 1985.

Design, selection, manufacture and case histories

1. D F Adams, *Test Methods for Composite Materials*, Technomic, Basel, 1990.
2. M F Ashby, *Materials Selection in Mechanical Design*, Pergamon, Oxford, 1992.
3. L A Carlsson and J W Gillespie, *Delaware Composites Design Encyclopaedia*, 6 Vols, Technomic, Basel, 1989–1991.
4. R J Crawford, *Plastics Engineering*, Pergamon, Oxford, 1987.
5. A Kelly and S T Mileiko (eds), *Fabrication of Composites*, Handbook of Composites, Vol 4, North Holland, Elsevier, Amsterdam, 1983.
6. S M Lee, *Dictionary of Composite Materials Technology*, Technomic, Basel, 1989.
7. P K Mallick and S Newman (eds), *Composites Materials Technology*, Hanser, London, 1990.
8. D H Middleton (ed.), *Composite Materials in Aircraft Structures*, Longman, Harlow, 1990.
9. L N Phillips (ed.), *Design with Advanced Composite Materials*, The Design Council, London.
10. *Replacement of Metals with Plastics, Plastics Processing*, EDC, National Economic Development Office, London, 1985.
11. R G Weatherhead, *FRP Technology: Fibre Reinforced Resin Systems*, Applied Science, London, 1980.

Joining of composites

1. R D Adams (ed.), *Structural Joints in Engineering*, Elsevier, New York, 1983.
2. A A Baker and R Jones (ed.), *Bonded Repair of Aircraft Structures*, Martinus Nijhoff, Amsterdam.
3. A J Kinloch (ed.), *Structural Adhesives: Developments in Resins and Primers*, Elsevier, London, 1986.
4. F L Matthews (ed.), *Joining Fibre Reinforced Plastics*, Elsevier, London, 1986.

Environmental effects

1. G S Springer (ed.), *Environmental Effects on Composite Materials*, 3 Vols, Technomic, Basel, 1981, 1984, 1987. *See also* STP 658 below.

Plastics technology

1. R W Meyer, *Handbook of Polyester Moulding Compounds and Moulding Technology*, Chapman and Hall, London, 1986.
2. J F Monk, *Thermosetting Plastics, Practical Moulding Technology*, George Godwin, London, 1981.

Specialist publication series

1. Edited volumes on composite materials as Special Technical Publications (STP) of American Society for Testing and Materials, AMSTM, Philadelphia, Pennsylvania, USA (selected titles).

STP452 Interfaces in Composites, 1969
STP617 Composites Materials Testing and Design, 1977
STP636 Fatigue of Filamentary Composite Materials, ed K L Reifsnider, L N Lauraitis, 1977
STP658 Advanced Composites: Environmental Aspects, ed J R Vinson, 1978
STP723 Fatigue of Fibrous Composite Materials, ed L N Lauraitis, 1981
STP749 Joining of Composite Materials, ed K T Kedward, 1981
STP775 Damage in Composite Materials, ed K L Reifsnider, 1982
STP787 Composites Materials Testing and Design, ed I M Daniel, 1982
STP876 Delamination and Debonding of Materials, 1985
STP937 Toughened Composites, ed N J Johnston, 1987

2. Major conference series
(i) Proceedings of the International Conference on Composite Materials: ICCM I 1976, ICCM II – Eds B Noton, R Signorelli, K Street, L Phillips, 1978, American Institute of Mining Metallurgical and Petroleum Engineers, New York.

ICCM III – Advances in Composite Materials, (2 volumes). Eds A R Bunsell et al, Pergamon, Oxford, 1980, C Bathias, A Martreuchar, D Menkes, G Verchery.
ICCM IV – Progress in Science and Engineering of Composites. Eds T Hayashi, K Kawata, S Umekawa, (2 volumes). Japan Society for Composite Materials, Tokyo, 1982.
ICCM V – ICCM V. Eds W C Harrigan Jr, J Strife, A K Dhingra, AIME, Metallurgical Soc, Warrendale, Pa.

ICCM VI – ICCM VI/ECCM 2. Eds F L Matthews, NCR Buskell, J M Hodgkinson, J Morton, (6 volumes). Elsevier App Sci, London, 1987.
ICCM VII – ICCM VII. Eds Y Wu, Z Gu, R Wu, International Academic Publishers, Beijing/Pergamon, Oxford, 1989.
ICCM VIII – Composites, Design, Manufacture and Application, (4 volumes). Eds S W Tsai, G S Springer, SAMPE, Covina, California, 1991.
ICCM 9 – ICCM 9. Ed A Miravete
  Vol 1 Metal Matrix Composites
  Vol 2 Ceramic Matrix Composites and other Systems
  Vol 3 Modelling and Processing
  Vol 4 Composites Design
  Vol 5 Composites Behaviour
  Vol 6 Composites Properties and Applications
  University of Zaragoza, Spain and Woodhead Publishing, Cambridge, 1993.

(ii) Proceedings of European Conference on Composites:
Developments in the Science and Technology of Composite Materials, European Association for Composite Materials, Bordeaux/Elsevier Applied Science, London.
ECCM 1   1985
ECCM 2   (see ICCM 1 above), 1987
ECCM 3   1989 (eds A R Bunsell et al)
ECCM 4   1990 (eds J Füller et al)
ECCM 5   1992 (eds A R Bunsell et al)

(iii) Proceedings of International Conference on Composite Interfaces (ICCI)
Elsevier, New York, 1986, 1988, 1990, 1991, (ed. H Ishida) (various titles).

(iv) International Conference on Interfacial Phenomena in Composite Materials
Ed. F R Jones, 1989; eds I Verpoest, F R Jones, 1991; eds F R Jones, T W Clyne, D C Phillips (Special Issue of Composites, June 1994), Butterworth-Heinemann, Guildford and Oxford.

(v) Proceedings of SPI Composites Institute Annual Conference, recent volumes available from Technomic, Westport CT and Basel.

(vi) Proceedings of Composites Structures
Ed. I H Marshall, Elsevier Applied Science, London, 1981, 1983, 1985, 1987, 1989, 1991, 1993.

Index of specific symbols

A = interfacial area 240
C = cure parameter 119
C = cycle ratio 312
C_θ = fibre orientation factor 273
C_1 = fibre reinforcing efficiency 283
C_2 = fibre orientation factor 283
CME = coefficient of moisture expansion 365
CTE = coefficient of thermal expansion 276, 365
D = needle diameter 341
D = moisture diffusion coefficient 368
D_∞ = true one-dimensional moisture diffusion coefficient 368
D_c = crystallite disorder 27
DFT = degree of fibre surface treatment 30
E = Young's modulus 23, 212–361
E_{ct} = composite modulus in transverse direction 249–53
E_f = fibre tensile or Young's modulus 212–361
E_f/ρ = specific modulus 2
E_m = matrix modulus 94, 212–361
E_H = modulus of hybrid laminate 227
E_1 = tensile modulus of longitudinal ply 261, 348
E_2 = tensile modulus of transverse ply 261, 348
EPEW = epoxy equivalent weight 88
F = pull-out force 243
F = load 249–53, 282
F = probability of transverse fracture 262
G_c = critical strain energy release rate 324, 326
G_i = shear modulus of interface 243
G_m = shear modulus of matrix 270
G^* = complex shear modulus 118
G' = real or storage shear modulus 118
G'' = imaginary or loss shear modulus 118
G_{Ic} = critical strain energy release rate in mode I crack opening 73, 100, 104, 105

G_{IIc} = critical strain energy release rate in mode II crack opening 73
ΔG = strain energy release rate range 326
HDT = heat distortion temperature 88, 132, 276
K_{Ic} = critical stress intensity factor for mode I crack growth 105, 384
ΔK = stress intensity range 309
L = beam length 255
L = stitch length 341
L = fibre length 235, 269
$L_{a\perp}$ = apparent crystallite width 27
$L_{a\|}$ = apparent crystallite length 27
L_c = fibre critical length 235, 273
L_c = stacking size of a crystallite 27
M = moments per unit width 349
M = molar mass 88
$\overline{M}_n$ = number average molar mass 118
$\overline{M}_W$ = weight or mass average molar mass 118
M = moisture content at time t 367
M_1, M_2 = moisture content at times t_1 and t_2 respectively 368
M_∞ = equilibrium or maximum moisture content 257, 367, 371
N = force per unit width 349
N = number of fatigue cycles 301–27
N_f = number of fatigue cycles to failure 301–27
P = applied load 239
P_m = maximum load 240
Q_{ij} = lamina principal reduced stiffness 348
R = $\sigma_{min}/\sigma_{max}$ 301–27
R = matrix radius in microdebond test and Cox model 240, 269
R_a = fibre aspect ratio 270
R_c, R_x, R_y = fibre aspect ratio 273
R_e = Reynolds number 188
R_t = fibre transfer ratio 270
RH = relative humidity 367, 373, 389
S = shear strength 297
S = peak stress in fatigue 301–27

Index of specific symbols

S = stress function 312
T_c = cure temperature 116
T_g = glass transition temperature 100, 103, 116, 147, 373
T_{g0} = glass transition temperature at cure time, $t_c = 0$ 116
$T_{g\infty}$ = glass transition temperature at cure time, $t_c = \infty$ 116
gel T_g = glass transition temperature at gelation, when $t_{c,gel} = t_{c,vit}$ 116
T_1 = stress free temperature 254
T_2 = service temperature 254
T_{CRIT} = critical temperature for thermally induced cracking of cross ply laminate 390
T_β = temperature of β-transition 120
U_1, U_3 = displacement of longitudinal ply 261
U_2 = displacement of transverse ply 261
V_f = fibre volume fraction 108, 133, 212–85
V_m = matrix volume fraction 212–85
V_{fi}, V_{fj} = volume fraction of fibres of length $l_i l_j$ 282
V_x, V_y = volume fractions of fibres of length x and y 273
W_d = weight or mass of a dry laminate 367
W_i = interfacial work of fracture 240
W_p = work done on pull-out per fibre 282
W_w = weight or mass of a wet laminate 367
W_{SL} = work of adhesion 44
X_c = compressive strength in fibre direction 297
X_t = tensile strength in fibre direction 297
Y_c = compressive strength perpendicular to fibres 297
Y_t = tensile strength perpendicular to fibres 297
Z = orientation parameter 28
a = crack length 240
a = delamination crack growth area 326
a = Weibull shape parameter 262
a_1 = thickness of longitudinal plies of a crossply laminate 261
a_2 = thickness of the transverse ply in a cross ply laminate 261
b = effective width of interface 243
b = thickness of shear transfer zone 261
b = laminate breadth in diffusion calculation 368
b = outerply thickness of a $0°/90°/0°$ laminate 249
c = laminate width 249
c = moisture concentration in Fick's second law of diffusion 369
c/2 = interlayer spacing 27
d = minimum opening in mixing head 188

d = semi-inner ply thickness of a $0°/90°/0°$ laminate 249
d = yarn diameter 341
d_c = critical yarn bending diameter 341
d_s = diameter of stitch yarn 343
h = laminate thickness in diffusion calculation 368
l = fibre length 280
l = laminate length in diffusion calculation 368
l_ϵ = characteristic fibre length at strain ϵ 279
l_c = critical fibre length 134, 279, 315
l_e = embedded length 237, 239–41, 243
$l_i l_j$ = individual fibre length 281
l_t = fibre transfer length 279
n_i = number of cycles of given stress range 326
p = course spacing 341
q = wale spacing 341
r = fibre radius 267, 269, 279
r_f = fibre radius 240, 243
s = crack spacing 221
t = fabric thickness 341
t = crack spacing 249
t_{av} = average crack spacing in the transverse ply of crossply laminate 250
t_c = cure time 116
$t_{c,gel}$ = cure time for gelation 116
$t_{c,vit}$ = cure time for vitrification 116
w_g = mass fraction of gel 118
w_s = mass fraction of sol 118
w_x, w_y = width of insertion yarns in 0° and 90° orientation 343
x = diffusion coordinate in Fick's second law of diffusion 369
x = distance from fibre end 235, 267, 270, 279
x = flaw depth 380
x = semi-chord length of an unbalanced $0°/90°$ beam 255
x = stress transfer coordinate of embedded single filament 235
y = stress transfer length 249
α = constant in pull-out test data analysis 243
α_f = linear expansion coefficient of fibres 254
α_l = linear expansion coefficient in the fibre direction of a unidirectional lamina 254, 372
α_m = linear expansion coefficient of matrix 254, 372
α_t = linear expansion coefficient transverse to fibres of unidirectional lamina 254, 372
β_l = moisture swelling coefficient of longitudinal ply 257

β_t = moisture swelling coefficient of transverse ply 257
γ = surface free energy 252
γ_L = surface free energy of a liquid 44
γ_S = surface free energy of a solid 44
γ_{SL} = interfacial free energy between solid and liquid 44
γ_t = surface free energy of transverse ply 252
γ_{12} = shear strain 297, 348
δ = displacement from flat of 0°/90° beam 255
δ = out-of-phase angle (in dynamic mechanical analysis) 118
ϵ = applied strain 262
ϵ_c = composite strain 326
ϵ_m = matrix strain 212–85
ϵ_o = Weibull scale parameter 262
ϵ_{lt}^{th} = thermal strain in the longitudinal ply in the transverse direction of a 0°/90°/0° laminate 254
ϵ_{lts}^{p} = Poisson strain for longitudinal splitting of longitudinal plies of 0°/90°/0° laminate 252, 259
ϵ_{ll}^{th} = thermal strain in longitudinal direction of longitudinal ply 258
ϵ_{lt}^{p} = Poisson strain in longitudinal ply in transverse direction of a 0°/90°/0° laminate 253–9
ϵ_{lls}^{min} = minimum applied strain for longitudinal splitting of longitudinal plies of a 0°/90°/0° laminate 253
ϵ_{lls} = applied composite strain for longitudinal splitting strain of longitudinal plies of a 0°/90°/0° laminate 253
ϵ_{tl}^{th} = thermal strain in the transverse ply in the longitudinal direction of a 0°/90°/0° laminate 252–7, 372
ϵ_{tt}^{th} = thermal strain in transverse direction in transverse ply 259
ϵ_{tl}^{s} = swelling strain in transverse ply in longitudinal direction 257
ϵ_{uf} = fibre failure strain 23
ϵ_{um} = matrix failure strain 94
ϵ_{ut} = failure strain of a transverse lamina 249–53
ϵ_{ut}^{min} = minimum failure strain of a constrained transverse ply in a 0°/90°/0° laminate 252
ϵ_{utl}^{min} = minimum cracking strain of a constrained transverse ply in a 0°/90°/0° laminate 252
ϵ_{utl}^{i} = cracking strain of inner ply of a crossply laminate 253
ϵ_{utl} = cracking strain of transverse ply in a 0°/90°/0° laminate 249–53
ϵ_x = dielectric constant 103
η = viscosity 118
η = fabric tightness factor 343

η = Halpin-Tsai geometry function 218
η_l = fibre length efficiency factor 273
η_o = fibre orientation constant 273
θ = angle of bias yarns 343
θ = contact angle 44
θ = fibre orientation angle in a lamina 348
λ_i = insertion yarn linear density 343
λ_s = stitch density 343
v = average stream velocity 188
v = frequency of incident light 264
v_f = Poisson ratio of fibre 257
v_l = Poisson ratio of longitudinal ply 259
v_m = Poisson ratio of matrix 240, 257
v_t = Poisson ratio of transverse ply 259
ρ = density 2, 188
ρ = flaw radius 380
ρ = radius of curvature of 0°/90° beam 255
σ = stress 23, 212, 361
σ = peak stress in fatigue 313
σ_a = applied stress 250, 324, 380
σ_f = laminate strength measured at equivalent testing rate to fatigue cycle 312
$\sigma_{f,max}$ = maximum fibre stress 236
σ_{max} = maximum stress 301–27
σ_{max} = maximum stress at the crack tip 380
σ_{min} = minimum stress 301–27
σ_R, σ_r = residual strength 312, 324
σ_{uf} = fibre strength 23, 218, 341
$\sigma_{ufr(Lc)}$ = fragment strength at critical length 236
σ_{uf}/ρ = specific strength 2
σ_{uc} = strength of composite 324
σ_{um} = matrix strength 94
σ_{utc} = transverse compressive strength 364
σ_1 = component of stress parallel to fibres in a lamina 297, 348
$\Delta\sigma$ = additional stress 249
σ_2 = component of stress perpendicular fibres in a lamina 297, 348
$\sigma_{2,x}$ = the tensile stress in the transverse ply as a function of distance x from the transverse crack 261
$\Delta\sigma_0$ = additional stress on 0° plies in plane of transverse crack 249
τ = interfacial shear stress or strength 240, 244, 279
τ_{av} = average interfacial shear stress 243
τ_{deb} = debonding shear stress 237
τ_i = interfacial shear stress or strength 235, 270, 273
τ_{max} = maximum shear stress 244, 262
τ_{12} = shear stress in plane of lamina 297, 348
ϕ = thread count number 343
ϕ = shear lag geometry factor 249
Δ = logarithmic decrement 118
Δ = Raman peak position 265

General index

0°/90°/0° laminates   245–67
A-stage   86
ABPO fibre   58
abrasion   156
absorbed moisture   362–79, 384–91
  see also moisture absorption
absorber   141
accelerator   122, 163, 378
acetylene chromene terminated resins   106
acetylene end caps   98
acicular inorganic particles   4–7
acicular magnesium hydroxide   7
acid anhydrides   90
acid catalyst   106
acidic environments   379–88
acoustic emission   237, 310
acoustic emission methods   355
activated resin   205
additives   102–15, 166, 183
adhesion assessment   230–45, 260
adhesion promotion   15, 29–33, 42–8
adhesives   104, 105
adsorbed water   30
advanced thermoplastics   69–74
AE see acoustic emission
aero-engines   98
aerospace applications   98, 100, 103, 135, 183, 292–6
air inhibition   125
aircraft interior panels   13
airframe construction   292–6
airframe structures   135, 292–6
Al_2O_3 filaments   10
aligned short fibres   280
alkaline environments   380
alkyd resins   168
allylnadic imides   98, 101
allylphenyl monomer   99
AlN   18
alternating copolymerization   129
α-alumina   8
alumina fibres   8–12
alumina hydrates   9

alumina trihydrate   182, 200
alumina whiskers   10
aluminium   17
aluminium, properties   82
aluminium acetate   9
aluminium alloys, properties   82
aluminium chloride   9
aluminium formate   9
aluminium nitrate   9
aluminium nitride   19
aluminosilicates   8
American Society for Testing and Materials (ASTM)   357
amines   90
amino resins   168
γ-aminopropyltriethoxysilane   46
ammonia   66
ammonium bicarbonate   29
amorphous interlayer bonding   149
amorphous thermoplastics   69, 146
amplitudes of the stress wave pulse   355
angle ply laminates   214, 245
anhydrite   6
anisotropic behaviour   222–5
anisotropic elasticity   223
anisotropy   81, 84–5, 212–25
anodic oxidation   29
antibuckling guides   360
anticorrosion applications   379
antioxidants   166
APC-2   73, 186
apparent crystallite length   27
apparent crystallite width   27
γ-APS   46
AR glass   38
aragonite   6
ARALL   225
aramids, aramid fibres   12–15, 58, 76, 181, 216, 264, 299
ARMAFLO   387
aromatic diamines   99
aromatic polyamides   299

aromatic polymer composites 74
  see also APC-2
asbestos 4
aspect ratio 273
ATLAC resins 131, 284
attenuation of ultrasonic wave 354
autoclave 63, 97, 107, 134, 138–44
autoclave moulding 138–44
automotive applications 109, 135
automotive panels 189
AVCO 62
average crack spacing 250
AVIMID 71, 100, 147
axial fibres 157

B-stage 86
B-staging 86
back pressure 166
backing films 201
backing resin 376
bacterial degradation 190
balanced laminates 245, 350
ballistic protection 13, 83
ballistics 328
barely visible impact damage (BVID) 329, 332
barrier layer 383
barrier resin 145
basal plane edge sites 31
basal planes 31
BCB resins 98, 104–5
BDMA 91
beetle resins 384
benzocyclobutene (BCB) resins 104–5
benzocyclobuteneimide resins 104–5
benzophenonetetracarboxylic acid dimethyl ester 97
benzyldimethylaniline 91
BF_3 complexes 92, 95–6
binder 198, 376
binder wet-out 198
biodegradeability 6
bis(benzocyclobutene imides) 98
bismaleimide resins 96–101
bismaleimide triazine 95
bismaleimides 90, 95, 96–101
bisoxazoline phenolic composites, properties 104
bisoxazoline-phenolic resins 103–4
bisoxazolines (PBOX) 103
bisphenol A 86–90, 124, 129–31
bleed pack 138
BLENDUR 105
blistering 376–9
BMC 41, 150–3, 168, 184
BMC, mechanical properties 152
BMC, typical composition 151
BMI 96–103
BMI composites, properties 100

BMI prepregs 100
BMI properties 100
BN 18
boats 376
boehemite 8
boehemite gels 8
borazine 18
boron carbide 18
boron fibres 15–18, 293
boron filaments 16
boron halide 16
boron hydride 16
boron nitride 18
Bowyer-Bader methodology 273
braided fabrics 49
braiding 85
British Standards Institute (BSI) 357
brittle composites 32
brittle matrix composites 219–22
brittle thermoset matrix 280
brucite 6
BTDE 97
bubbles 377
buckling 308
bulk moulding compound 41, 150–3, 168, 184
butyrolactone 95

C-glass 38
C-glass veil 387
C-scan technique 139, 354
C-stage 86
C^{13}NMR 46
calcium carbonate 5, 111, 151, 200, 202
calcium carbonate, aragonitic 5
calcium metasilicate 5
calcium sodium phosphate 6
calcium sulphate 5, 385
calorific values 185
caprolactam 191
carbon coating 64
carbon fibre 19–38, 76, 216, 230, 254, 264, 278, 295, 305, 334, 338
carbon fibre, crystal orientation 24–29, 34
carbon fibre, degree of interlayer (intracrystallite) disorder 27
carbon fibre, HM 21, 24
carbon fibre, HS 21, 24
carbon fibre, IM 24
carbon fibre, modulus 21, 34
carbon fibre, PAN based 19–34
carbon fibre, pitch based 34–8
carbon fibre, Reynolds and Sharp mechanism 26
carbon fibre, strength 21, 26, 34
carbon fibre, structure 24–9, 34–8
carbon fibre, Type I 21
carbon fibre, Type II 21
carbon fibre, Type A 21

General index

carbon fibre epoxy 293
carbon fibre preparation 19–24, 34–48
carbon–carbon composites 24, 49
carbonization 21
carbosilane polymer 63
carboxylic acid groups 30
catalyst 86, 102, 163, 206
catalyst injection 207
catalytic graphitization 22, 24
cellulose, regenerated 9
centrifugal casting 144, 387
ceramic 8–12, 18–19, 62–8
ceramic fibres 8, 18, 62–8, 176, 264
ceramic fibres, properties 67
ceramic matrices 62
ceramic matrix composites 63, 176
ceramic, polymer precursor 9
ceramics, polymer precursor route 62–8
ceramics, polymeric precursors 18, 66–8
CRFP 300, 305, 310, 327, 364
characteristic damage state 247, 286
characteristic strength 297
characterization of blistering 377
chemical bonding theory 46
chemical plant 130, 379, 388
chemical resistance 94, 124, 381
chemical vapour deposition 16, 62
chemically resistant polyester 384
chemisorbed 46
chemosets 86
china clay 111
chlorosilane 66
chopped glass fibres 41, 176, 278
chopped glass strands 41, 151
chopped rovings 205
chopped strand 111, 200
chopped strand mat (CSM) 39, 43, 133, 163, 278, 377, 387
chrysotile 4
CIBA GEIGY 913/914 87
clamp breathe 171
clamping force 187
CMC 63
cobalt soap 122, 129
coefficient of moisture expansion (CME) 257, 365
coefficient of thermal expansion 75, 82, 108, 254, 257, 365
coefficient of variation 218
coin tap test 353
comparative evaluation of interfaces 241
compatibility of fibres with resins 29–34, 42–8, 210
competitive materials, properties 82, 126
COMPIMIDE 99–100
complex modulus 118
composite, laminate properties 347
composite, modulus of 134, 215–18, 223, 227

composite, properties of 73, 82, 83, 85, 100, 104, 114, 126, 133, 152, 178, 186, 205, 210, 217, 223, 276, 347
composite, strength of 134, 218–222, 223, 228
composite micromechanics 212–85
composite properties, phenolics 107
composite technology 1–4
compression after impact (CAI) 104
compression modulus 347
compression moulding 97, 99, 106, 108, 112, 134, 152, 204, 278
compression strength 36, 114, 305, 347
compression test rig 359
compression testing 359
compressive creep 114
compressive deformation 60
computer aided engineering 194
condensation 45
condensation mechanism 170
conductive fibres 76
conductivity 213
conformability 49
conjugated π-electron system 13
consolidation 164
constrained cracking 247
contact radiography 352
contaminants 377
continuous fibre laminates 212–30, 245–64
continuous fibre mat 41, 181
continuous fibre reinforced thermoplastic composites 73, 80, 145
continuous fibre reinforcement 222
continuous fibre reinforcement, strength 218
continuous fibre reinforcement, Young's moduli 215
continuous random mats 39, 41
continuously impregnated compound 151
COPNA resin 105
copper alloys 82
core–sheath structure 29
corrosion products 385
corrosion resistance 129, 130, 179
corrosion resistant resins 131, 145
corundum 8
cotton flock 111
coupling agent 42
coupling in laminates 245, 349
Cox model 269, 281
crack propagation 240
crack spacing 221
crack stopping 168
creep resistance 4
critical crack size 26
critical fibre length 134, 235, 279
critical fibre volume fraction 220
critical temperature for thermal cycling 390

critical yarn diameter 341
crosshead extruder 80
crosslink density, heterogeneity 122
crosslink density 122
crosslinking in polyethene 56
crossply cracking 247–63
crossply laminates 231, 245–63, 321
crystallinity 70
crystallite orientation 28, 34, 59
crystallization pressure 385
CRYSTIC 131, 384
CSM see chopped strand mat
cumulative damage 326
cure advancement 141
cure cycle 139, 141
cure kinetics 117, 182
cure temperature 115
cure time 115
cure–property diagrams 119
cure–temperature property 120
curing 90, 115, 122, 160, 164, 170, 200
curing agents 122, 376
curing shrinkage 125
curing strain 259
CVD 16, 62
cyanate ester 101
cyanate ester resins 102
cyanurate resins 90
cycle time 166, 179
cyclic loading 323 see also fatigue
cyclic oligomer 45
cycling frequency 315
cyclization 20
cyclopentadiene 191

damage 249, 310, 351
damage accumulation 286
damage growth laws 324
damage tolerance 229
DAP 168
Darcy equation 196
dawsonite 6–7
DDS 92
debond 238
debond length 233
debonding 237, 244, 351
decaborane 19
decohesion 238
deformable layer hypothesis 46
degradation 13, 115
degree of fibre surface treatment 30
delamination 108, 286, 289, 317, 318, 328, 331, 335
delayed brittle fracture 379
delivery eye 154
depolymerization 192
DERAKANE 129, 384
DESBIMID resin 99
design 81, 305, 317

devitrification 115
DFT 30, 33
DGEBA 86–90, 129
diallyl phthalate 169
diaminodiphenylsulphone 92
4,4'-diaminodiphenylmethane 97
diaphragm forming 148
dicyandiamide (DICY) 92
dielectric constant 102, 105, 351
Diels-Alder reaction 99
diethylene glycol 124
diffusion 366–75, 380
diffusion control 117
diglycidyl ether of bisphenol A 86–90, 120
dimensional random fibre 283
dimensional stability 114
dimethylaniline accelerator 129
dimethyldichlorosilane 63
dip impregnation 156
direct measure of interfacial adhesion 238
direct net shape knitting 52
direct rovings 39
discontinuous fibre 74, 134, 269, 274–85
discontinuous fibre reinforced thermoplastics, mechanical properties 75
discontinuous fibre reinforced thermosets, mechanical properties 284
diuron 92
DMC, typical composition 151
DMC see dough moulding compound
doctor blade 157
dodecamethylcyclohexasilane 63
dolomite 200
double belt lamination 148
double cantilever beam 360
dough moulding 128
dough moulding compound 39, 134, 150, 171, 184, 278, 284
dropweight 328
dropweight impact 329, 331
dry jet wet spinning 12
dry spinning 67
ductile matrix composites 219
durability 381
dwell 141
dye penetrant methods 351
dynamic mechanical analysis 353
dynamic response 334

E-glass 2, 38–48, 76, 80, 274, 379
see also glass fibre
E-glass chopped strands 202
ease of fabrication 135
economics of composite manufacture 133–7
ECR glass 38, 380
edge effects 357
edges of basal planes 31

elastic constants 83, 347
electrochemical treatment 29
electronic speckle pattern interferometry
    (ESPI) 351
elongation at break 178
    see also failure strain
embedded fibre length 233, 243
emulsion 42
emulsion bound mat 39
encapsulation 169
end notched beam 361
end tabs 358
energy absorbing capacity 338
energy absorption 138
energy management 179
engineering compounds 76
engineering materials 82
engineering phenolics 114
engineering thermoplastics 69, 75, 167
enhanced moisture absorption 372,
    388-91
environmental effects 362-91
environmental stress corrosion
    cracking 145, 379-87
EPIC resins 105
epichlorohydrin 86
epikote 88
EPIKURE 91
EPON 88
epoxides see epoxy resins
epoxy bathmasters 199
epoxy equivalent weight 90
epoxy novolak 130
epoxy resin, properties 94
epoxy resins 21, 31, 86-96, 100, 105, 116,
    129, 168, 182, 191, 211, 280, 365
    see also diglycidyl ether of bisphenol A
        and DGEBA
equilibrium moisture content 257, 362-71
ESCC 145, 379-87
ESCC failure modes 382
ethylene glycol 124
eutectic 99
event counting 355
exotherm 92, 115, 164, 168, 187, 189
extensometers 357
extrusion 74, 76, 112, 171, 176

fabrication 133-211
fabrication cycles 84, 138
factors affecting blistering formation
    377-8
failure modes in fatigue 318
fast wet-out 210
fatigue 179, 299-326
fatigue, aramid fibre composites 299-304
fatigue, aramid fibres 300-1
fatigue, carbon fibre composites 305-9
fatigue, damage accumulation 326-7

fatigue, damage assessment 323
fatigue, glass fibre composites 309-16
fatigue, glass fibres 315
fatigue, life prediction 317-27
fatigue, residual strength curve 311
fatigue, S-N plots 301, 305, 306, 307, 308,
    312, 313, 318
fatigue, strain control model 302
fatigue, strain-life plot 314-20
fatigue strength 321
fatigue testing 357
FIBERAMIC 66
FIBERITE 87, 102
FIBEROD 80
fibre alignment 39, 172
fibre bands 154
fibre breakage 274, 286, 318
fibre bunching 212
fibre bundles 274, 377
fibre deformation 265
fibre distribution 213
fibre failure 296
fibre fibrillation 14
FIBRE FP 10
fibre interface 213, 230-45
fibre length 77, 134, 150
fibre length distribution 269
fibre management 168, 171-6
fibre matrix debonding 253
    see also debonding, decohesion
fibre orientation 77, 82, 134, 152, 171, 177,
    188, 278, 341
fibre orientation distribution 269
fibre orientation factor 283
fibre pattern 163
fibre preforms 48-54, 196
fibre pull-out 168, 238, 242, 282
fibre reinforced thermoplastics 41, 69-86,
    145-50, 219
fibre reinforced thermosets 41, 278-85
fibre reinforcement 198
fibre reinforcing efficiency 134, 273, 279
fibre spinning 9, 12, 19, 54, 58
fibre strength 1, 14, 23, 37, 38, 56, 59, 67,
    218
fibre surface treatment 30, 242
fibre tension 154-5
fibre toxicity 4
fibre volume fraction 134, 155, 189, 280,
    341
fibre-matrix adhesion 15, 29, 42, 230-45
fibre-matrix bond strength 230-45, 264
fibre-matrix debonding 244, 318, 381
fibre-matrix debonds 247
fibre-matrix interaction 31, 192
fibre-matrix interface 64, 221, 222,
    230-45, 279
fibres, AlN 18
fibres, B_4C 18

fibres, BN 18
fibres, discontinuous 74, 269–85
fibres, gel spun 55
fibres, high performance 15
fibres, melt spun 55, 63
fibres, modulus of 1, 14, 23, 37, 38, 55, 59,67
fibres, polyethylene 54
fibres, short 74, 269–85
fibres, SiCO 64
fibres, SiCNO 65
fibres, SiC 62, 65, 66
fibres, SiCNO 66
fibres, Si_3N_4 18, 65, 66
fibres, strength of *see* fibre strength
fibrillar macrostructure 25
fibrillar structure 14
fibrillation 302
filament guides 154
filament strength 42
   see also fibre strength
filament winding 99, 102, 108, 147, 154–60
filament wound pipes 145, 385, 387
filament wound vessels 387
filler 76, 150, 168, 183, 186, 195, 200, 202
fillers, acicular 4–8
film former 156
finish, glass fibres 39, 42–46
finite element analysis 261, 284
fire resistance 105, 107, 109
fire retardance 124
fire retardants 130, 183
flame retardance 109
flammability 103, 200
flaws 1, 24
flexible thermoset matrix 282
flexural modulus 83, 84, 85, 100, 103, 114, 178
flexural strength 4, 83, 100, 103, 114, 178, 205
flexure tests 360
flow controlled prepreg 92
flow moulding 178
fluorinated monomer 98
Fokker debond tester 354
foreign object impact 327
formaldehyde 106, 109
fracture 224
fracture mechanics 240, 286
fracture processes 337
fracture toughness 99, 104, 130, 145, 383
fracture toughness, effect of moisture 384
fracture toughness testing 360
fragmentation test 231, 235–8
FRANKLIN fibre 5
free carbon 67
free edge 286
free radical chemistry 127
free radical mechanism 170, 200
free radical polymerization 129

free radical scavenger 125
free radicals 122
free volume 94
frequency in fatigue 315
friction of debonded fibres 240
frictional forces 155, 244
FRTP 41, 69–86, 145–50, 219
FRTS 41, 278–285
fuel saving 183
fuel tanks 159
full cure 120
fumarate unsaturation 123
fumaric acid 122

G_{Ic} 72, 73, 100
G_{IIc} 73
G_c 324
gas gun 328
gel, alumina hydrate 8
gel coats 43
gel glass 116
gel point 118
gel spinning 55–57
gel spun fibres 55
gelation 115
gelation to cure time 198
gelcoat 160, 206, 376, 381
gelcoat application 163
geodesic path 154
GLARE 225
glass composition 38
glass fibre 38, 128, 133, 150, 181, 209, 278, 280, 193, 338, 380
glass fibre composites, mechanical properties of 82, 85, 108, 114, 126, 133, 152, 205, 276, 284
glass fibre composites, thermal strains 254
glass fibre finish 42–8, 376
glass fibre mat 38–42, 376
glass fibre modulus 38
glass fibre-polypropylene, mechanical properties 178
glass fibre reinforced polyester 126, 200, 376
glass fibre sizing 42–8, 107, 152, 201
glass fibre strength 38
glass flake 191, 387
glass formation 115
glass mat reinforced thermoplastic 134, 176
glass transition 115–120
glass transition temperature 69, 70–1, 87, 95, 103, 107, 115–20, 128, 374
   see also heat distortion temperature and heat deflection temperature
glass-polyamide 74, 83
glass-polypropylene 83
glass-polypropylene composite 82
glycidyl amines 87

General index

GMT 134, 176
graded sand fillers 387
granular polyester 168
GRAPHIL 23
graphite 28
graphite lattice 28
graphitization 21
graphitized pitch 24
green stage 206
Griffith flaws 1
GRP, durability 43, 376–388
GRP 39, 133, 300, 310, 351, 376
  see also glass fibre composites

Halpin-Tsai equations 217
hand lay 134, 143, 160–5, 206
handleability of fibres 21, 39, 42
hard size 152, 201
hardeners 86
health and safety 84
heat deflection temperature 4, 205
heat distortion temperature 82, 94, 114, 128, 132, 205
helical winding 157, 159
heterogeneity 213
hexagonal close packing 212
hexamethyldisilazane 66
hexamethylene tetramine 169
hexamine 106, 110, 169
high density linear polyethylene 56
high modulus carbon fibre 22, 29
high modulus polyethylene fibres 56
high performance composite laminates 138
high performance composites 86
high service temperatures 107
high strength carbon fibre 1, 19, 21, 29
high strength polyethylene fibres 55
high temperature composites 12, 95, 96, 101
high temperature fibres 18
high temperature matrices 101, 109
high temperature performance 130
high temperature polymers 96–108
high temperature safety 84
higher energy impacts 335
holography 357
honeycombs 143
hoop tension 157
Hopkinson bar 328
hot press moulding 150, 200, 204
hot pressing 15
hybrid composites 60, 225, 301
hybrid configurations 226
hybrid effect 225–30
hybrid fibre 81
hybrid laminate 245
hybrid laminate, modulus 227
hydridopolysilazane (HPZ) 66

hydro-rubber forming 148
hydrochloric acid 385
hydrofluorosilic acid 9
hydrogen bonds 14
hydrogen gas 62
hydrolysis 102, 376, 380
hydrolytic resistance 190
hydroquinone 125
hydrothermal stability 294
hydroxy terminated polyester 189
hydroxycarbonate, sodium aluminium 7
hydroxyl groups 30, 43, 90
hydroxyl-isocyanate reaction 189–92
hygrothermal conditions 362
hygrothermal stability 117
hyperperoxides 122

imaginary or loss modulus 118
imide oligomers 104
impact damage 305
impact performance 135, 327–41
impact performance, residual compression strength 330–4
impact resistance 81, 83, 334, 347
impact strength 4, 33, 114, 178, 205
impregnation of fibres 79, 86, 154, 156, 196, 198
impregnation bath 158
improvements 309
in-house recycling 185
in-mould coating 191
in-plane random fibres 283
incineration 183–4
individual filament dispersion 274
induction heating 148
induction period 125
ingress of solvents 70–1
inhibitor 202
initiator 378
injection moulding 74, 76, 106, 112, 134, 150, 152, 165, 168, 171, 204, 278, 284
injection moulding compounds 274
injection pressure 167
injection/compression moulding 113
inlay yarn 344
instrumented falling weight impact 335
interface 31, 80, 81, 243, 269, 306, 311, 363
interfacial bond strength determination 230–45
interfacial shear strength 230–45, 270
interfacial shear stress 267, 281
interfacial yielding 240
interlaminar fracture toughness 336
interlaminar shear strength 31, 33, 72
interlaminar shear stress 57
interlaminar strength 336
interlaminar stresses 286, 289
intermediate modulus (IM) fibres 22
intermingled fibres 226

internal and external mixing   207
internal stress   376
interphase   31, 46, 80, 381
interply delamination   83
interweaving   71
intracrystallite disorder   27
Ioscipescu test   231, 360
isochrones   195
isocyanate   105, 211
isoimide   98
isomerization   122
isophthalic acid   123, 377
isoviscous contours   116
itaconic acid   19, 122
item finishing   164
item removal from the tool   164

J-polymer   71, 147
jamming   343
joints   163

K_{Ic} 384
KAPTON   96
Kelly-Tyson model   272, 281
KERIMID   101
KEVLAR   58, 76, 265, 278, 299, 300, 363
KFRP   300
kinetic control   117
kink bands   300
kneader moulding compound   151
knife penetration   83
knit composites   341
knits   49
knitted fabric   347
knitted fabric composites, properties   341–6
knitted structures   50

L_c 235, 278
ladder polymers   20
laminate construction   83, 84, 386
laminate theory   82, 214, 347
laminated plate theory   247, 286
laminates   82, 101, 107, 138, 214
laminates, 0°/90°/0°   243–67
laminates; crossply, angle ply   245
landfill   184
LaRC   98
laser holographic methods   351
laser induced shock   328
latent catalysts   105
latent cure   102
latent hardeners   86
law of mixtures   215, 225
leaching   380
legislation   183
life prediction   317, 327
limiting strain for stress corrosion   384
linear elastic fracture mechanics   287

liquid crystal polymer   75, 176
liquid crystal polymer fibres   176
liquid crystal solution   34, 58
load-displacement curve   239
long glass fibre reinforced thermoplastic   145–50, 176–80
longitudinal cracking of fibres   14
longitudinal splitting   252
longitudinal stiffness   215
longitudinal strength   218
loop length   341
Lotus Elan   199
low cost composites   79
low energy impacts   335
low pressure SMC   204
low profile   126
low profile additive   151, 203
low profile resins   126
low smoke generation   109
low styrene emission resins   125
lubricant   42

macroscopic shear   172
MAGNAMITE   23
magnesium alloys   82
magnesium hydroxide   6
magnesium oxide   202
magnesium oxysulphate   6
maleic acid   122
maleic anhydride   122
management of reinforcing fibres   171–6
mandrel   154
manufacturing technology   133
matched metal moulding   85
material cure   164
material/process selection   135
MATRIMID   99
matrix cracking   128, 237, 247, 249, 286, 290, 296, 318
matrix cracks in single embedded filament fragmentation   233
matrix failure strain   221
matrix fibres   71
matrix modulus   70, 75, 128
matrix properties, thermosets   88, 94, 100, 103, 126, 128, 132, 210
matrix properties, thermoplastics   70, 75, 276
matrix swelling   257
matrix-rich regions   212
maximum strain failure criterion   297
maximum stress failure criterion   297
maximum work theory   225
MDA (4,4′-diaminodiphenylmethane)   97
mean strength of fibres   218
mechanical abrasion   302
mechanical degradation   185
mechanical properties   104

mechanical properties, bisoxazoline-
    phenolic composite 104
mechanical properties, cyanate resins 103
mechanical properties, epoxy
    composites 104
mechanical properties, epoxy resins 94
mechanical properties, glass fibre
    thermosets 108, 126, 284
mechanical properties, matrices 70
mechanical properties, phenolic
    composites 108, 114
mechanical properties, phthalonitrile
    resin 105
mechanical properties, polyester
    composites 126
mechanical properties, polyimide
    composites 100
mechanical properties, polyimide
    resins 100
mechanical properties, thermoplastic
    composites 73, 82, 83, 85, 178, 186,
    276
mechanical properties, thermoset
    composites 100, 104, 108, 114, 126,
    152, 210, 284
mechanical properties, urethane
    methacrylate resins 128
mechanical properties, vinyl ester
    resins 132
mechanism of blister formation 376
melamine formaldehyde 169
melt impregnation 72
melt spinning 55–7, 67
melt spun fibres 55, 57, 63
melting point 75
mesophase 34
metal 62
metal matrix 176
metal matrix composites 63
metal stearates 111
metal/composite laminates 225
metathesis 191
γ-methacryloxypropyltrimethoxysilane 46
methane 62
methyl acrylate 19
methyl methacrylate 127, 211
methylol phenols 109
methyltrichlorosilane 62
mica 169
micro-indentation test 231
microchemistry 30
microcracking 98, 100
microdebond test 238, 243
microdefects 307
micromechanics 212–85
micronized hardener 93
micropores 31
microstructural effects 251
microstructure 323

microvoids 300, 376
mild steel 82
milled glass fibres 4, 41, 191
Miner's law 326–7
mineral fillers 4–8, 111
mineral thickening 202
minimum cracking strain 252
MMC 63
MODAR 127, 182, 191
mode of failure 229, 317, 382
modified acrylate 127, 182, 191
modified acrylic resins 127, 182, 191
modulus, composite theory 215
modulus, composites see mechanical
    properties, thermosetting composites
    and mechanical properties,
    thermoplastic composites
modulus, fibre see fibres, modulus of
modulus, matrix see matrix properties
    thermosets and matrix properties
    thermoplastics
moiré fringes 357
moisture, entrapment 107
moisture absorption 257, 362–79, 384, 389
moisture absorption, fibres 15, 300
moisture absorption, resins 87, 95,
    372–4
moisture in fatigue 300, 315
moisture induced property
    degradation 315, 362, 376
moisture induced swelling 257, 363
molecular composites 61
molecular deformation 61
molecular order 34, 59
molecular orientation 20, 34
monitoring cure 117
monuron 92
mould shrinkage 153, 168
moulding compounds 110, 114
moulding cycle 170
γ-MPS 46
multiaxial warp knits (MWK) 53, 342
multilive feed 172–6
multiple cracking of fibres 219
multiple cracking theories 249
multiple impression tooling 199
multiple matrix cracking 221, 280
multistream systems 195
MY720 87
MY750 86

nadicmethylenetetrahydrophthalic-
    anhydride 91
NARMCO 5245C 389
NDE 351–6
near net shape preforms 211
net shape 153
net shaped preforms 196
network formation 190

NICALON 66
NICALON fibres 63
nickel faced tooling 199
NMA 91
NOMEX 12, 143
non-destructive evaluation of composites 351–6
non-linear stress-strain curves 284
non-woven fibres 49
nonylphenol 102
5-norbornene-2,3-dicarboxylic acid monomethyl ester 97
normalized tensile strength 82
novolak 106, 110, 151
novolak resin 106
nylon 6.6 75, 114
NyRIM 189

off-axis fibres 267
off-axis performance 222
off-axis properties of unidirectional laminae 222
optical inspection 351
organic peroxides 122
organometallic polymers 18, 64, 66
organosilane 44
orientation dependence of tensile strength 223
orientation dependence of Young's modulus 223
orientation of fibres 82
orthophthalic resins 123, 377, 387
osmosis 376
oxidation 20
oxidative stability 59, 64
oxidative stabilization 34
oxygen plasma treatment 57

PAN 19, 24, 29
paper-making technologies 178
paraffin wax 125
Paris law 309
partial crystallization 203
particles, acicular 4–8
particles, reinforcement 4
particulate reinforcement 4
payout eye 154, 159
PBO fibre 58
PBT fibre 58
PBT thermoplastic 114
PEEK 70, 71, 146, 186, 331, 338, 365
PEI 71, 100, 146
PEKK 71
pendulum impact 328
penetrant enhanced X-radiography 291
penetration 337
performance index for the interface 241
peroxide catalysts 129
PES 70, 71, 87, 146

PET 75, 114, 121, 177
phase separation 115, 116
phenol 106, 109
phenolic composite, thermal expansion coefficient 108
phenolic resin 103–104, 106–115, 117, 151, 168, 169, 182, 280, 293
phenolic resin composites, mechanical properties 108, 113–15
phenolic SMC 201
phenoxy resins 87
phosphate fibre 6
phosphoric acid 385
photoelasticity 357
phthalic anhydride 123
phthalocyanines 105
phthalonitrile resins 105
physisorbed oligomers 46
piezoelectric transducers 232
pigments 166, 183
pitch 34
planar aromatic structure 20
planar ribbon-like form 20
planetary winders 159
plasma oxidation 30
plasma treatment 57
plastic deformation 244
plasticization 363
plunger 166
plunger type injection moulding 166
PLYTRON 80
PMR 96
PMR-15 97, 105
PMR-15 lamination 142
PMR-II-50 98
Poisson contraction 216, 217, 281
Poisson ratio 75, 217, 234, 253, 348
Poisson strain 252
poly(2,5(6)-benzoxazole(ABPBO) 58–62
poly(amide ether) 103
poly(p-phenylene benzobisoxazole) (PBO) 58–62
poly(p-phenylene benzobisthiazole) (PBT) 58–62
poly(p-phenylene terephthalamide) see aramid
polyacrylate 43
polyacrylonitrile 19, 24, 29
polyamides 5, 80, 90
polyaromatics 34
polyarylene ether 100
polyborodiphenylsiloxane 63
polybutylene terephthalate (PBT) 114
polycarbonate/PBT alloy 177
polycarbosilane 63
polyester copolymer molecular structure 91
polyester fibre veil 387

General index

polyester moulding compounds 150, 200
  see also BMC, DMC, SMC
polyester resin 121–7, 182, 206, 280, 293, 376
polyesterol 189
polyethene fibres 54
polyether 189
polyether structure 91
polyetheretherketone 70, 71, 146, 186, 331, 338, 365
polyetherimide 71, 100, 146
polyetherketoneketone 71
polyetherol 189
polyethersulphone 70, 71, 87, 146
polyethylene 54, 57
polyethylene fibres 54, 244
polyethylene terephthalate 75, 121
polyimide foam 144
polyimide resins 95, 102, 295
polyimides 96, 295
polymer binder 39
polymer foam 144
polymer powders 71
polymer precursor 9, 18, 62
polymer precursor route (PPR) 62
polymeric binders 42
polymeric fibre properties 14, 59
polymeric precursors 9, 18, 62, 66
polymerization 189
polymerization of monomeric reactants 96
polymers, rigid rod 58
polymethacrylimide 144
polynuclear aromatics 105
polyphenylene sulphide 71, 114
polypropylene 75, 80, 171, 173, 177, 184, 334
polysilanes 63
polystyrene 43, 126
polystyrene crosslinks 130
polystyrylpyridine (PSP) 105
polytitanocarbosilane 65
polyureas 189
polyurethane 43, 187, 211
polyvinyl acetate 43, 126, 162, 203
polyvinyl alcohol 162
post-curing 94
potassium 385
powder bound CSM 39, 387
powder bound mat 39, 387
powder coating 80
powder impregnation 81
powders, phenolic moulding, performance of 114
PPS 71, 114
pre-accelerated resin 163, 207
pre-impregnated fibres 138
pre-impregnated moulding materials 200
pre-impregnated sheet 70
precatalysed resin 207

precracks 377
precursors 18, 19
prediction of fatigue life 317–27
preform 187, 193, 209, 341
preformability/drapeability 210
preparation of resin and reinforcement for hand lay 163
prepreg 70, 80, 86, 93, 97, 102, 104, 138, 146, 214, 245
prepreg tape 159
prepreg technology 134
pressclave 138
pressing 84
pressure 167
pressure vessels 159
PRIMASET 102
principal moduli 347
principal reduced stiffnesses 348
printed circuit board 100, 103
probabilistic theory 279
process aids 183
process controllers 168
processing speed in pultrusion 180
propenyl monomer 99
properties of BMC 152
property modifiers 183
propylene glycol 124
protective polymeric size 42
PT resin 102
pulforming 181
pull-out 231
pull-out test data 241
pulse video thermography (PVT) 353
pultruded composite tapes 79
pultruded composites 79
pultruded products 108
pultrusion 75, 76, 85, 107, 128, 134, 147, 180
pulwinding 181
PVA 162
PVAc 43, 126, 162, 203
pyrolysis 186

quasi-isotropy 214

R-curve 290
R-glass 38
R-ratio 302
RADEL 71, 147
radome applications 296
rail shear 360
Raman spectroscopy 61, 264, 270
Raschel Tricot 51
rate of manufacture 134
reaction injection moulding 187–92, 209–11
reactivity ratio 123
real or storage modulus 118
reconstitution 185

recyclability   183
recyclable   179, 192
recycling   79, 85, 183
refractory metal filament   16
reinforced reaction injection moulding   41, 186–92
reinforced resin injection moulding   41
reinforced thermoplastic sheet   176
reinforcement efficiency   134, 273, 280
reinforcement type   206
relative costs   135
relative costs of materials of similar rigidity   135
relative costs of materials of similar strength   135
β-relaxation   119
release agent   162, 182, 202
release cloths   138
release films   138
relic processes   9
residual compression strength   330
residual radial thermal stresses   243
residual strain   252, 254–60
residual strength   331
residual strength models   324
residual stress   233, 376
resilience   190
resin cracking   311, 351
resin cure   193
resin gelation   198
resin injection moulding   187, 193
resin injection or transfer   134
resin microcracks   204
resin rich areas   212
resin transfer   108
resin transfer moulding see RTM
resin/glass ratio   163
resins see individual types   86, 102, 106, 121, 127
resistance welding   148
resit   86
resitol   86
resole   86, 106, 110, 151
resolution   352
restrained layer hypothesis   46
reversible hydrolytic bonding mechanism   46
rheology   9, 182
rigid rod polymer fibres   58, 264
rigid rod polymers, mechanical properties   59
RIM   187–92, 209–11
ring-down counting   355
ring-opening polymerization   191
RK carbon   24
robots   154
rocket   159
ROHACELL   144
roller impregnation   157

rotational casting   144
rovings   39, 43, 154, 181, 208
   see also glass fibres
RPM pipe system   387
RRIM   41, 186, 187–92
RTM   39, 99, 102–5, 128, 134, 187, 192–200, 209
RTM technology   209
RTS   176
rubber   183
rubber block processing   148
rubber block stamping   85
rubber modified epoxies   116
rule of mixture   82, 225, 279
RYTON   71, 147
RYTON PPS   71, 147

S-2 glass   2
S-glass   2, 38
S-N plot   301, 305
sacrificial polymer binders   176
SAF   19
SAFFIL fibres   11
SAFFIL staple fibre   10
SAFIMAX   10
sandwich fibre orientation   277
satellite structures   296
saturated acids   123
saturated polyesters   121
saturation fragment length   235
scanning electron microscopy   25
scanning secondary ion mass spectrometry   46
Schapery equations   257
SCOREX   171
SCORIM   171
SCORTEC   171
scrap plastics   184
screw back speed   167
screw injection moulding   166
secant modulus   361
self-cleaning impingement mix-head   188
self-induced thermal polymerization   20
SEM   25
semi-interpenetrating network   46, 98
semicrystalline polymers   69, 146
semipermeable membrane   376
service durability   317
shape parameter   262
shaping thermoplastic composites   72
shear buckling   332
shear controlled orientation   171
shear controlled orientation technology   171
shear cracking   335
shear in injection moulding   171
shear lag analysis   243, 247–54, 261, 269
shear lag   249
shear lag model   266

shear lag theory 281
shear modulus 83, 243, 347, 348
  see also matrix properties and composites, properties of
shear properties 302
shear strength, matrix 224
shear strength, composite 224
shear stress 221, 243
shear transfer zone 261
shears 171
sheet moulding compound 41, 134, 150, 184, 200, 278, 284
sheet stamping 84
shelf life 84, 125
short beam shear strength 73, 100, 104, 231
short beam shear test 231
short E-glass fibres 191
short fibre reinforced thermoplastics 75, 185, 274
short fibre reinforced thermosets 278–85
short fibre reinforcement 269–78
short fibre thermoplastic composites 75, 185, 274–8
short fibres 274
short glass fibres 173
shredding 186
shrinkage 164
shrinkage control 201
shrinkage control additive 150, 202, 203
shrinkage on cure 125
Si_3N_4 fibre 18
SiC fibre 17, 18, 62, 66
SiCNO fibre 67
SIGMA fibre 62
silane coupling agent 5, 42–8
silica 380
silicon carbide 16, 62
silicon carbide fibres 17, 18, 62, 66
silicon carbonitride fibres 66
silicon carbonitrides 66
silicon nitride 62, 66
single fibre pull-out 243
single filament pull-out test 238
sintering 15
size formulations 43, 81
sizing 31, 38, 42–8, 81
sizing resin 21, 43, 81, 100
skin-core heterogeneity 25
skin-core structure 78, 279
SMC 41, 134, 150, 184, 200, 278, 284
SMC, mechanical properties 205
SMC, typical composition 202
smoke and toxic fumes 107
sodium aluminium sulphate 385
sodium hydroxide 386
sodium thiocyanate 19
softening point, epoxy resin 89
softening point, glass 38

sol glass 116
solid state thickening 203
solvent assisted impregnation 81
solvent attack 71
special acrylic fibre 19, 22
specific modulus 1, 13, 135, 293
specific stiffness 13
specific strength 1, 13, 135, 293
SPECTRA fibres 57
spin-stretch factor 19
spinning 34
spiral cracking 380
splitting of fibres 14
splitting of laminates 286
spray deposition 205
spun pipe 144
square close packing 212
SRIM 39, 95, 128, 134, 187, 209–11
stacking sequence 138, 289
stacking size 27
stainless steel 82
stamping 134, 148, 178
static fatigue 379
statistics of fibre failure 228
statistics of fibre strength 233
STATOIL 80
step-addition polymerization 45, 189
step-growth polymerization 45, 189
stick-slip 244
stiffness 213, 317
  see also modulus
stiffness reduction models 324
stitch length 341
stoichiometric concentrations 90, 188
stoichiometry 90, 188
straight-chain polymer structure 13
strain control model of composite 302
strain energy release rate 288, 290, 324
  see also G_C and G_{IC}
Strain free temperature 256
strain gauges 357, 358
strain life diagrams 319
strand integrity 42
strands 39, 274
strength, fibre 14, 17, 37, 55, 59, 218
  see also fibre strength
strength, carbon fibres 1, 23
strength, composite 82, 108, 134, 213, 218, 228
  see also composites, properties of
strength, glass fibres 38
strength, matrix 100, 128
  see also matrix properties
strength 1, 36, 70, 135, 213, 284, 317, 363
strength after impact 331
stress concentration 238, 316
stress corrosion 300
stress distribution at the interface 240
stress free temperature 256

stress intensity factor, K 287
stress intensity range, ΔK 309
stress on the fibre 221
stress relaxation 255
stress transfer 269
stress wave amplitudes 355
stress whitening 351
stress-life curves 299–327
stress/strain relationships for
    laminates 347
structural changes during cure 118
structural reaction injection moulding *see*
    SRIM
structural thermoplastic composites
    69–74, 79–86, 176–9
styrene 99, 122, 129, 150, 203
styrene emission 125
subcritical damage 296
sulphuric acid 12, 380
surface finish 150, 204
surface free energy 44, 252
surface microstructure 30
surface oxidation 21, 29–33
surface oxygen concentration 30
surface tissue 377
surface treatment 5, 20, 29–33, 42–7, 57
surface wettability hypothesis 46
susceptibility to damage 294
    see also damage
swimming-pools 376
swirl mat 181
symmetric laminates 214, 245, 249, 350
synergism 363, 380
synergistic effects 363, 380

T_g 70, 71, 87, 95, 100, 103, 107, 115–20, 128, 374
T_g temperature-property diagram 119
T-flange 52
T-joints 52
tack 86, 138
tacticity 20
tangent modulus 361
tape laying 143
tape placement 85, 147
tape winding 85
TDI 189
TEM 22, 25
temperature, strain free 256
temperature, stress free 256
TENAX 23
TENFOR 57
tensile modulus, laminate 178, 215–18, 223
tensile strength 232
tensile strength, fibre *see* fibre strength
tensile strength, laminate 114, 178, 205, 280, 347, 363

tensile strength of unidirectional
    laminae 218
tensile strength perpendicular to the
    fibres 222
tensile testing 357
terephthalic acid 123
terminal methacrylate groups 127
terminal unsaturated vinyl groups 129
tertiary aromatic amines 122
tertiary butyl perbenzoate 151, 202
tetrabrominated bisphenol A 130
tetrafunctional epoxy resins 87
tetraglycidyl amine 87
tetrahydrofuran 95
tetrahydrophthalic anhydride 94
tex 39
textile acrylic precursor 24
theories of multiple microcracking 251
thermal conductivity 75, 108
thermal cycling 98, 257, 388
thermal degradation 14
thermal expansion 213
thermal expansion coefficient 75, 82, 108, 178, 254–7, 365
thermal imaging 352
thermal internal stress 228
thermal spiking 388
thermal stability 8, 14, 17, 59, 62, 64
thermal strain 252, 254–7, 390
thermoelastic stress analysis 353
thermoforming bend angles in
    composites 83
thermoforming of composites 83–4
thermoplastic additive 125, 203
thermoplastic composites 138, 244, 274
thermoplastic matrices 69, 74, 79
thermoplastic matrices, properties 75
thermoplastic matrix composites 71, 145, 171, 176, 182, 186
thermoplastic matrix composites, elastic
    constants 83
thermoplastic matrix composites,
    mechanical performance 73
thermoplastic matrix composites,
    properties 70–3, 82, 85, 186, 274–8
thermoplastic modifiers 87
thermoplastic polyesters 80
thermoplastic polymers 69
thermoplastic polyurethanes 80
thermoplastic prepregs 146
thermoplastic resins 294
thermoplastic toughening 100
thermoplastics 74, 165, 280, 331
thermoset matrices 96–132, 180, 278
thermoset matrix composite,
    properties 100, 104, 108, 114, 126, 152, 205, 210, 284
thermosets 86, 106, 126, 150, 168, 185, 294
thermosetting modifiers 90

General index

thermosetting phenolic matrix
  composites 106–15
thermosetting polyester matrix
  composites 121–7, 171
thermosetting polyimides 96–101
thermosetting resins 86–132, 138, 200, 209
thick moulding compound 151
thick ply behaviour 248
thickener 202
thickening 201
thicker glass fibres 230
thickness of gelcoat 377
thin ply 248
THORNEL 23
three-dimensional network
  polysiloxane 46
three-dimensional random fibres 283
through thickness fibre reinforcement 54
time dependent mechanical properties 364
time dependent phenomenon 310
time temperature transformation curing
  diagrams 94
time-temperature transformation 115
time-temperature-transformation
  diagrams 115–21
titanium alkoxide 65
ToF SIMS 48
toluene diisocyanate 189
tool design 160
tool design and manufacture 160
tool preparation 162
tooling 143
TORAYCA 23
TORLON 71, 147
torsional braid analysis 117
tough composites 32
toughened epoxy 331, 333
toughened polyimide resins 99
toughness 72, 73, 100, 104, 106, 130, 145, 168, 324, 384
tow 39
toxicity 4, 98, 102
transfer moulding 112
transformed reduced stiffnesses 349
transmission electron microscopy 22, 25
transverse compressive strength 364
transverse cracking 212, 247, 264
transverse cracking strain 250
transverse cracks 212, 247, 261
transverse interface strength 234, 260
transverse loading 231
transverse stiffness 216
transverse strength 222, 385
transverse stress 216
treatment of fibre surface 29–34, 42–7
β-triamino-N-tris(trimethylsilyl)borazine 18
triazine 102
trichlorosilane 66

tricot stitch 342
triethylaluminium 19
triethylene tetramine 94
tristrialkylaminoboranes 18
Tsai-Hill criterion 224, 298
Tsai-Wu criterion 298
TTT diagrams 115–21
tubes 154
tumble winders 159
tungsten 16
tungsten boride 17
tungsten wire 17
turbostatic graphite 28
turbostatic graphite 36
TWARON 12
  see also aramids, aramid fibres
twin pot 207
twist 39
twisting 155
Type A carbon fibre 22, 29
Type I carbon fibre 22, 29
Type II carbon fibre 21, 29
TYRANNO 66
TYRANNO fibres 63

UHMW polyethylene (UHMWPE) 180
ULTEM 96
ultrasonic techniques 354
ultrasonic welding 148
ultraviolet light 13
ultraviolet light degradation,
  self-screening 13
unbalanced 0°/90° beam 255
unidirectional composite 213–15
unsaturated polyester resin 117, 121, 127, 129, 144, 150, 163, 182, 191, 200, 211, 379
urea formaldehyde 169
urethane crosslinking 203
urethane methylacrylate 127

V_{crit} 219, 221
vacuum assisted resin injection 195, 199
vacuum bag 139, 148
vacuum forming 84
Van der Waals' interactions 14
VARI 195, 199
variational approach 261
very short fibres 187
VICTREX 71, 147
vinyl copolymerization 121
vinyl ester resin 127, 129–32, 144, 182, 379, 384
vinyl ester SMC 200
vinylsilane 67
viscosity 86, 99, 102, 127, 138, 150, 155, 168, 170, 187, 194, 202, 208
viscosity modifiers 86
visible light 13

visual inspection 351
vitrification 115
void content 107
void-free laminate 138
voids 142
volatile 105
volume fraction of fibre 133, 212–22
volume fraction of matrix 216

warp knit 49, 344
water absorption 91, 114, 192, 256–8, 362–79
wax 162
wear-out 310
weaving 84
weft knit 49, 344
weft knit composites 344
Weibull distribution 262
Weibull scale parameter 262
Weibull statistical methods 237
weld lines 153
welding 71
wet lay-up 278
wet lay-up of CSM 278
wet slurry process 177
wet-out 39, 43, 128, 154
whirling arm winders 159

whiskers 5, 8
winding 157
wollastonite 4, 5, 111
wood fibres 363
wood-flour 109, 110, 169
work of adhesion 44
work of fracture 282
work of fracture, interfacial 240
working time 125
woven cloth 49, 134, 226
woven rovings 39, 133, 163, 387

X-ray diffraction 22, 27
X-ray inspection 352
X-ray microradiographs 173
XPS 48

Yajima 63, 66
yarn diameter 341
yielding 244, 334
Young equation 44
Young's modulus 205, 215, 347
 see also modulus

Z-blade compounding 151
zero bleed 139
zinc stearate 151, 202